The History of British Birds

The History of British Birds

D. W. Yalden
with
U. Albarella

UNIVERSITY PRESS

OXFORD
UNIVERSITY PRESS

Great Clarendon Street, Oxford OX2 6DP

Oxford University Press is a department of the University of Oxford.
It furthers the University's objective of excellence in research, scholarship,
and education by publishing worldwide in

Oxford New York

Auckland Cape Town Dar es Salaam Hong Kong Karachi
Kuala Lumpur Madrid Melbourne Mexico City Nairobi
New Delhi Shanghai Taipei Toronto

With offices in

Argentina Austria Brazil Chile Czech Republic France Greece
Guatemala Hungary Italy Japan Poland Portugal Singapore
South Korea Switzerland Thailand Turkey Ukraine Vietnam

Oxford is a registered trade mark of Oxford University Press
in the UK and in certain other countries

Published in the United States
by Oxford University Press Inc., New York

© D.W. Yalden and U. Albarella 2009

The moral rights of the authors have been asserted
Database right Oxford University Press (maker)

First published 2009

All rights reserved. No part of this publication may be reproduced,
stored in a retrieval system, or transmitted, in any form or by any means,
without the prior permission in writing of Oxford University Press,
or as expressly permitted by law, or under terms agreed with the appropriate
reprographics rights organization. Enquiries concerning reproduction
outside the scope of the above should be sent to the Rights Department,
Oxford University Press, at the address above

You must not circulate this book in any other binding or cover
and you must impose the same condition on any acquirer

British Library Cataloguing in Publication Data

Data available

Library of Congress Cataloging in Publication Data

Data available

Typeset by Newgen Imaging Systems (P) Ltd., Chennai, India
Printed in Great Britain
on acid-free paper by
CPI Antony Rowe, Chippenham, Wiltshire

ISBN 978–0–19–921751–9

1 3 5 7 9 10 8 6 4 2

Contents

	Introduction	1
1.	The bird in the hand...	3
	Identifying bird bones	3
	What to identify?	3
	Problems of identification	10
	Problems of dating	17
	Sources of bones	21
	Conclusions	23
2.	The early history of birds in Britain and Europe	25
	Archaeopteryx	25
	Cretaceous birds	27
	Cretaceous-Tertiary transition	29
	Tertiary birds	31
	Pleistocene birds	33
	The Last Glaciation	37
	Continental Europe	39
	Conclusions	47
3.	Coming in from the cold	49
	Late Glacial birds	50
	Younger Dryas	55
	Mesolithic birds	56
	Reconstructing the Mesolithic bird fauna	60
	Birds of open country	68
	Conclusions	70
4.	Farmland and fenland	73
	Neolithic birds	73
	Bronze Age	78
	Fenland	79
	Conclusions	92

5.	**Veni, Vidi, Vici**	95
	Iron Age Britain	95
	Early domestication	97
	Domestic Fowl	97
	Domestic geese	102
	Domestic Duck	104
	Domestic Dove	105
	Other Roman introductions	106
	Wild birds in Roman Britain	108
	Conclusions	113
6.	**Monks, monarchs, and mysteries**	115
	Birds in placenames	115
	Archaeological Saxon birds	130
	Norman birds – castles, feasts, and falconry	134
	Falconry in archaeology	135
	Cranes, Ernes, Brewes, and other Mediaeval birds	139
	Birds in early literature and art	149
	Conclusions	151
7.	**From Elizabeth to Victoria**	153
	Assembling a list of British birds	153
	Birds lost and gained	159
	Great Bustard	163
	Great Auk	167
	Capercaillie	170
	Raptors	170
	Conclusions	174
8.	**Now and hereafter**	175
	Birds in the twentieth century	175
	Changing attitudes	176
	The balance of the bird fauna now	180
	The bird fauna in the future	188
	The future of predators	197
	Appendix	203
	References	231
Index		257

Introduction

In writing this account of the history of birds in the British Isles, it was intended to draw to the attention of the ornithological world the extensive archaeological information that is available, and to draw to the attention of the archaeological world the wider importance and interest of the results that are so often hidden away in supplements and appendices to accounts of archaeological excavations. There is an obvious parallel here to *The History of British Mammals*. However, the historical constraints are less severe on birds, for which the isolation of Great Britain from continental Europe, of Ireland or Man from Great Britain, and the isolation of the northern and western islands do not assume the importance that they do for mammals; nor does intentional or accidental Human introduction. As a consequence, this account does not adhere so strictly to a historical layout, though the march of time is an underlying theme.

A major problem in writing it is the obscure and scattered nature of much of the archaeological record. 'Grey literature', unpublished but publicly available reports to those commissioning excavations (especially English Heritage reports), are a particular problem. In an effort to overcome this, a partial set of records was supplemented by a more systematic trawl of the literature available in the John Rylands University Library, University of Manchester. The Leverhulme Trust granted D. W. Yalden a research grant that allowed Rob Carthy to be employed for 6 months specifically to garner much of the information summarized here. He set up the database of archaeological sites and bird records, as well as an EndNote database of the relevant literature. I am very grateful to both him and the Trust for their invaluable support. It is intended that the database will be made freely available to the scientific community once this account is published. These immediately available sources were supplemented by records that Umberto Albarella had compiled for central England, and a similar archive for the north of England assembled by Keith Dobney. A number of others have generously helped with commenting on parts of this account, supplied extra records (sometimes as yet unpublished) or sent reprints on other sites; among them are Sheila Hamilton-Dyer (Southampton), the late Colin Harrison (London), Gil Jones (Leeds), Roger Jones (Hertfordshire), Matthew Rogers (Bristol), Cecile Mourer-Chauviré (Lyon), Dale Serjeantson (Southampton), Catherine Smith (Perth), Sue Stallibrass (Liverpool), John Stewart (London), and Tommy Tyrberg (Kimstad, Sweden).

A separate line of relevance concerns placenames, and we are grateful for the advice offered by, among others, Richard Coates (Sussex), Margaret Gelling (Birmingham), Carole Hough (Glasgow), and Peter Kitson (Birmingham).

A major contribution came from various undergraduates who conducted B.Sc. projects in their third year under my supervision; it was a mutually advantageous partnership, and I am grateful for their enthusiastic contribution to this account, even if they did not realize that I would end up exploiting their efforts. They include Simon Boisseau (placenames for ravens, raptors, and cranes), Steven Bond (extinction rates of some raptors in the nineteenth century), Rajith Dissanayake (passerine humeri), John Heath (identification of passerine bones), Christopher John (archaeological record of birds), Iain Pickles (placenames for domestic birds), Richard Preston (passerine tarsometatarsi), James Whittaker (eagle placename), and David Younger (variability of bird bones). Between them, they established what might be possible and interesting in this field.

The account was initially written in 2004–05 by D.W. Yalden. It was then sent to U. Albarella, who had intended to be a co-author, in January 2006. Other work then diverted our attention until late 2007. U. Albarella has read the whole text, adding comments and contributions throughout, but in the event had been left little to do. The form of authorship on the title page is intended to reflect this. D.W. Yalden takes full responsibility for opinions and errors contained here, and is grateful to have been spared worse errors by U. Albarella's additions.

K. Dobney was also to have been a co-author. In the event, pressure of other work prevented his full participation, but we thank him for his thoughts on the project, and his contributions to the data-base.

The value of Dr A. J. Morton's DMAP programme for generating the distribution maps is gratefully acknowledged.

During the period 1966–90, D. W. Yalden had the good fortune to join the Peakland Archaeological Society in excavating Foxhole Cave in the Peak District, under the direction of the late Don Bramwell. At a time when very few others had an interest in bird bone identification, he had developed an expertise, and the reference collection to support it, which was sought by many other archaeologists (as the reference list makes clear). We shared many zoological stories as we attempted to extract rodent bones and identify larger mammal bones, and he passed on much accumulated wisdom and knowledge to me. He had been working on a book of his own at one time, and I inherited many of his notes. I hope this account bears some comparison with what he would have written, and I acknowledge his friendship and tutelage. Other members of the Peakland Archaeological Society, including Roger Jones, the late Ken Holt, Sonia Holt, and the late Norman Davenport are also fondly remembered and thanked. Foxhole was a cool and draughty classroom at times, but some of its lessons reappear here.

D.W. Yalden
24 December 2007

1
The bird in the hand...

...is worth two in the bush, says the old saying, generalizing the advice to a hunter of earlier times to concentrate efforts on the reliability of the catch already made. For the historical ornithologist, the equivalent reliable catch is a well-dated specimen of a well-identified bone. It should provide a firm indication that some particular species of bird, perhaps one now locally extinct, occurred at some particular site and time. It is important, though, to examine the uncertainties surrounding both 'well identified bone' and 'well-dated specimen'.

Identifying bird bones

Mammals are readily identified from their teeth, skulls, jaws, and other bones. The popular perception is that birds' bones are so similar that they cannot be reliably identified. They are also much less robust than mammalian bones, leading to the equally popular perception that there is no valuable or significant subfossil or fossil record of birds. One of our main aims is to demonstrate that both perceptions are quite wrong. To do so, we must first discuss the bones of the avian skeleton, to concentrate on those of most value, and also address the genuine problems of identifying these bones in a group that contains many more species. Roughly, there are 9,500 species of bird in the world, as against 4,300 mammals; more parochially, there are about 200 breeding birds in Britain, but only 60 mammals (admittedly, these totals include the seabirds that nest on British cliffs, but omit the dolphins in the surrounding seas; the former are common in archaeological sites, whereas the latter are rather rare, and can only be loosely described as British).

What to identify?

The bird skeleton (Figure 1.1) is a highly modified version of a small dinosaur skeleton, and the very distinctive bones can readily be identified as bird bones, not easily confused with those of mammals. Even bats' bones, which might be expected to resemble bird bones, are very different, because their wings are anatomically very distinct. A bird's shoulder girdle forms a strong well-braced hoop, with a tall, pillar-like coracoid bone (absent in most mammals) running from the big sternum, with its distinctive keel, to the shoulder joint (Figure 1.2). The coracoid and scapula (a broad 'shoulder blade' in mammals, including bats, but a thin, more knife-like, blade in birds) together provide at their junction the socket (glenoid joint) for the humerus, the upper arm bone. The furcula, or wish bone (fused collar bones), is a V-shaped bone lying in front of the coracoids, often springy but sometimes

4 | The bird in the hand...

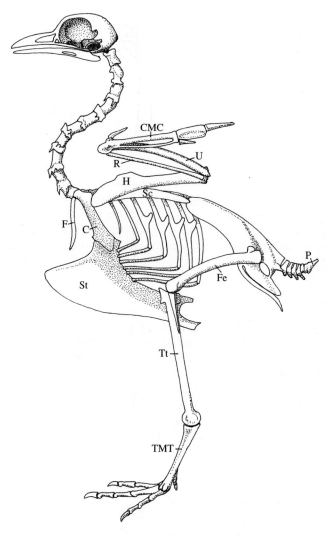

Fig. 1.1 Bird skeleton, with some of the more important bones (for archaeological identification) identified: C Coracoid; CMC Carpometacarpal; F Furcula (= wishbone, fused clavicles); Fe Femur; P Pygostyle; R Radius; Sc Scapula; St Sternum; TMT Tarsometatarsus; Tt Tibiotarsus; U Ulna.

forming a very solid 'V'. The humerus is one of the most distinctive bones and, being one of the most robust bones in the avian skeleton, one of the most useful to the practising archaeologist. Its head, the shoulder joint, bears a complex articular surface, with depressions on the dorsal side where wing-folding muscles insert (Figure 1.3). The prominent deltopectoral crest, where the main flight muscles attach, runs about a third or halfway along the anterior side, and there is a complex elbow joint at the distal end, all of which provide identification characters. It is gently curved in most groups, but has a very straight shaft in passerines. Of the two bones in the forearm, the radius is a thin straight bone, not very distinctive, but the

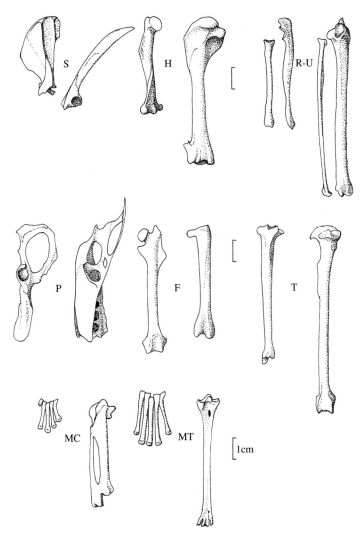

Fig. 1.2 Bird bones (right of each pair) compared with equivalent mammal bones (Grey Squirrel and Carrion Crow, both about 600 g; scale bars 1 cm). Top row: S Scapula; H Humerus; R-U Radius and Ulna. Middle row: P Pelvis; F Femur; T Tibia/Tibiotarsus. Bottom row MC Metacarpals/Carpometacarpal; MT Metatarsals/Tarsometatarsus.

ulna is a stout, gently curved bone that supports the secondary flight feathers, and often has small bumps (nodes) that indicate the position of each feather. Birds have only two small wrist bones (we have seven), but then a bone in the hand, the carpometacarpal bone, which is another of the very useful bones for identification purposes. It represents three metacarpals (equivalent to three of the five bones of our palm), fused together in a distinctive manner with another wrist bone; it supports in life most of the primary flight feathers (those that make up the tip of the wing) (Figure 1.4). There are only four finger bones (phalanges) in

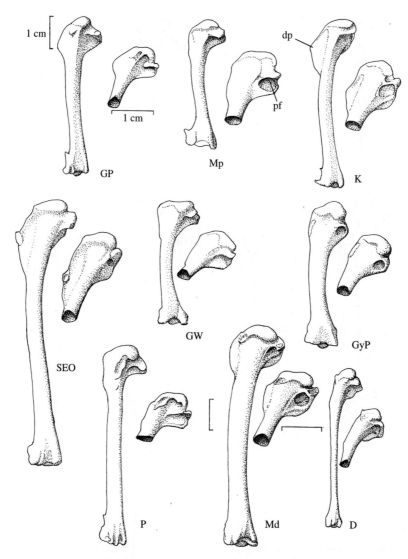

Fig. 1.3 Bird humeri, dorsal surface, to show variation, with oblique view of head, slightly enlarged. Selected to be of similar size, around 200 g (but owl and duck larger, about 300 g). GP Golden Plover; Mp Magpie; K Kestrel; SEO Short-eared Owl; GW Green Woodpecker; GyP Grey Partridge; P Puffin; Md Mandarin; D Dabchick. The deltopectoral crest (dp) is particularly prominent in Falconiformes (K); the shaft is bowed in Galliformes (GyP), Strigiformes (SEO) and Falconiformes (K) but rather straight in Charadriiformes (GP, P). The pneumatic fossa (pf) is particularly well developed, sometimes double, in Passeriformes (Mp).

each wing, compared with the 14 that we have in each hand; they support the remaining wing tip feathers, but are tiny bones, no use for identification purposes.

The main bones of the hind limb are equally distinctive and diagnostic. The pelvic girdle, the hip bone, is a wide thin bone, fused to the vertebrae dorsally but wide open ventrally,

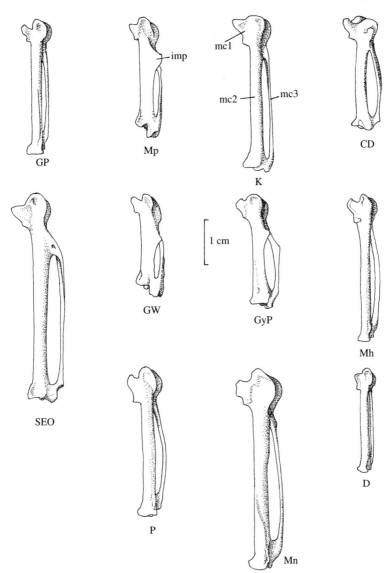

Fig. 1.4 Bird carpometacarpi, to show variation. Species as Fig. 1.3, with addition of CD Collared Dove; Mh Moorhen. The bone results from fusion of metacarpals 1 (mc1), 2 (mc2, the stoutest) and 3 (mc3) with a carpal bone at the proximal (upper) end. An intermetacarpal process (imp) is present in Galliformes (GyP), and in Passeriformes (Mp) and Piciformes (GW) it is fused to mc3. A strongly bowed mc3 and consequently wider intermetacarpal gap is characteristic of some orders.

quite unlike the equivalent bone in mammals. One argument, though surely only part of the truth, is that it is open ventrally because birds lay such large eggs, relative to their body size, and those eggs have to pass out between the arms of the pelvic girdle. It is, however, a very thin sheet of bone, so rather fragile, and not much used in practical identification, except that the bowl (acetabulum) that forms the hip joint is quite robust, and has some diagnostic value. The femur, the thigh bone, is perhaps the bird bone most like its mammal equivalent. However, it is very short, has a rather cylindrical head (more globular in mammals), and lacks the wide groove for the knee cap at the knee end – birds do not have a separate knee cap, though there is a narrow groove for the tendon from the equivalent muscle. The tibia, or shin bone (strictly, the tibiotarsus) is a very elongate bone with an irregular triangular-shaped proximal (knee) end and a sharply keeled pulley-shaped distal (ankle) end. Birds do not have separate ankle bones (they are fused to their neighbours), but they have a very distinctive foot bone, the tarsometatarsus, which is the equivalent of the cannon bone in a horse or cow. It represents two or three ankle bones (tarsals) fused to three elongated foot bones (metatarsals) also fused together; the three separate pulleys at the distal end, for the toe bones, show its derivation. Birds run (or hop) on their toes, and what is commonly referred to as their 'knee' is in fact their ankle (we never see their true knee, which is enclosed in the muscles and feathers of the body). The toe bones are rarely of much use for practical identification, but while most birds have one short toe pointing back and three longer ones pointing forward, some have lost the hind toe, others have two toes forward and two back, yet others can move one of the toes forward or back. These differences affect the shape of the distal end of the tarsometatarsus, giving it added diagnostic value (Figure 1.5).

Passerines have the most distinctive tarsometatarsi, as a group: the three condyles at the distal end are small, evenly sized and evenly spaced alongside each other. In most birds, the condyles for the side toes are placed higher (more proximally) than the central one, though in raptors and owls the very large condyle for the inner (second) toe is aligned with the middle (third) toe, and the outer (fourth) toe has a smaller, more proximal, condyle; they have a facet for the hind (first) toe, which is strongly marked, because it is, of course, a large toe forming an important part of the grasping mechanism. Seen end on, the condyles form almost a semi-circle, with the outer and inner toes almost facing each other, as part of their prey-grasping mechanism. Passerines too show a distinct facet for the hind toe, an essential part of their perching mechanism. Waders overlap in size with passerines, and are equally well represented in many archaeological sites. Their tarsometatarsi tend, of course, to be long for their size, and have distinctively large, projecting, middle condyles. The condyle for the inner toe is displaced backwards relative to the other toes, and the front face of the bone is concave, grooved, for much of its length. Their relatives the auks and gulls have rather similar, though shorter, stouter bones, those of auks being widened, and flat-fronted, as part of their swimming function. Game birds by contrast have short, sturdy tarsometatarsi, with strong condyles for all three main toes; the condyle for the inner toe is distinctively bilobed in side view. Ducks and geese also have very short broad tarsometatarsi, but they are stout, rather flat, and have a distinctive hypotarsus carrying the tendons across the ankle joint (Figure 1.5).

Skulls, especially the beaks, are of course very diagnostic, just as are the skulls and teeth of mammals, because many groups are distinguished by their diet. Bird skulls are, however, so fragile compared with mammalian skulls, that they have only a limited practical value to the archaeologist. Perhaps paradoxically, palaeontologists, looking at much older specimens, are more often able to use them, for some deposits contain beautifully preserved complete skeletons, whereas the skeletons in archaeological sites are usually isolated bones – or fragments of them.

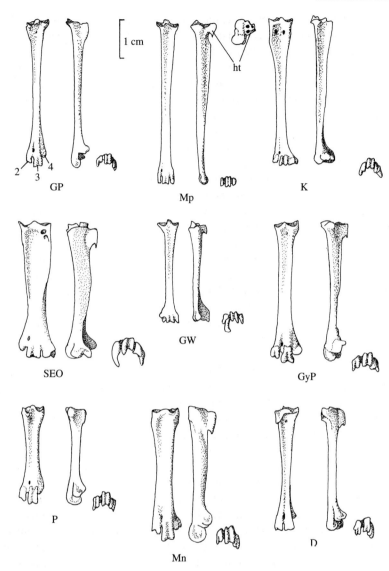

Fig. 1.5 Various types of tarsometatarsi associated with different types of feet. Species as Fig. 1.3. Right tarsometatarsi in dorsal (anterior), lateral and distal views. The three pulleys (trochleae) for the three main toes (2, 3, 4) are characteristically of even length and alignment in Passeriformes (Mp), and the well-developed hypotarsus (ht) is penetrated by 4–6 tendons. Note the strong base for the 1st toe in Piciformes (GW), and the curved appearance of the trochleae in distal view in Falconiformes (K) and Strigiformes (SEO).

Among archaeological specimens, the two eagles have bills of very different shapes, much deeper in White-tailed than Golden, and Raven skulls, for instance, sometimes survive in archaeological sites.

In summary, the humerus, metacarpus, tibiotarsus, and tarsometatarsus are all quite robust bones, with a variety of anatomical features that are diagnostic to the ordinal level, at least. Other bones can be identified, but are either more fragile, so less likely to be preserved in archaeological sites (remembering that they have to survive not only initial burial but also archaeological excavation), or are too similar between bird groups to be useful.

Problems of identification

Given an example of one of the more distinctive bones, reasonably well preserved, from an archaeological site, how easy is it to identify it to species? Species differ particularly in size, though of course the distinctions can be obscured by individual and, in many species, by sexual variations. Groups of species – genera and families – differ also in minor morphological details. Orders differ very substantially in morphology, as is well illustrated by Cohen (1986) and by Gilbert *et al.* (1996). Thus identification tends to be a matter of assigning bones to Order, on morphology, and then using a combination of assessing size and checking the detailed morphological features to get close to a species identification. Fortunately, many of the important or interesting species from archaeological sites are either taxonomically isolated in Europe (e.g. Gannet, Crane), or combine distinctive morphology and size (e.g. Raven, Great Auk). In groups with few species which are very different in size (for instance, Cormorant, Shag, and Pygmy Cormorant), an identification of a decent bone, say a humerus or metatarsal, can be firmly made. The auks make a similar graded size series, from Great Auk, Guillemot, Razorbill, Puffin, Black Guillemot to Little Auk, with little overlap (Figure 1.6). In other cases, like the ducks, the morphological distinctions of at least some bones, like the metatarsals, allow ready separation of diving ducks (*Aythya* and *Bucephala*) from dabbling ducks (*Anas*), but the species within each of these groups are so similar that a firm identification is less likely. The Mallard is appreciably larger than other species of *Anas*, but close to Pintail, which overlaps Wigeon, then Gadwall and Shoveler; distinguishing these is difficult, though there are small morphological differences in some bones (Woelfle, 1967). Teal is distinctively smaller than all of them, but barely distinguishable from Garganey. The enormous variability added to this mix by the various breeds of domestic duck, all descendants of Mallard, merely adds to the confusion. Geese present a similar very common problem area. Pink-feet are distinctively smaller than Greylag, but the intermediate sized Bean Geese overlap both of them, and the slightly smaller White-fronted Goose overlaps extensively the Pink-foot in size. The fact that female geese are somewhat smaller than their mates adds to the variation, and therefore confusion. In a few cases, DNA extracted from the bones has been used to confirm their identity (Dobney *et al.*, 2007). The Whooper and Mute Swan also overlap extensively in size, but are usually morphologically distinguishable. For instance, the more terrestrial feeding activity of Whooper Swans is reflected in a broader distal end to the metatarsal bone, which is somewhat longer but more slender, on average, and the sternum, in particular, is readily distinguished by the cavity for the extended trachea, reflecting their ability to make trumpeting calls (Figure 1.7).

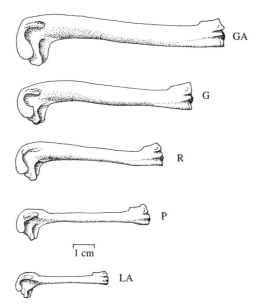

Fig. 1.6 Range of auk humeri, to show how sizes differ between relatives. GA Great Auk; G Guillemot; R Razorbill; P Puffin; LA Little Auk.

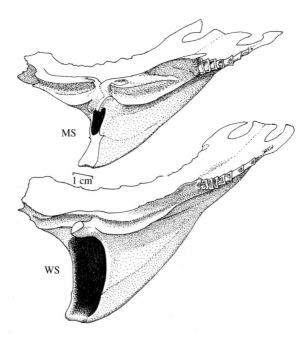

Fig. 1.7 Mute (MS) and Whooper (WS) Swan sterna, in oblique anterior view, to show the excavated keel that houses the enlarged trachea (associated with its trumpeting call) in the Whooper Swan.

The archaeological record of birds is dominated by the bones of domestic species, ducks, geese and, especially, Domestic Fowl. Domestic Fowl are 'gamebirds' – both a legal and a taxonomic term – i.e. members of the order Galliformes. Close relatives (grouped in the family Phasianidae) include the Peacock, Pheasant, and Grey Partridge, also Guineafowl and Turkey. Not quite so close are their relatives in the grouse family Tetraonidae, including Red Grouse, Ptarmigan, Black Grouse, and Capercaillie, as well as the Hazel Hen in Europe. All of these are important food species for humans and other predators, with robust bones that occur regularly. They are important also archaeologically and historically. Many of the Phasianidae have been introduced to the British Isles, therefore should give evidence for or comply with dating. The Tetraonidae show important climatic, ecological, and geographical replacement, from northernmost Ptarmigan, through dwarf shrub/scrub Red Grouse, woodland edge Black Grouse, conifer forest Capercaillie, and deciduous woodland Hazel Hen, and therefore give evidence of climatic and ecological changes during postglacial times. This is evidently an important group from which to derive historical data and, as a group, well represented archaeologically. How easy is it to identify the various species? Fortunately, there are now a number of manuals that help, though they are not readily available. Domestic Fowl and Pheasant are quite close in size and morphology, so distinguishing their bones has attracted attention over many years (Lowe, 1933; Erbersdobler, 1968). Even when their sizes match (and breeds of Fowl are so variable that size is not a very reliable character), there are morphological characters that allow most major bones to be discriminated, so long as they are reasonably complete. The sternum, for instance, a major bone because it carries so much meat, has a distinctively shaped rostrum or anterior spine, differently shaped precostal processes, and numerous other differences of shape (Figure 1.8). The grouse are even more different in shape, though Black Grouse and Capercaillie are rather similar to each other. The cock Capercaillie is much bigger, though the hen is nearer to Black Cock in size – and occasionally, the two hybridize, just to add to the confusion. Capercaillie are more likely to be confused, on size, with Peacock or Turkey, but these two also differ in shape from it, and from each other. The smaller game birds are harder to differentiate reliably. For instance, Ptarmigan are closely related to Red Grouse, therefore morphologically similar, and although they are a little smaller, they do overlap in size. Though their metatarsi can be separated – Red Grouse are bigger – their humeri overlap in size, so that a large one will be Red Grouse, a small one Ptarmigan, but some in the middle of the size range will not be distinguishable. The partridges *Perdix* and (in southern Europe) *Alectoris* are close enough in size to these grouse to need careful scrutiny, though differences in shape do separate the families more readily. The Hazel Hen also falls into this size range, and is particularly close in size to Grey Partridge, though it differs in detailed shape (Kraft, 1972).

Among other species of particular interest for the study of British birds, Golden Eagle and White-tailed Eagle are not particularly closely related, so that in addition to size differences (White-tailed has much longer wings, and therefore wing bones), most bones can be distinguished morphologically. The metacarpal carries a spiral groove in the Golden Eagle, whereas the equivalent is straight in the White-tailed Eagle, the articular surface for the fourth (outer) toe is flat in the Golden but rounded and extends further distally in the White-tailed Eagle, and the coracoid has a much wider anteroventral corner in the White-tailed Eagle. One detailed difference is remarkable – some of the toe bones of the White-tailed Eagle are fused, a very distinctive feature (Figure 1.9). The various falcons form a graded

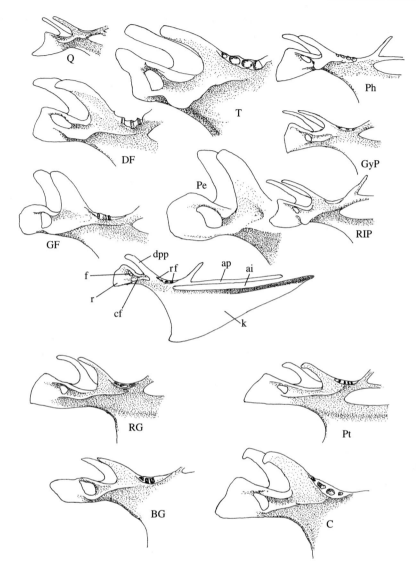

Fig. 1.8 Galliform sterna, to show diagnostic differences between species. Relevant anatomical features are labelled on the small drawing of the Red Grouse sternum in the centre: ai abdominal incision; ap abdominal process; cf coracoid facet; dpp dorsal precostal process; f foramen; k keel; r rostrum (manubrium); rf rib facets. The foramen in the rostrum is a characteristic galliform character. The enlarged oblique views of the rostrum and dorsal precostal processes highlight their different shapes and lengths; in Phasianidae (above) the processes are usually longer relative to the rostrum than in Tetraonidae (below), but are more upright in Guinea Fowl and Peacock. Q Quail; DF Domestic Fowl; GF Guineafowl; T Turkey; Pe Peacock; Ph Pheasant; GyP Grey Partridge; RlP Red-legged Partridge. RG Red Grouse; BG Black Grouse; Pt Ptarmigan; C Capercaillie.

14 | The bird in the hand...

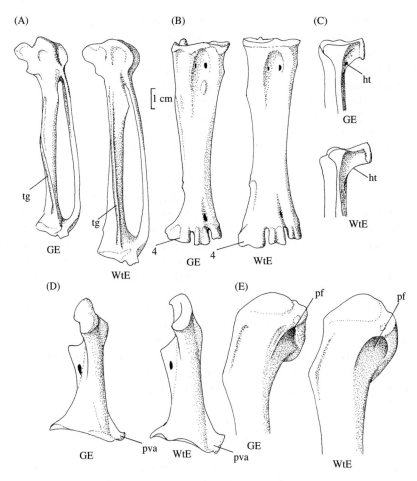

Fig. 1.9 Eagle bones compared. A carpometacarpus: Golden Eagle (GE) has a spiral tendon groove (tg) on metacarpal 3 which is relatively straight in White-tailed Eagle (WtE). B, C, tarsometatarsus: more slender, curved laterally, with a smaller trochlea 4 and (C) a rhomboidal, but smaller, hypotarsus (ht), with a foramen in it, in Golden Eagle. D coracoid with a much smaller postero-ventral angle (pva) in Golden Eagle. E humerus: has a much deeper but narrower pneumatic fossa (pf) in Golden Eagle.

series, from the large Gyr, through Peregrine to Kestrel and Hobby and the small Merlin, which can be separated on size reasonably well, though Hobby and Kestrel overlap. The possible importation of exotic species for falconry could cause further confusion. The broad-winged Accipitridae are more difficult, because of extensive overlap in size between, for instance, kites and Hen Harriers, but morphological as well as size differences are demonstrated by Otto (1981) for the fore skeleton and Schmidt-Burger (1982) for the hind limb bones.

Waders form an interesting though minor (numerically) group of food species, of which Woodcock is much the most frequent. Because of its distinctive size, it can usually be identified readily. By contrast, distinguishing Golden Plovers from Grey Plovers is barely possible, though the former is much more abundant, more widespread (especially in winter), slightly smaller, and was subject to specific hunting techniques, so is surely the plover present in most sites. The Lapwing, which seems to be a similar sized bird, is actually appreciably larger, in most limb bones, and can be recognized quite readily.

The most difficult group, inevitably, is the passerines. Identifying them as passerines is fairly straightforward. The humerus, for example, has a complex arrangement of depressions, fossae, at its proximal end on the dorsal side, which among other things hold the wing-folding muscles. The shaft is straight (not bowed as in, for instance, similar-sized small waders), and the complex of condyles distally is also distinctive. The metatarsus ends in three very evenly sized trochleae, whereas in many birds the middle trochlea is larger and extends well beyond the two side-toes. The difficulty is in distinguishing passerines from each other. Even the Corvidae, which are so much larger than most passerines, pose problems in that Rook and Crow overlap substantially in size, as do Jackdaw and Magpie, though the detailed guide by Tomek & Bocheński (2000) allows better discrimination, on morphological characters and bone proportions. At least the Raven has a very distinctive size and morphology, as does Jay at the opposite end of the size range (so long as Nutcracker can be plausibly ignored; Azure-winged Magpie and Siberian Jay are much smaller). Among the smaller passerines, distinguishing, say Song Thrush from Redwing, or Blackbird from Fieldfare or Ring Ouzel is likely to be impossible, or only possible with the best of specimens and a good comparative collection or some additional skill, such as DNA typing. This would rarely be available, or worthwhile, in an archaeological context. It is though remarkable how quite subtle size differences can sometimes be detected. Most birdwatchers would not expect to tell Meadow Pipit from Tree Pipit on size, yet the Tree Pipit humerus is perceptibly larger. However, even such an osteologically distinctive species as Swallow overlaps substantially in size with House Martin.

In summary, distinguishing bones of passerines or waders to family level is usually possible, genera are usually separable, but specific identification is very tricky; reciprocally, identifications offered in bone reports should be treated circumspectly (Figure 1.10).

One point inevitably emerges from all these discussions: a reference collection is essential. Even a partial collection helps considerably to ensure at least the correct assignment of a specimen to its order. Hence anyone actively involved in identifying bird bones, whether from archaeological sites or owl pellets, soon finds themselves scavenging corpses from roads and beaches, asking zoos, bird hospitals, or vets for dead specimens, and hiding decaying specimens in pots to recover bones later. If a reasonably representative collection can be assembled, it is often possible to eliminate obviously wrong answers and get close to a right one. More likely identifications can then be checked at a national museum collection, perhaps at Tring (where the British Museum (Natural History) collection is housed), at the Royal Scottish Museum, Edinburgh, or in Cardiff, Dublin, or Belfast. Some local museums (e.g. Sheffield) and university departments (e.g. Archaeology, Southampton) also have good collections.

A couple of examples will illuminate this discussion. Back in the 1970s, an excavation at Abingdon yielded a collection of bird bones that were sent to Don Bramwell, then one of the few people with experience of identifying bird bones in Great Britain. It included two smallish metacarpi with a very distinctive anatomy – that is, they were potentially identifiable – but which fitted nothing in his quite extensive reference collection. Persevering, he realized

16 | The bird in the hand...

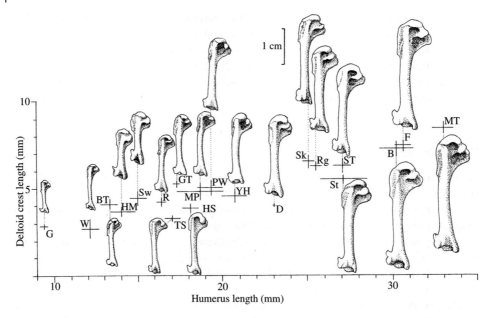

Fig. 1.10 Size range of a selection of passerine humeri, to illustrate the difficulties of separating related species. The crosses indicate one standard deviation of the sizes in modest samples (mostly 5–10) of reference skeletons (except only 1 Dipper (D)). Passerines readily divide into a smaller group (warblers, chats, finches, etc) of 5–25 g body mass, and a larger group (thrushes, larks, starlings) of 50–110 g, with few species (e.g. Dipper) in between. Within these groups, hirundines (House Martin HM, Swallow Sw) have short, stout humeri, but overlap. Seed-eaters (Tree Sparrow TS, House Sparrow HS, Yellowhammer YH) have stouter humeri than insectivores (Great Tit GT, Meadow Pipit MP, Pied Wagtail, PW). In the larger group, Skylark (Sk) and Redwing (Rg) overlap in size, but the pneumatic fossae are very shallow in larks, deeper in thrushes. Song Thrush (ST) and Starling (St) are similar lengths but differ in morphology. Note the complete overlap of Blackbird (B) and Fieldfare (F) (and Ring Ouzel). Mistle Thrush is distinctively larger. At the opposite end, Goldcrest (G) is indistinguishable from Firecrest, and Blue Tit (BT) is likely to overlap other small tits. Wren (W) overlaps Willow Warbler and Chiffchaff (after Dissaranayake 1992).

that they were very similar to Cormorant and Shag, but much smaller, and he suggested that they would prove to belong to Pygmy Cormorant; they were sent to Tring, and so it proved (Bramwell & Wilson, 1979; Cowles, 1981; see p. 92). A more personal example actually involved an owl pellet analysis, of some Tawny Owl pellets from near Macclesfield. One of them included a large intact tarsometatarsus, much larger than the usual thrushes or Starling, which was evidently from a bird weighing about 200 g. What birds of that size might a Tawny Owl eat? One of the corvids, perhaps Magpie or Jackdaw, seemed most likely, but it was clearly not a passerine bone. A Black-headed Gull, Lapwing, or Golden Plover? No, it didn't match any of them either. The site in question was a wet marshy one, and Water Rail (too small) or Moorhen (too big) were considered, and on morphology not a Rail anyway. The problem was left on one side for some weeks, but then a thought occurred. A Kestrel weighs about 200 g, and sure enough the morphology when checked was exactly right.

Throughout the compilation of records that underlies this book, we have, for the most part, had to accept the identifications offered by the original identifiers. A few collections have been reviewed, most notably by the late Colin Harrison (see especially Harrison, 1980a, 1987a), but many are lost, or at least untraceable, and in any case there are far too many for all to be re-examined. Some groups would benefit from restudy, and there are some good PhD research topics suggested by our account. If we can present our review as a working hypothesis of what we think is known, and stimulate others to challenge it by further study, we will have succeeded in a major objective.

Problems of dating

Dating can be absolute or relative, direct or indirect. Absolute dates (years in the historical record) can come from documents and artefacts, or from the specimens themselves. Annual rings in trees and layers in lake deposits give absolute dates, and they are direct dates of the layers in question. Bones lying in a lake deposit might be dated indirectly from the layers in which they lay, though it would be a remarkable event to be able to do so. The best known way of obtaining direct dates of organic materials, including bones, is radiocarbon dating. Plants incorporate a small amount of radioactive carbon (^{14}C) in the carbon dioxide they use to synthesize material, and animals eat these plants. The minute amount of radioactive carbon decays with time, such that the amount left after 5,560 years is halved. The rate of radioactive decay is not affected by temperature, pressure, chemical, or biological changes, so the rate of decay gives a direct measure of the time that has passed since the plant absorbed the carbon dioxide. Because the amount of radioactive carbon is so small, it effectively vanishes after about 40,000 years and the technique cannot be used on older material, but that is quite long enough to estimate time for the archaeological period, the last 15,000 years or so, that interests us here. There is a further complication, that the dates provided do not exactly match calendar years as one goes back in time; at the end of the Last Glaciation, about 10,200 years ago according to the radiocarbon clock, careful analysis of tree rings and other sources of direct dates suggests that the correct date was more like 11,700 years ago. Radiocarbon years are usually quoted as 'years b.p.' (before present), while absolute calendar dates, usually termed calibrated dates, are quoted as 'years BP' or even 'years BC' (for which, take off 1950 years, because radiocarbon dating works from a baseline of AD 1950). We have quoted the radiocarbon dates, as presented by the original accounts, throughout this account.

There is a further problem with radiocarbon dating, that it is expensive (currently, about £100 per date), so more frequently relative dates are used. If bones occur in obvious archaeological contexts, a twelfth century castle, perhaps, or a Bronze Age barrow, it is often sufficient to assign them that appropriate cultural date. In practice, most bones are dated this way. There are some obvious pitfalls to doing this. Bones might have been dropped into a ditch that was dug into earlier layers, conferring on them apparent dates that are much too early. Some deposits, particularly loose scree in cave sites, are rather 'porous', so that bones work their way into earlier layers, or are carried there by burrowing animals such as Badgers and Foxes. However, for the great bulk of bird bones from conventional archaeological sites, relative dating works well. For earlier archaeological periods, the time spans are greater, because cultures changed more slowly (lasted a longer time) (Table 1.1), but the more recent times, with such very datable artefacts as coins and jewellery, even direct documentary evidence, give more precise dates.

Table 1.1 Geological and archaeological periods.

GEOLOGICAL PERIODS	AGE Ma	
PLEISTOCENE	2	
PLIOCENE	5	
MIOCENE	23	
OLIGOCENE	38	
EOCENE	54	*Lithornis*
PALAEOCENE	65	
CRETACEOUS	135	*Hesperornis* *Enaliornis* *Ambiortus* *Sinornis*
JURASSIC	194	*Archaeopteryx*

PERIOD	AGE ka	SITES
FLANDRIAN	10	Star Carr, Thatcham
LATE DEVENSIAN	40	Pinhole Cave
MIDDLE DEVENSIAN		Kent's Cavern
EARLY DEVENSIAN	120	Tornewton Cave
IPSWICHIAN	130	Tornewton Cave
WOLSTONIAN	186	Tornewton Cave
PRE-IPSWICHIAN	245	
?		
PRE-IPSWICHIAN		
?		
HOXNIAN	400	Swanscombe
ANGLIAN	450	
?		Boxgrove, Westbury
CROMERIAN	500	West Runton
?		
?		
PASTONIAN	1800	

The geological timescale used to date fossil birds is reasonably familiar, at least in general. Birds evolved from small bipedal dinosaurs in the Jurassic period, 194 to 135 million years ago (Ma). The earliest certain bird, *Archaeopteryx*, dates to the Upper Jurassic, about 150 Ma (Chapter 2). Birds from the succeeding Cretaceous, 135 to 65 Ma, are known from Spain, China, Mongolia, and the USA, but are scarce in Britain. Nor are there Palaeocene (65 to 55 Ma) fossil birds from Britain. However, in the Eocene London Clays, about 54–47 Ma, a substantial avifauna of some 55 or more species has been found (Feduccia, 1996). There is then another gap in the British fossil bird fauna, through the Oligocene, Miocene, and most of the Pliocene, until a few specimens turn up at the top of the Pliocene, about

Table 1.1 (*Continued*)

ARCHAEOLOGICAL PERIODS	POLLEN ZONE NAMES	AGE years b.p.	SITES
POST-MEDIAEVAL	SUB-ATLANTIC		
MEDIAEVAL			
NORMAN		1000	Stafford Castle
SAXON			West Stow, Hamwic
ROMAN		2000	Colchester, Barnsley Park, Wroxeter
IRON AGE		2700	Glastonbury, Meare
BRONZE AGE		3500	Burwell Fen
NEOLITHIC	SUB-BOREAL	5500	Knap of Howar, Isbister, Quanterness Dowel Cave
MESOLITHIC	ATLANTIC	7000	Port Eynon Cave
	BOREAL	9000	
	PRE-BOREAL	10000	Star Carr, Thatcham
LATE PALAEOLITHIC	YOUNGER DRYAS	11000	Chelm's Combe, Ossom's Cave
	WINDERMERE	14000	Robin Hoods Cave, Goughs Cave
	OLDER DRYAS	15000	

2 Ma. The Pleistocene, the period of the ice-ages, covers the last 1.8 Ma. Because of the frequent cycles of glacial and interglacial times, a strict chronology is hard to apply in our latitudes; succeeding ice sheets wiped out the traces of earlier ones, while deposits tend to be confined to individual sites, and hard to correlate across the country, let alone to elsewhere in the world. Deep sea cores, which retain a complete record, suggest as many as nine glacial and nine interglacial periods (Shackleton, 1977; Shackleton *et al.*, 1991) but it is hard to recognize as many as four of each in Britain. A simplified system of Anglian, Wolstonian, and Devensian (Last) Glaciations, but Cromerian, Hoxnian, pre-Ipswichian, and Ipswichian (Last) Interglacials separating them, plus the mild Flandrian or Postglacial period in which

we are living, gives a loose template against which to present the accumulating knowledge of our early bird faunas (cf. Stuart, 1982; Yalden, 1999).

In Britain, and for this book, the last 15,000 years form the period of most interest, because the maximum spreading of the ice sheet in the Last (Devensian) Glaciation at about 20–18,000 years ago (20–18 ka) wiped out most biological activity in this country. Our present fauna and flora has arrived since then (Chapter 3). Initially, as the ice retreated about 15,000 years ago, in the Late Glacial period, a flora of open-ground species, wormwood, grasses, sedges and herbs, was able to colonize. By about 12,000 b.p., birch scrub covered much of southern Britain. Human hunters, of the Upper Palaeolithic (Old Stone Age) culture, spread into what is now Great Britain, leaving their food remains and stone tools in caves in places such as the Gower Peninsula in South Wales, the Mendip Hills in Somerset, and Creswell Crags on the Derbyshire/Nottinghamshire border. However, the climate then deteriorated again for a short period. Ice caps formed on the Scottish mountains again, and spread as far as Loch Lomond, so geologists call this period the Loch Lomond Readvance; it is better known by the archaeologists' term, the Younger Dryas (because Mountain Avens *Dryas octopetalla* is a plant commonly preserved in sites of this age). At about 10,200 b.p. (probably about 11,700 BP), the climate suddenly improved, to herald the warm Postglacial period, also known as the Flandrian or Holocene, in which we are fortunate to find ourselves. This climate change was very rapid: about an 8°C rise in mean summer temperatures in 50 years or less. It took the forest vegetation some 2,000 years to spread back into Britain, but animals reacted much more quickly. Beetles provide the best documentation of this, but what information we have of birds, mammals, and indeed humans matches the evidence from insects. The humans who returned were still hunters, using stone tools, but of a new culture, the Mesolithic (Middle Stone Age). Their encampment at Star Carr near Scarborough is one of the first post-glacial sites that is both well dated and informative about the birds and mammals then living in Britain. As woodland spread back across the landscape, lowland Britain may have become too thickly wooded to provide easy hunting conditions, though archaeologists and ecologists are still arguing about this. There must have been clearings along river valleys and the coast, perhaps more widely. Either way, the fine flint arrowheads and tiny flakes that Mesolithic people used to barb spears are frequently found in the uplands, in the Pennines, for example, and they may have hunted deer and Aurochsen (wild cattle) in the more open glades and woodland edges that surrounded the less tree-covered uplands. They clearly used coastal sites to gather fish and molluscs, as well as birds and seals, for instance on Oronsay. Their only domestic animal was the dog, already domesticated from the Wolf.

As the ice caps had melted in the post-glacial period, so sea level correspondingly rose to drown the Doggerland that formerly extended across to Germany and Denmark. Probably this happened by around 8,000 b.p., drowning much coastal foraging habitat and many Mesolithic sites in the process. About 5,500 b.p., however, the New Stone Age (Neolithic) culture spread into what were by then the British Isles. This culture originated in the Middle East, about 9,000 years ago, and spread more quickly westwards through the Mediterranean areas of southern Europe than northwards. However, it certainly reached the Atlantic and North Sea coasts by around 6,000 b.p.. We do not know much about the ships used by these people, but they evidently were competent sailors, carrying not only themselves but their domesticated livestock, sheep, goats, cattle, and pigs, as well as cereals and other plants to sustain an agricultural existence. They arrived in both Ireland and Great Britain about 5,800 b.p. (4,600 BC), and the Mesolithic way of life died out very suddenly. Both the mammal remains at early Neolithic sites (Yalden, 1999) and analysis of the carbon isotopes in human bones (which indicate the difference between terrestrial and marine diets) show that these ancient Britons quickly gave up their hunter-gatherer existence and exploited instead the new crops and livestock (Richards *et al.*, 2003). These new farmers slowly cleared some of the forest, so providing open habitats for farmland birds, creating both open pasture, especially on the downlands of southern Britain, and cereal fields. By 4,500 b.p., they were creating large monuments such as Stonehenge in essentially open countryside, more appropriate for Skylarks than Chaffinches. Their tools though were still made of bone, antler, and especially flint, as mined for example at Grime's Graves, in Norfolk, using antler picks. Metal tools, initially copper and then bronze, were added to their armoury around 4,100 b.p, (about 2,500 BC) and then iron tools appeared about 2,700 b.p. (i.e. about 880 BC). The Celtic peoples, the Ancient Britons, using these iron tools were invaded by Roman peoples temporarily in 55 and 54 BC, under Julius Caesar, and then more permanently under Claudius in AD 43. The Romans in turn retreated as their capital was threatened around AD 410, leaving a Romano-British culture threatened, then displaced in England, at least, by Anglo-Saxon invasions from northern Germany and Denmark. The Anglo-Saxon society that emerged from the Dark Ages was itself threatened by Viking invasions in the period AD 800–1000, before being subsumed by those Vikings who had settled in Normandy, the Normans, after AD 1066. The Mediaeval period, covering the thirteenth to sixteenth centuries, and the Post-mediaeval seventeenth to twentieth centuries, complete the cultural sequence. In Ireland, which the Anglo-Saxons never settled, the Christian Celtic cultures survived through, despite Viking invasions and settlement, while in Scotland the interactions of Anglo-Saxons (in the south), Picts, Scots (Celts invading from Ireland), and Vikings (especially in the islands) produce a more complex chronology than in England. Never-the-less, for the purposes of this account of the bird life of these islands, the succession of cultural periods, Upper Palaeolithic, Mesolithic, Neolithic, Bronze Age, Iron Age, Roman, Anglo-Norman, Mediaeval, and Post-mediaeval, provides us with the broad timescale we use here to describe and evaluate the changes in bird faunas. Most archaeological sites, and the bird remains contained in them, can be allocated at least to these broad periods.

Sources of bones

Bones are poorly preserved in acid sands or peats, better preserved in limestone caves or the silts of flood plains. Some early bird specimens come from maritime clays, and some of the earliest Pleistocene archaeological sites (Boxgrove, Swanscombe) are in coastal or riverine gravels. Most of the Late Glacial sites are caves in limestones, particularly in Carboniferous Limestones of the Mendip Hills of Somerset, the equivalent outcrops of Devon and South Wales, including the Gower Peninsula, and in the Peak District, shared between Derbyshire and Staffordshire. Permian (Magnesian) Limestones at Creswell Crags on the Derbyshire/Nottinghamshire border have also yielded important evidence. Post-glacial history is more usually represented at conventional archaeological sites, such as the Mesolithic camp site at Star Carr, the Iron Age village at Glastonbury, and the famous Irish eighth century site of

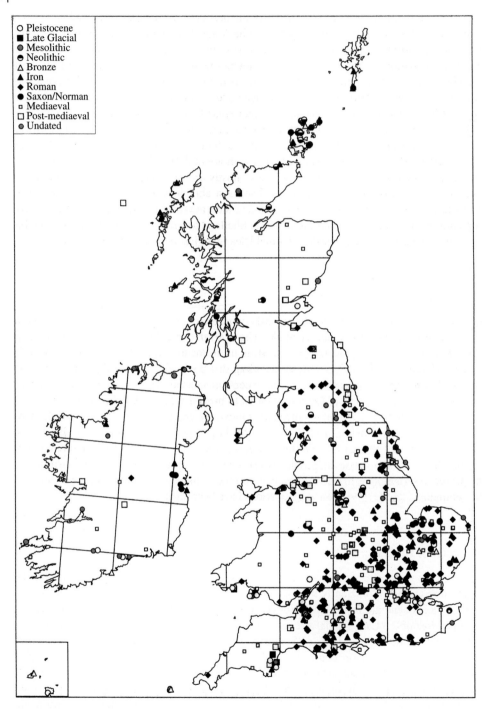

Fig. 1.11 Map of archaeological sites yielding bird bones: heavily clustered in England, thinly sampled in Ireland, Man, Scotland and Wales, but note the strong sample of Orkney sites. Older (Pleistocene/Late Glacial) sites are mostly cave sites, so clustered in limestone areas.

Lagore. With the Roman settlement, many of the best faunas come from excavations of conventional castles, villas, and other buildings. The Anglo-Saxons seem to have abandoned the cities that the Romans built, living at least initially in small farmsteads such as West Stow. As their population increased, they too developed towns, and the excavations of Ludenwic (London) and Hamwic (Southampton), not to mention the important series of excavations at Eoforwic/Jorvik (York), have given us much useful information about bird life of those times. With the Norman invasion came another episode of castle-building, and excavations of, for example, Launceston, Stafford, and Wakefield Castles have also provided extensive bird faunas. In analysing these faunas, we have a data base of over 9000 records (8,953 as of 17 March 2004, when we started writing, with about 200 added since). These identify a species at a site and age/layer, from 740 sites, mostly archaeological sites but including the Pleistocene sites from gravels and caves. The most abundant record comes from England (594 sites), because most archaeological and cave sites are there, but Ireland (27), Man (four), Scotland (80, including 19 on Orkney, two on Shetland and nine in the Hebrides) and Wales (28) are also represented, as are the Channel Isles (with just four sites) (Figure 1.11).

Conclusions

Most larger bones of larger species of birds can be recognized reliably, though access to a good reference collection is invaluable. The relevant manuals are also important aids. Species in more diverse groups are more difficult to identify reliably; conversely, identifications offered in published literature, including this book, need to be accepted with some caution. For the most part, we have had to accept the identifications offered by the original describers, there being simply too many for us to have checked them all. Dating is usually achieved by reference to the archaeological context, which in turn requires that the excavation was carefully conducted. In many cases, only larger bones were extracted and identified. On the one hand this is convenient, as the larger species are also more readily identified, but on the other, the result is a double bias against the record of the smaller species, particularly the passerines: these are difficult to identify, and their remains are only reliably recovered if the sediments from archaeological excavations have been sieved.

2
The early history of birds in Britain and Europe

Archaeopteryx

Despite all the fossil collecting that has gone on since *Archaeopteryx lithographica* was first found, in 1861, it is, still, a very clear candidate for the title 'the earliest bird'. This famous example of a non-missing link, from the Sölnhofen area of southern Germany, is now known from nine specimens, all dated to the Upper Jurassic period about 150 Ma (million years ago). The first skeleton to be described, the one in the Natural History Museum, London, is the type specimen, and shows much of the plumage but has a broken skull and a partially scattered skeleton. The Berlin specimen, the second one (1877), is a more complete skeleton, albeit somewhat crushed, and has an even better preserved plumage. The little fifth skeleton, known as the Eichstatt specimen (the town nearest to where it was found, and where it is now housed) barely shows any feathers, but has a better preserved skull. The fine-grained limestones in which *Archaeopteryx* was preserved show the feathers of the wings, including the clearly distinct asymmetrical primary feathers of the hand and the more symmetrical secondary feathers of the arm, much as in modern birds. However, the tail, an elongate dinosaur-like organ, also carries feathers, arranged in pairs down its length and quite unlike the shortened fan of modern birds. Also unlike modern birds are the claws on the three fingers, the teeth in the beak, and the much less specialized skeleton (free metacarpals in the hand, simple ribs, a short dinosaur-like coracoid). It has been suggested that *Archaeopteryx* would have been described, from its skeleton, as a small dinosaur were it not that the imprints of the feathers were also preserved. This is perhaps a slight exaggeration. The hind toe, for instance, is turned back to oppose the three longer front toes, allowing it to perch in the way modern birds do. No dinosaur has such an arrangement. And while the skull bears teeth, it has the enlarged brain case and slender jaws of a bird, not the heavy skull of a dinosaur. The pelvic girdle is also very distinctive, with a very odd, two-pronged, ischium, not much like that of modern birds but not exactly like the equivalent dinosaur bones either (Figure 2.1).

While its anatomy is well described (Elzanowski, 2002a), much debate surrounds the life style of *Archaeopteryx* and its significance for interpreting both the ancestry of birds and their subsequent evolution. Though its humerus is much longer than that of a Magpie, its wingspan is similar, about 55 cm for the Berlin specimen, and its body length suggests a similar size to a large Brown Rat *Rattus norvegicus*, so it probably weighed about 250–300 g (Yalden, 1984; Elzanowski, 2002a). (The different fossils are themselves different sizes; the bones of the London specimen are about 10% longer than the Berlin example, and it

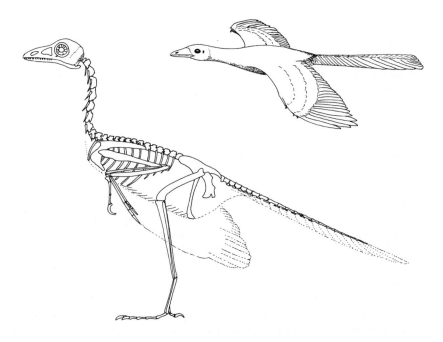

Fig. 2.1 *Archaeopteryx*. Reconstructed skeleton (after Elzanowski 2002, Yalden 1984) and an impression of *Archaeopteryx* in gliding flight. Note teeth, clawed fingers, long bony tail, opposable hind toe.

perhaps weighed 470 g, about the size of a Rook.) Not only are the primary and secondary feathers distinct, but they are asymmetrical, distal vanes narrower than proximal ones, and the feather shafts are curved, characters that only make sense if the wings were used in flapping flight. The asymmetrical vanes mean that the feathers close against each other on the downstroke, and open up on the upstroke, while the curved shafts produce the same effect (Norberg, 1985). However, the details of the rather simple rib cage and pectoral girdle indicate a much less refined muscular and respiratory system than in modern birds, so it is equally sure that flight was neither as prolonged or as manoeuvrable as in modern birds. Probably, it was adequate to allow *Archaeopteryx* to scramble and flap away from predators, in the way that young gamebirds can use their wings to escape predators long before they can fly properly (Elzanowski, 2002b). The claws on its hand are very narrow and sharp, like the claws of woodpeckers, and it could have used them to scramble up tree trunks or rocks (Yalden, 1985). Its hind claws were also quite sharp, though less so than the claws on the hand, and while they could also have been used for climbing, it seems possible that *Archaeopteryx* spent some of its time foraging for insects on the ground or among rocks, as well as in bushes. Its long hind limbs certainly suggest some ground foraging, though it could not run fast enough to take off from the ground without flapping, so probably had to scramble up to a height and then fly, gaining flying speed in the way that many birds and bats do, by dropping away from a branch or small cliff (Elzanowski, 2002b).

Cretaceous birds

No other birds are known, from Europe or anywhere else, in the Jurassic, but a range of Lower Cretaceous birds has now been described, from China, Mongolia, and Spain. Collectively covering about 30 million years, from about 140 to 110 Ma, these birds show a variety of advances in their structure, compared with *Archaeopteryx*. Most of them have tails shortened into a pygostyle, suitable for carrying a fan of feathers. The claws on their hands are reduced to vestiges or lost completely. Their coracoids become taller, robust bones as in modern birds, and the large sternum, with the keel to carry the flight muscles, becomes evident. They lose their teeth, and so acquire the toothless beak that is so characteristic of all modern birds (and nearly all fossil birds, too). The different genera of Lower Cretaceous birds show these various modern characteristics appearing in an irregular or mosaic pattern, as though several different lineages of birds were evolving better flight mechanisms in parallel. For example, *Jeholornis* from north-east China still has a long tail, and a very *Archaeopteryx*-like pelvic girdle, but has a pillar-like coracoid and very few teeth; it seems to have been an early seed-eating bird, given that over 50 ovules of a plant called *Carpolithus* are preserved in its stomach. The contemporaneous *Sinornis*, also from China, and the slightly later *Iberomesornis* from Spain, a tiny bird about the size of a Great Tit, have a pillar-like coracoid and an odd, elongate pygostyle in combination with a rather primitive pelvic girdle, remnant finger claws and teeth (Figure 2.2). Another Chinese bird from the Early Cretaceous, *Confuciusornis*, is more advanced in having a toothless beak, like modern birds, but retains the long clawed fingers of *Archaeopteryx*, and has an elongate pygostyle. All these share the opposable hind toe of perching birds, but another contemporary Chinese bird, *Chaoyungia*, while retaining a toothed beak, has a reduced hind toe, like modern wading birds, and shows the earliest keeled sternum. Its forelimb and pectoral girdle were essentially modern, albeit with a reduced claw on at least the third (longest) finger. The recently described *Hongshanornis* appears to be a very early relative of the modern birds; coming from the Early Cretaceous of Inner Mongolia, it has a toothless beak, remnant claws on its fingers, but long hind legs that suggest a wader-like ecology (Zhou & Zhang, 2005). It is evident that the early Cretaceous birds show a remarkable diversity of advanced and primitive characters, with much parallel evolution towards a modern flight apparatus. *Ambiortus*, from the Early Cretaceous of Mongolia, about 130 Ma, is probably the oldest bird known to have a modern flight skeleton, with fused carpometacarpus, large sternum and keel, and an extended coracoid with the pulley system for the wing-raising muscles. Interestingly, it retains a claw on its third finger (Kurochkin, 1985). Its skull is not known.

There is a scatter of bird fossils throughout the Cretaceous, though not enough to give us a coherent story, yet, of the evolution of modern birds. The earliest fossil bird from Britain is from the earliest period of the Upper Cretaceous, from the Greensand at the base of the Chalk in Cambridgeshire, and about 100 Ma ago. Named *Enaliornis* by Seeley in 1876, it is known from a scatter of bones, possibly from different sites, including three brain cases, part of a pelvic girdle, femora, tibiotarsi, and tarsometatarsi (Galton & Martin, 2002). While insufficient to provide a complete description, they indicate a bird about the size of a pigeon, but, from the hind limb bones, evidently a seabird related to the later, and much better known, *Hesperornis*. *Hesperornis* is one of two famous toothed birds from the Upper Cretaceous, about 80 Ma, the other being *Ichthyornis*. They were described in 1880 from

28 | The early history of birds in Britain and Europe

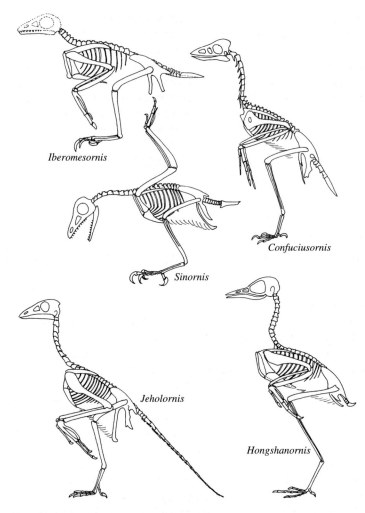

Fig. 2.2 A selection of Lower Cretaceous birds. *Iberomesornis* comes from Spain, *Hongshanornis* from Inner Mongolia, and the others from N.E China. Note the mosaic nature of evolutionary change represented here. *Jeholornis* retains a long tail, *Hongshanornis* has a modern-looking pygostyle, and the other three have an odd-looking elongate pygostyle. *Iberomesornis* has lost its finger claws, but retains teeth, as does *Sinornis*. *Iberomesornis, Sinornis* and *Hongshanornis* have a tall, modern-type, coracoid (after Hou *et al.* 1996, Sereno & Cheggang 1992, Zhou & Zhang 2002, 2005).

remarkably complete skeletons found in the Niobara Chalk, a marine deposit in Kansas, which also produced such reptiles as mosasaurs, ichthyosaurs, and plesiosaurs. *Hesperornis regalis* was a large bird, the size of an Emperor Penguin, and clearly flightless, as indicated by the tiny wings and flat, keel-less, sternum. However, it had large hind limbs with flattened, streamlined, foot bones that probably, in life, bore lobes like grebes (rather than webs like penguins). Unable to stand upright, it must have slid along the ground, like divers. The lower bill carried about 30 small recurved teeth along its length, while the upper bill had the front half toothless and about 15 teeth further back, in the maxilla. A number of other genera and species of Hesperornithidae have been described from Late Cretaceous deposits in North America, Sweden, Russia, and even Antarctica, though not (so far) from Britain. *Ichthyornis* was a much smaller bird from Kansas, about the size of a tern, with a powerful flight musculature, as indicated by the large keel on the sternum and the prominent crest on the humerus, where the flight muscles attach. Like *Hesperornis*, it had teeth the length of the lower bill, but the upper bill is not fully known so it is uncertain whether it too was partially toothless. Since its original discovery, it too has been reported widely from North America, from Antarctica, and from Belgium (Dyke *et al.*, 2002). These toothed birds clearly belonged to widespread, successful groups of fish-eaters.

Cretaceous-Tertiary transition

However, the end of the Cretaceous, marked by the disappearance of mosasaurs, dinosaurs, pterosaurs, ichthyosaurs, and ammonites, also saw the demise of these toothed birds. Indeed, there is very little Cretaceous evidence of the birds that did survive into the Tertiary. Some possible 'transitional shorebirds' have been described from the Lance Formation, the latest Cretaceous in North America. These are specimens, usually isolated bones, that seem to share characteristics of Stone Curlews with those of some much later Eocene fossils that show a mixture of wader (charadriiform), duck (anseriform), and ibis (ciconiiform) characteristics (Feduccia, 1995, 1996)). Some possible primitive relatives of the ratites and tinamous are also present (ratites include the flightless Ostrich, Rhea, Emu, and Cassowary, which share a flat, raft-like sternum; their primitive sort of palate anatomy, termed palaeognathous, is shared with the tinamous of South America). It seems likely that these two groups provided the avian survivors of the Cretaceous extinctions, but they are not well preserved. This is, however, a subject of much current controversy. While the palaeontological record is characterized by a scarcity of evidence for the birds that must have been present during the transition from Cretaceous to early Tertiary times, the molecular evidence strongly suggests that many lineages of modern birds were already present. Thus palaeontologists strongly believe that there was a very rapid, explosive, evolution of birds in the Palaeocene, as the few surviving lineages (ancestral types of wader/duck and ratite/tinamou) evolved into the ancestors of the modern orders within the earliest 10 Ma or so of Palaeocene time (e.g. Feduccia, 1995; Benton, 1999). Molecular ornithologists contrarily argue that the extensive differences between the genes of modern birds, coupled with the time that was likely to be needed to evolve those differences, mean that the modern orders must have been founded anything between 100 and 160 Ma, well back in the early Cretaceous (e.g. Cooper & Penny, 1997). On this interpretation, several lineages must have been present in the Late Cretaceous, and survived the period of extinction at 65 Ma, yet have left no fossil evidence of their Cretaceous existence. This controversy is worth some further examination.

Genetic evidence has been invaluable in discerning genetic relationships between bird species and groups. Basically the reasoning is very simple. The DNA sequence from any species differs by some small percentage from that of any other species. If three species are compared, for the same gene, the two with fewest differences are, obviously, more closely related than either is to the third species. Moreover, a bigger difference has presumably evolved over a longer period. Genetic evidence has shown, for instance, that the various seabirds recognized as penguins, shearwaters, albatrosses, and divers (loons) are more closely related to each other than to other orders of birds, and, more surprisingly, are actually relatively recent, more derived groups. Traditionally, they had been regarded as more primitive, that is, an early radiation. Even traditional, morphologically and palaeontologically based, ornithologists are willing to accept this new judgement of their relationships. The controversy starts with trying to assign dates to the times of divergences. It is not possible to do so without reference to some palaeontological time marker. It is also necessary to assume that the rate of genetic change has been constant over some time period since a divergence. Palaeontological time markers are usually the earliest distinct fossil of one or other lineage: considering say anseriform–galliform divergence, the earliest definite duck or gamebird would indicate that these two had diverged from their presumptive common ancestor. If swans and geese were to differ from ducks by 5% of their DNA codes, and we think the earliest fossil swan is 10 million years old, then a difference of 50% between ducks and hens would imply that they diverged 100 million years ago. An alternative time marker is sometimes used, the dates by which continents, carrying particular bird groups, are known to have separated, by continental drift. For example, kiwis only occur in New Zealand, which is believed to have separated from Australia in the Late Cretaceous about 82 Ma, so the 18.4% difference between kiwis and emus, their nearest living relatives, must have developed over about 82 Ma.

In assigning dates to their phylogeny, Cooper & Penny (1997) take a date of 70 Ma as the occurrence of the earliest diver, but this identification is considered very doubtful by others (Feduccia 1996); the relevant specimen may well be a hesperornithid. In that case, the earliest diver is probably *Colymboides anglicus* from the Eocene London Clays of the Thames estuary, only about 53 Ma. Because it is accepted that this seabird group was a late divergence in bird phylogeny, taking such an early date for this divergence as 70 Ma obviously pushes the dates of divergence of other, even earlier evolving, bird groups well back into the Cretaceous. There are similar problems with using the movements of continents to estimate divergence times. Implicit in the suggestion that kiwis and emus diverged 82 Ma is the assumption that their common ancestor was already flightless, so had to be carried on the diverging continents as they drifted apart. Other evidence shows, however, that flightlessness, and the anatomical features that characterize it, can evolve very quickly in birds. If a flighted kiwi ancestor flew to New Zealand, and then lost its ability to fly, it could have done so much more recently than 80 Ma. A different approach to the apparent absence of early fossils of modern bird groups is taken by Benton (1999), trying to reconcile the palaeontological and genetic evidence. He points out that the gaps in the known record of fossil bird lineages in the Tertiary can be used to estimate the likely size of the gap before the earliest known occurrence of that lineage. On this basis, for example, the earliest known examples of swift (55 Ma), nightjar (55 Ma), and owl (58 Ma) lineages, in combination with the subsequent time gaps observed in their lineages, indicate likely earliest dates (with 95% probability) of 62 Ma, 67 Ma, and 63 Ma respectively. In other words, nightjars, with the biggest gaps in their subsequent history, might have evolved in the latest Cretaceous, and thus survived through the Cretaceous–Tertiary extinction of 65 Ma, but probably did not, and it is very unlikely that the other two groups did so. A more direct analysis of the Mesozoic (mostly Cretaceous) bird record (Fountaine *et al.*, 2005) demonstrates that it is a perfectly good one – enough species are known, and many of the specimens themselves are complete enough – to demonstrate that these known birds really do not belong to any of the 'modern' bird groups. Nor are any decent fossils of modern (neornithine) groups known from the Late Cretaceous. Perhaps, as the molecular scientists argue, the relevant species existed only in parts of the world from which there is little or no fossil record. More likely, as the palaeontologists argue, they don't exist because modern groups had mostly not yet evolved; what fossils there are, and their Palaeocene descendants, seem to have been waterbirds of some sort (Dyke *et al.*, 2007a). Palaeontologists also argue that the assumption of a molecular clock running at a constant rate does not apply to the early stages of modern bird radiation. It is evident that both molecular and direct fossil evidence have much to contribute still to this debate, and it will be an active area of research and discussion for the next decade.

Tertiary birds

Not only are there scant remains of modern type birds in the Late Cretaceous, but nor are there many bird specimens from the succeeding 10 million years of the Palaeocene. However, in the succeeding Eocene, important and diverse faunas from five sites (and a scatter of specimens from additional sites) give us early glimpses of the radiation of modern birds (Mayr, 2005). The earliest of these major faunas, from the Fur Formation in Denmark, dates to just above the Palaeocene/Eocene boundary, about 54 Ma. This is a small fauna, but contains some exquisitely preserved bird skeletons, many of articulated bones preserved 'in the round', including even soft tissues and feathers. The fauna includes about 30 species, though many are not yet described or named (Lindow & Dyke, 2006). There are primitive gamebirds, waders, parrots, mousebirds, trogons, and swifts, as well as possible owls, rails, and coraciiforms (roller/kingfisher/hoopoe relatives). Among the best preserved is *Lithornis*, a flying bird with the primitive (palaeognathous) palate seen in ratites and tinamous. The next fauna in age, also from the Lower Eocene at about 53 Ma, comes from the London Clay around the Thames Estuary, at various sites including the Isle of Sheppey and The Naze, in Essex, and equally in the Hampshire Basin along the south coast from Dorset to Sussex and on the Isle of Wight (Harrison & Walker, 1977; Steadman, 1981; Dyke, 2001). More than 50 species are recorded. Mostly these are represented by isolated bones, but they too are preserved uncrushed. Many of them are referable to modern families, though others belong to extinct families that show characters intermediate between two modern ones, and a few belong to entirely extinct groups. Primitive members of the nightjar, stone curlew, falcon, hawk, duck, bustard, owl, roller, wood-hoopoe, cuckoo, turaco, and mousebird families are present, along with, for example, a galliform *Paraortygoides radagasti* that cannot be placed in any of the four modern galliform families (Megapodidae, Phasianidae, Numididae, or Cracidae (Dyke & Gulas, 2002)). Mostly the birds are small or very small, and the fauna as a whole seems a strange mixture of what we would now regard as tropical forest birds (wood-hoopoes, mousebirds, trogons, turacoes, parrots) and more likely inhabitants of a European landscape (divers, petrels, ospreys, auks). One major group is significantly absent: although small birds are preserved, there seem to be no passerines. One partial metacarpus, *Primoscens*, was named as such (Harrison & Walker, 1977), but it is difficult to distinguish passerine metacarpals from those of woodpeckers (Benton & Cook, 2005), and primoscenids turn out to have zygodactyl feet (two toes opposing two toes) like woodpeckers (Lindow & Dyke, 2006). The London Clay fossils are mostly small fragmented specimens, sometimes a few associated bones and usually nicely preserved in the round, but hard to interpret. The faunas from the slightly later sites of the Green River in Utah (about 50 Ma) and the famous Messel oil shales near Darmstadt in Germany (about 49 Ma) are very similar, though their fossilization is very different. The Green River fauna, laid down in fine silt, contains birds that are well-preserved, often complete, skeletons. It includes at least 39 species from 14 or 15 orders, including an early frigate bird *Limnofregata*, an owl, a swift, an oil-bird, various rail relatives, mousebirds, coraciiforms (kingfisher/bee-eater/roller relatives) and, best known, the wading duck *Presbyornis* (Feduccia, 1996). The Messel fauna is remarkable because whole body fossils are preserved in oil shales, sometimes with plumage and gut contents, albeit the skeletons tend to be crushed flat. Among about 30 species described so far are a swift *Scaniacypselus*, a primitive hoopoe *Messelirrisor*, kingfisher *Quasisyndactylus*, parrot *Psittacopes*, nightjars, rails, a galliform *Paraortygoides* like that from the London Clay, woodpeckers, and mousebirds. A predator, *Messelastur*, is closer to owls, but probably also related to Falconiformes. Surprisingly, in a European context, is a swift-like hummingbird ancestor *Parargornis* (Mayr, 2000, 2005). Both the variety of these Eocene faunas and the genera present are very similar across all four sites, and detailed studies are often enhanced by comparing, say, flattened whole body specimens from Messel with broken but uncrushed specimens from the London Clay. However, the primitive *Lithornis* seems to be absent by the Middle Eocene (e.g. at Messel), and may be a Cretaceous survivor that finally died out as modern groups stated to evolve.

The biggest fauna of all is the famous fauna from the quarries in the phosphorites at Quercy in south-west France, rather later in time, ranging from Upper Eocene to Upper Oligocene, or from about 40 to 35 Ma. Specimens have been collected from those quarries since at least the 1860s, and are still being collected. Some 90 or more species are reported, though many of the names are old, and much needed revision may reduce their number, but at least 25 families are represented. The fauna shares some features of the earlier Eocene faunas, including some genera; there are numerous coraciiforms, nightjars, swifts, and owls, some trogons, mousebirds, cuckoos, rail relatives, a heron, hawks, cathartid (now American) vultures, waders, and gamebirds. The later faunas include a more modern element as well – some phasianid gamebirds, and perhaps one extant genus (an avocet *Recurvirostra sanctaeneboulae*); most importantly, from the Upper Oligocene, about 25 Ma, comes the first passerine reported from Europe. In the succeeding Miocene, passerines become more diverse although, being mostly small, rarely numerous.

The place and timing of the appearance of passerines has been a topic of much discussion and controversy. In the modern fauna, some 60% of the 9,500 bird species are passerines, and their current diversity has led some to question the reality of their earlier absence. Such a diversity 'must' have required a long time to evolve, is the argument. Both on their anatomy and their molecules, passerines are certainly a distinctive group (Slack *et al.*, 2007). Characteristically, they are small; the Raven and the Australian Superb Lyre-bird *Menura novaehollandiae* are exceptionally large, about 1 kg in weight, but most are in the range of 10–100 g. Among their anatomical characteristics is a perching foot, with a strong hallux (big toe or hind toe) opposing the three main toes; many have the complex syrinx, which is correlated with them being songbirds. It seems very likely that they originated in the southern continents, perhaps in Australia, as the earliest fossil passerine comes from Eocene deposits there (Boles, 1995). Moreover, molecular analysis of modern passerines confirms that the more primitive passerines come from southern continents, the relics of the giant southern continent Gondwanaland, which split up in the Cretaceous. The most isolated and genetically primitive are the few members of the riflemen, family Acanthisittidae, confined to New Zealand (Ericson *et al.*, 2002; Slack *et al.*, 2007). The Suboscines, the passerines that are not also songbirds (Oscines), occur principally in South America and South-east Asia (India was also part of Gondwanaland) (e.g. tyrant flycatchers Tyrannidae, broadbills Eurylaimidae, pittas Pittidae). The more primitive of the Oscines (including the lyre-birds) are also found in Australia. It looks as though the more advanced, oscine, passerines only extended their range to northern continents during the Oligocene (Barker *et al.*, 2002). The beautifully preserved Messel birds, often at least as small as modern passerines, leave little doubt that passerines were absent from Eocene Europe. The London Clay and Quercy faunas emphasize that conclusion (Blondel & Mourer-Chauvire, 1998). The earliest passerines from Europe come from the Early Oligocene of Germany and France, though not yet fully described. The analysis of some partial tarsometatarsals from the Middle Miocene of France and Germany adds some convincing detail (Manegold *et al.*, 2004). These are preserved well enough to reveal details of the canals for the tendons of the foot as they pass the ankle, through a bony bridge called the hypotarsus. Most modern passerines show six enclosed canals in the hypotarsus, but the New Zealand Acanthisittidae have only two enclosed canals; one of these Miocene fossils has only one canal, another has three. These indicate that the passerines then in Europe did do not belong to any of the modern families of European passerines (Manegold *et al.*, 2004).

Pleistocene birds

As cooler climates developed, from the Late Oligocene about 30 Ma, the tropical-looking birds characteristic of the Eocene and earlier Oligocene European faunas retreated to Africa. Most of Britain was submerged under shallow seas through the Miocene, so bird (and mammal) faunas are absent. From elsewhere in Europe, Miocene avifaunas show that passerines were becoming the dominant land-birds, and many extant genera appear. The fossil record of birds in Britain recommences in the later Pliocene, about 2 Ma, and in the Pleistocene, roughly the last 1.8 million years. This was a period in which successive, increasingly more severe, cold (glacial) periods, interrupted by briefer warm (interglacial) periods, determined the fauna and flora of northern latitudes, including our islands. From the end of the Pliocene, we have a very few bird bones from the Red Crags, clays laid down under shallow marine conditions now exposed in a few places along the Suffolk coast. Most notable is the remnant of an albatross, originally named *Diomedea anglica*, for albatrosses are usually thought to be southern hemisphere birds (Harrison & Walker, 1978b). A tarsometatarsal from Foxhall, Suffolk with one associated toe bone constitutes the type specimen, now in Ipswich Museum. A partial right ulna from the earlier Pliocene Coralline Crag of Orford, Suffolk and a partial tibiotarsus from Florida were the only other known elements, but recently another, complete, right ulna and partial humerus were recovered from the Norwich Crag near Coverhithe

(Dyke *et al.*, 2007b). In size they are all close to Royal Albatross *D. epomophora*, but about 5% smaller; in shape they are closest to the Short-tailed Albatross *D. albatrus* of the Pacific, though appreciably larger. The availability now of a complete wing bone and a complete leg bone gives an index of their relative lengths. The legs are relatively rather longer than in typical *Diomedea*, and along with *D. albatrus* and its relatives, *D. anglica* is now assigned to a different genus, *Phoebastria*, containing the Short-tailed Albatrosses of the North Pacific (Dyke *et al.*, 2007b). The genus has evidently a very shrunken modern range. Along with the albatross is *Cepphus storeri*, related to the Black Guillemot *C. grylle* and its Pacific relatives *C. columba* and *C. carbo*, and perhaps ancestral to all three (Harrison, 1985).

Another gap, of perhaps a million years, elapses before the next glimpse of our bird fauna, from the Cromer Forest Beds, exposed at numerous coastal sites such as West Runton, Norfolk. By this time, however, modern species are present. From deposits of the Pastonian interglacial, a temperate period about 400 ka (thousand years ago) comes a small fauna, including Bewick's Swan, Mallard, Buzzard *Buteo* sp. (Common *B. buteo* or Rough-legged *B. lagopus*?), Guillemot, and Razorbill (Harrison, 1985). This could be a present-day wintering bird fauna, but one additional species would be a very unlikely visitor now to East Anglia: the first record of Eagle Owl *Bubo bubo* from Britain also comes from these deposits. It indicates a smaller form than the current northern European Eagle Owl, comparable in size with the present North African race *B. b. ascalaphus*.

A glacial period with no recorded bird fossils separates this from the fauna of the next, Cromerian, interglacial of about 350 ka. The Upper Freshwater Beds along the north Norfolk coast, especially at West Runton, have yielded an extensive fauna (unsurprisingly, given their name) of aquatic birds, including Cormorant, Whooper Swan, and Greylag Goose. Moorhen and Green Sandpiper are present. Dabbling ducks (Mallard, Teal, and Wigeon), diving ducks (Red-crested Pochard, Goldeneye, Tufted Duck, and Pochard), and saw-bills (Smew, Red-breasted Merganser) have been identified, along with a thick-legged Eider that is perhaps an extinct species (*Somateria gravipes*). Somewhat improbably, there also appears to be Mandarin, an Oriental species of oak woodland (Harrison, 1985). Other species that show disjunct distributions, notably the Azure-winged Magpie (found now in southern Spain and Portugal but otherwise in China and Japan), indicate that this is not an impossible occurrence. Some genera of mammals inhabiting oak woodlands, such as hedgehogs *Erinaceus* and wood mice *Apodemus*, show similar disjunct distributions. One may suspect that there was, in earlier times, a continuous belt of deciduous woodland stretched across the Palaearctic from Atlantic to Pacific, a belt that is now broken by the intervention of the arid interior of central Asia. The Cromerian avifauna is not confined to wetland species. Passerines of oak woodland include Blackbird, Song Thrush or Redwing, Nuthatch, Starling, and Jay. This could almost be the tick-list from a wetland in temperate oak woodlands in East Anglia today, though the Mandarin, Red-crested Pochard, and the Eider would look rather out of place. Another small fauna from Ostend, Norfolk, also includes Red-crested Pochard, along with Pochard and Common Scoter, and a fourth species that is another exotic eastern element, a Junglefowl *Gallus europaeus* (Harrison, 1978). There are other species of *Gallus* recorded elsewhere in Europe in earlier times; *Gallus beremendensis* from the Late Pliocene or Early Pleistocene of Hungary (Janossy, 1986), and another from the Pliocene of France (Mourer-Chauvire, 1993), but these are much earlier in time than the British specimens, and it is not clear whether they are directly related, even the same.

The cooling indicating the end of the Cromerian Interglacial and the approach of the next (Anglian) Glacial is marked, perhaps, by the presence of Red-throated Diver and Common Scoter at Mundesley, further south round the Norfolk coast (Harrison, 1985).

Another small but significant avifauna has been described from the important archaeological site of Boxgrove. This site reveals the earliest evidence yet for human habitation of Britain. It is now some 12 km north of the Sussex coast, inland from Bognor, but was then much closer to the sea. The open campsite was sited just in the shelter of a low chalk cliff-line, and the humans were represented not only by abundant stone tools, hand axes made from the local flint, but a single leg bone (tibia) as well. The birds present have been identified, sometimes tentatively, as cf. Whooper Swan, Greylag Goose, Mallard, Widgeon, cf. Garganey, Teal, Tufted Duck, Goldeneye, Grey Partridge, Moorhen, a medium-sized wader, perhaps Woodcock or Golden Plover, Black-headed Gull, cf. Kittiwake, Great Auk, Tawny Owl, Swift, cf. Robin, cf. Hedge Sparrow and Starling (Harrison & Stewart, 1999). Most of these are only represented by one or two bones, hence the tentative identifications, though the Mallard is well represented, by 38 or so bones. In particular, the Great Auk, perhaps the most notable in this list, is only indicated by the proximal end of a right humerus, but this is one of the most distinctive bones of a very distinctive species. It is significant, too, as probably the earliest record of this unfortunate species anywhere in the world. It is not too easy to detect either the environment or the manner in which these bones arrived on site, but wetlands nearby seem to have been a good hunting ground for the early Britons. They were certainly hunting the larger mammals at this site (horse, rhinoceros, deer, elephant), whose bones bear the cut marks of their flint tools, but there are no such direct clues on the bird bones to show that the birds too were hunted (Roberts & Parfitt, 1999). Perhaps they were – it is difficult to see how else the Great Auk might have arrived – but they might just have been using the freshwater lakes that would undoubtedly have been important to both the humans and their prey. The Swift was probably nesting in crevices in the chalk cliffs and hunting over the water.

The evidence from the small mammals, in particular, suggests that Boxgrove represents a later phase, another interglacial, than the Cromerian, but earlier than the Hoxnian or Great Interglacial (Yalden, 1999). There is no direct evidence from the vertebrate faunas of the intervening glacials that must have separated these interglacials, and the next bird faunas come from Swanscombe in Kent, East Farm, Barnham, in Suffolk, Cudmore Grove, Essex, and Hoxne itself, all of Hoxnian age. Swanscombe produces the largest bird fauna of this interglacial (Harrison, 1979, 1985; Parry, 1996). It includes Cormorant, Shoveler, Common Scoter, Goldeneye, Red-breasted Merganser, Osprey, an Eagle Owl as large as the modern European form, Wood Pigeon, Garden Warbler, and Serin. From Barnham, Stewart (1998) records unspecified dabbling duck *Anas* sp. and other ducks, probable Wood Pigeon, and Redwing/Song Thrush. Hoxne itself seems to have yielded only the remains of a duck; the rather larger fauna from Cudmore Grove has not yet been described (J. Stewart, pers. comm.).

From the Wolstonian Glacial, or perhaps the very end of the Hoxnian, as the climate got colder, at Swanscombe, Harrison (1979, 1985) lists possible White-fronted and Barnacle Geese, Common Scoter, and Capercaillie, probably a hen. A much larger fauna comes from the lower levels, the Glutton Stratum, of Tornewton Cave in Devon, 10 km inland from Torquay. This includes Black Stork, Shelduck, Goosander, Kestrel, and a large Eagle Owl.

Particularly interesting is a Crossbill *Loxia* sp., which could be Scottish *L. scotica*, Parrot *L. ptyoptsittacus* or, quite likely, the common ancestor of both of them. Somewhat more incongruous is apparently a red-legged partridge, described as a new species *Alectoris sutcliffei* because it is smaller than the extant species of *Alectoris*, and was perhaps adapted to cooler conditions.

There is some difficulty in assessing bird faunas of the next interglacials, because what has traditionally been supposed to be one interglacial, the Ipswichian, is now considered to be two or three such warm periods (Currant, 1989; Yalden, 1999). One, which might be termed informally the Pre-Ipswichian, is characterized by a lack of Hippopotamus: the Bear Stratum and Otter Stratum of Tornewton Cave, with a bird fauna including Shelduck, Brent Goose, and Goosander, might well belong here. White-tailed Eagle was also present in the Bear Stratum, and was probably the predator responsible for bringing the waterfowl into the cave (Harrison, 1987b; Stewart, 2002a).

The true Ipswichian Interglacial, characterized by the presence of hippopotamus in its mammal faunas, is known from sites such as Trafalgar Square, Peckham, and Brentford in London; Barrington, Cambridge; Victoria Cave near Settle, Yorkshire; Joint Mitnor Cave and the Hyaena Stratum of Tornewton Cave, Devon. Only the last of these has an avifauna as well. It includes Brent Goose, Ruddy and Common Shelduck, Wigeon, Kestrel, Skylark, Tree Pipit, Starling, and Raven (Harrison, 1980b). Two other caves, on the Gower Peninsula of South Wales, also yield bird fossils of this age. Bacon Hole has Cory's Shearwater, Bean Goose, Red Kite, Hobby, Turnstone, Golden Plover, Dunlin, Razorbill, Skylark, Swallow, Wheatear, Blackbird/Ring Ouzel, Starling, and Carrion Crow; Minchin Hole nearby confirms Dunlin, Razorbill, Skylark, and Starling (Harrison, 1987b). Assuming that the waders were wintering birds from the nearby shore, the Red Kite, Hobby, and particularly nesting Cory's Shearwater indicate a somewhat southern fauna, commensurate with the well-known occurrences of such southern mammals as Hippopotamus, Fallow Deer, and Spotted Hyaena in faunas of this age. There are a few bird fossils of this age from the London area as well, though not from the classic mammalian sites: Smew, Junglefowl, and Coot from Crayford, Kent; Cormorant, Mute/Whooper Swan, Greylag, and Red-breasted Goose from Gray's, Essex; Mute/Whooper Swan, White-fronted and Greylag Goose, Mallard, and Crane from Ilford, Essex; Mallard from Uphall, Essex, and Gadwall from Waterhall Farm, Hertfordshire (Harrison & Walker, 1977). Most of these species seem unremarkable, but the Junglefowl is either an eastern species or an intrusive specimen of much later date. Only the distal end of a radius, not the most distinctive of bones at the specific level, was recorded. The Crane from Ilford also deserves discussion. Harrison & Cowles (1977) identify this as a representative of a now extinct European Crane, *Grus primigenia*, larger than the Common Crane *Grus grus*, closer in size to, though distinguishable anatomically from, the Sarus Crane *G. antigone* of India. This species has been reported from various sites of Late and Post Pleistocene date, including Iron Age Glastonbury and King's Cave, Loch Tarbet on Jura, as well as sites in France, Germany, and Mallorca, and has been much discussed (e.g. Harrison & Cowles, 1977; Northcote & Mourer-Chauviré, 1988). One notion is that it filled a now vacant niche, of a larger crane alongside a smaller one, seen elsewhere in the world (e.g. the Whooping and Sandhill Cranes *G. americana* and *G. canadensis* in North America). However, it is more likely that the size range of *G. grus* has been underestimated – males are anyway bigger than females, and the species was apparently larger in the past (Stewart, 2007a; Driesch, 1999).

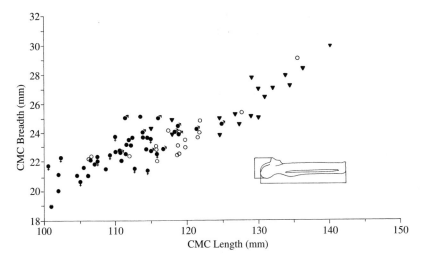

Fig. 2.3 Size of the carpometacarpal (inset) in the Common Crane *Grus grus* (solid dots), the Asian Sarus Crane *G. antigone* (triangles) and fossil specimens from Europe (circles) which include the putative extinct European Crane *G. primigenia*. The size ranges of the two modern species overlap, and the fossil specimens range across the two. Where the sex of modern specimens is known, it is indicated; note that males are larger than females, and it is likely that the large fossil specimens were large males of Common Crane (after Driesch 1999, Stewart 2007a).

Driesch (1999) and Stewart (2007a) plotted the lengths of the few available bones of archaeological *Grus* from western Europe against the available modern specimens of *G. grus* and *G. antigone* (Figure 2.3). There is in any case a substantial difference between females and larger males, which makes the species very variable. The plots do indeed suggest a larger size range for archaeological specimens, overlapping both modern *G. grus* at the lower end of the size range and *G. antigone* at the upper end. In that case, *G. primigenia* cannot readily be distinguished from, and should probably be included within, *G. grus*. The few Pleistocene specimens, in particular, are larger than the material from archaeological sites, and, just as for many mammals (Davis, 1981), it looks as though its size declined as the climate ameliorated. Larger size is well known to confer advantages to many species in severe climates, so long as adequate food sources are available; it allows accumulation of greater fat reserves, a wider size range in food and a relatively lower rate of heat loss (through a relatively lower surface/mass ratio). An alternative possibility, with some evidence to support it, is that cranes in former southern breeding populations, now extinct, may have been larger, or at least included larger birds (Stewart, 2007a).

The Last Glaciation

The most recent glacial period is known as the Devensian in Great Britain, the Weichselian in northern Europe, the Würmian in Alpine Europe, and the Wisconsinan in North America, to avoid the assumption that these are necessarily the same period in all these places.

Table 2.1. The birds listed from the Late Devensian of Pinhole Cave, Creswell Crags, Derbyshire (Jenkinson 1984, Bramwell 1984). Both dating and identification would be worth checking for many of these records, though some (e.g. Demoiselle Crane, Alpine Swift) have been confirmed.

Black-throated Diver	Demoiselle Crane	Woodlark	Great Tit
Grey Heron	Moorhen	Crag Martin	Long-tailed Tit
White Stork	Lapwing	Swallow	Nuthatch
Bewick's Swan	Ringed Plover	Meadow Pipit	House Sparrow
Brent Goose	Grey Plover	Starling	Tree Sparrow
Barnacle Goose	Golden Plover	Waxwing	Chaffinch
Greylag Goose	Turnstone	Jay	Brambling
White-fronted Goose	Snipe	Magpie	Bullfinch
Pink-footed Goose	Curlew	Nutcracker	Hawfinch
Mallard	Whimbrel	Jackdaw	Greenfinch
Wigeon	Greenshank	Rook	Linnet
Teal	Knot	Crow	Pine Grosbeak?
Garganey	Skua sp.	Raven	Crossbill
Ruddy Shelduck	Common Gull	Dipper	Rose-breasted Grosbeak
Tufted Duck	Black Guillemot	Wren	Corn Bunting
Common Scoter	Puffin	Hedge Sparrow	Snow Bunting
Goosander	Stock Dove	Blackcap	
Golden Eagle	Wood Pigeon	Wheatear	
Rough-legged Buzzard	Tengmalm's Owl	Whinchat	
Goshawk	Short-eared Owl	Redstart	
Osprey	Tawny Owl	Robin	
Merlin	Barn Owl	Ring Ouzel	
Kestrel	Hawk Owl	Blackbird	
Red Grouse	Alpine Swift	Redwing	
Ptarmigan	Kingfisher	Song Thrush	
Black Grouse	Lesser Spotted Woodpecker	Mistle Thrush	
Grey Partridge	Skylark	Fieldfare	

Nevertheless, it is increasingly clear that they are contemporary, different names for the same event, which began about 70,000 years ago. As the Last Glaciation, there is more evidence for faunas of this age than for earlier glaciations, evidence of which has often been obscured by the later ones. Moreover, there is enough detail to show that it was a period of fluctuating climate, not uniformly cold, but colder in some phases than others.

The earliest evidence of cooling comes from sites such as Banwell Bone Cave, Somerset and the Reindeer Stratum of Tornewton Cave (assigned by Currant & Jacobi (2001) to a Banwell Bone Cave mammal assemblage). Birds from Tornewton include Teal, Willow/Red Grouse, Ptarmigan, Little Bustard, Skylark, Fieldfare, Starling, and Carrion Crow (Harrison, 1980b). This is clearly the fauna of an open unwooded countryside. The Little Bustard might seem out of place to those who expect to see it on birdwatching trips to Spain or Portugal. However, its modern range extends far into the southern steppes of Russia, a cold open environment in winter, suggesting that it was a perfectly appropriate species for this fauna.

A slightly warmer phase, an interstadial within the Devensian Glaciation, is well recognized from its mammal faunas, often characterized by the presence of Spotted Hyaena – many cave sites seem to have been hyena dens, and they may well have accumulated bones of other species in the caves. Currant & Jacobi (2001) select the Lower Cave Earth of Pin Hole Cave, Creswell Crags, on the Derbyshire/Nottinghamshire border, as representative of this period. Dated hyena bones, from 42 ka to 23 ka, just within the range datable by ^{14}C, suggest a time frame for this period. Extensive bird faunas that probably belong in this phase include White-fronted Goose, Mallard, Goosander, Common Scoter, Ptarmigan, Goshawk, Rough-legged Buzzard, Demoiselle Crane, Turnstone, Snipe, and Raven. However, there is an enormous bird fauna recorded from this cave, including many other waders and passerines, 98 species in all (table 2.1, from Jenkinson, 1984). Unfortunately, although they were excavated very carefully in the 1920s by Armstrong (1928), the complex nature of climatic changes during the Devensian were not then fully appreciated, and it is hard to interpret the species list in ecological terms. For instance, there are obviously northern species such as Hawk Owl, Tengmalm's Owl, Nutcracker, and Waxwing, as well as those already listed, mixed with such obviously southern species as White Stork, Alpine Swift, and Kingfisher (Jenkinson, 1984; Bramwell, 1984). It is not easy to conceive of these all being members of the same fauna, even allowing for the southward migration of northern species in winter. It seems likely that the birds represent at least two subfaunas combined, one from a warmer phase when waters did not freeze in winter, the other from a colder phase.

Perhaps contemporary cold faunas come from Kent's Cavern, Devon, including Shag, White-fronted Goose, and Snowy Owl (Harrison, 1987b), and Windmill Cave, Brixham, with Common Shelduck and Common/Rough-legged Buzzard (Harrison, 1980b). The small mammals of this time are well known to include both Collared and Norway Lemmings (*Dicrostonyx*, *Lemmus*) as well as the northern voles *Microtus oeconomus* and *M. gregalis*. Thus the presence of raptors and owls is highly appropriate, but it is odd that this seems to be the only record of Snowy Owl for Britain, as the species is well recorded, even abundant, in Europe, for instance in France (Mourer-Chauviré, 1993) and Hungary (Janossy, 1986), during the Last Glaciation.

The coldest phase of the Devensian, when the ice cap extended as far south as the Gower coast in the west and the north coast of Norfolk in the east, was a time when there was little biological activity even in southern Britain. A few mammal bones have been dated to this period, about 20–15,000 years ago, for example the Arctic Fox from Castlepook Cave, County Cork at 19,950 b.p. and the Woolly Mammoth at 20,380 b.p. and Collared Lemming at 20,300 b.p. in the same cave (Woodman *et al*, 1997). There are probably no bird fossils from this period, and one might imagine that the bird fauna would have been a sparse one – perhaps resident Ptarmigan, breeding Snowy Owls, with Little Auks, geese, and northern waders in the summer. Essentially, little or none of the present breeding bird fauna (or indeed mammal fauna) of Britain is likely to have survived here then, so providing a clean slate for recolonization in the Late and Post-Glacial periods.

Continental Europe

If it is true that Britain, and other northern parts of western Europe (especially Scandinavia, which was also covered under an ice sheet), had little or none of their present bird fauna during the glacial maximum, that fauna must have been pushed south into warmer latitudes.

Classical theory has supposed that Iberia, southern Italy, and perhaps the Balkan peninsula acted as refuges for more northern species during the glacial maxima. Given that there were several glacial–interglacial cycles during the Pleistocene, the repeated retreat to and expansion from these refuges should also have played a part in the speciation of modern birds (and other animals and plants). For example, it has been suggested that Hooded Crows *Corvus cornix* represent a population that retreated to and differentiated in Iberia, while Carrion Crows *C. corone* retreated to the Balkans. There are several quite complex arguments wrapped up in this apparently simple and very plausible notion. The account just given of birds in Britain during the Pleistocene is notable for the increasing familiarity of most of the species and their combinations as faunas. A few genuinely extinct species certainly were present in the Pliocene and early Pleistocene, and this is much more evident elsewhere in Europe (e.g. in France and Hungary; Mourer-Chauviré, 1993, Janossy, 1986) than in Britain, where equivalent faunas are missing. These earlier forms seem to be ancestral to modern species, for instance precursors of Raven (*Corvus antecorax*), Black Grouse (*Tetrao partium*), Hazel Hen (*Bonasia preaebonasia*), and Capercaillie (*Tetrao praeurogallus*) (Mourer-Chauviré, 1993). By the middle Pleistocene, there are perhaps the early *Gallus*, *Grus primigenia* if that is real, *Alectoris sutcliffei* and the Thick-legged Eider *Somateria gravipes* but most birds seem to belong to modern species; at the very least, they cannot be readily distinguished from modern species (Stewart, 2002b). Birds do not seem to show the relatively rapid evolutionary changes that the mammals, particularly the voles and lemmings, show during the late Pleistocene as they adapted to the more severe conditions. One species does seem to show some change: the Eagle Owl of the earlier period was smaller, and seems to have evolved into the larger form now found in Europe by the Devensian. It has been suggested that owls are relatively sedentary, and becoming larger would be an appropriate strategy for a large owl, but that most birds would respond by migrating rather than evolving (Harrison, 1987b). Grouse too are largely sedentary, and it has been suggested that they likewise evolved a larger size, and an ability to live on a coarse diet (conifer needles, heather, bilberry, buds, and catkins of deciduous shrubs), during the Pleistocene, as a response to increasingly severe weather and the appearance of boreal habitats (Drovetski, 2003). But if most birds from earlier in the Pleistocene were of familiar species, it is hard to argue that isolation in southern Europe during the last glaciation contributed to their differentiation. They must have evolved rather earlier, perhaps during the Late Miocene and Pliocene.

Molecular evidence, of the sort already discussed in relation to the origins of the modern orders of birds, has also been applied to the question of when modern species evolved. This too sometimes suggests that modern species split from their common ancestors much further back than the Last Glacial. However, the black and white flycatchers, *Ficedula hypoleuca*, *F. albicollis*, *F. semitorquata*, and *F. speculigera*, differ from each other by about 3% of their mitochondrial DNA; on the basis of the molecular clocks suggested earlier, this does imply a separation about 70 ka (Saetre *et al.*, 2001). This indicates separation at the beginning of the Devensian glaciation, which would certainly have restricted these woodland species to southern refuges. The three European species have overlapping breeding distributions, but are ranged north-west to south-east, Pied Flycatcher *F. hypoleuca* in the west and north, Collared *F. albicollis* across the centre, and Semicollared *F. semitorquata* ranging eastwards from the Greece to the Caspian Sea and beyond (Figure 2.4).

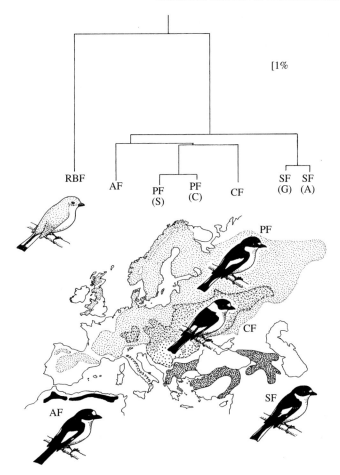

Fig. 2.4 *Ficedula* flycatcher distribution and phylogeny. The Red-breasted Flycatcher (RBF) is used to root the phylogeny of the black/white flycatchers. The Semi-collared Flycatcher (SF) is the most distinct, samples from Greece (G) and Armenia (A) clustering together. The geographically isolated Atlas Flycatcher (AF) is slightly more distinct from the Pied Flycatcher (PF, from both Spain (S) and Czechoslovakia (C)) than the Collared Flycatcher (CF) whose ranges overlap (based on Saetre *et al.* 2001).

More importantly, only *F. hypoleuca* occurs in the Iberian peninsula, while *F. albicollis* is the only species in Italy, and *F. semitorquata* is the only one in the Balkans and the Caucasus. It seems likely that these represent their Devensian refuges. *F. speculigera*, generally regarded as a local population of the Pied Flycatcher, is genetically as distinct from it as is the Collared Flycatcher, and deserves recognition as the Atlas Flycatcher. It presumably has been confined to North Africa, its current and perhaps past range, by the existence of its relatives to the north.

Another interesting and converse example is offered by the Crossbills *Loxia* sp. Their identification has been confused and much discussed. The smallest species, the Two-barred

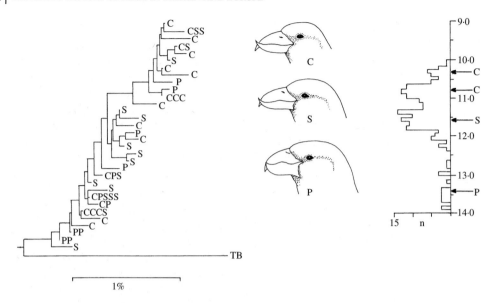

Fig. 2.5 Crossbill distribution and phylogeny. Ecologically, there are clearly 3 forms (species?) in Scotland, and peaks in their bill sizes (right) are well matched to two populations of larch-feeding Common Crossbills (C) and pine-feeding Scottish Crossbills (S), with a large tail of bigger specimens (but no peak) that match Parrot Crossbills (P) elsewhere in Europe. However, the molecular phylogeny (left, based on mitochondrial DNA, which evolves quickly and usually reveals relationships between close relatives) is a complete mix which fails to show any difference between these three, though the Two-barred Crossbill (TB) is well separated (after Piertney *et al.* 2001, Marquiss & Rae 2002).

Crossbill *L. leucoptera*, which specializes on feeding on larch cones, a relatively soft food source, is not taxonomically difficult nor controversial, but the taxonomy of the larger Red Crossbills, which feed on the harder cones of spruces and various pines, is more problematical. In Europe, the smaller Common Crossbill, which feeds particularly on spruce, is usually recognized as *L. curvirostra* while the larger Parrot-billed Crossbill, concentrating on pines, is usually recognized as *L. pytyopsittacus*. However, populations of supposed *L. curvirostra* Crossbills in southern Europe, for instance in Mallorca and North Africa, feed on pines. Importantly, so do the resident Crossbills in Scotland, of intermediate size, recently considered to be a full species, *L. scotica*, and therefore the only endemic bird species in Britain. An alternative taxonomic view is that these larger-billed southern isolates are actually forms of *pytyopsittacus*, not *curvirostra* or *scotica*. The fossil records, reviewed by Tyrberg (1991b), show Crossbills in southern Europe from the Middle Pleistocene, including the record from the Wolstonian of Tornewton Cave (?*curvirostra*) already mentioned, and from Grotte de Lazeret in southern France (?*pytyopsittacus*). In the Last Glacial, pine was certainly confined in Europe to south of the Alps and Pyrenees, and there are records of Crossbills (identified as both *curvirostra* and *pytyopsittacus*) in southern Europe at this time. Spruce was probably confined further east, in the Carpathians or around the Black Sea, and perhaps *curvirostra* was in fact restricted there, or further east, as well. As Tyrberg remarks, it is possible to envisage two scenarios for the present occurrence of Crossbills in Europe.

One would suggest that each species occurred in separate southern refuges during the Last Glacial, and spread north with the spread respectively of pine and spruce, *pytyopsittacus* from Iberia or Italy, *curvirostra* from the Balkans or Caucasus, meeting in Scandinavia, but leaving large-billed forms isolated in North Africa, Mallorca, and Scotland. The alternative is that one variable species occurred throughout southern Europe during the Last Glaciation, but competition increased as spruce became more common and favoured the smaller *curvirostra* forms. The larger billed forms would have been replaced in much of Europe, leaving large-billed isolates of *pytyopsittacus* stock in various places where pine woods survive, and also pushing the development of even larger bills in Scandinavian *pytyopsittacus*. Tyrberg predicted that molecular evidence should be able to resolve these hypotheses. If the first is true, *pytyopsittacus*, *scotica*, and the Mediterranean forms should be closely related to each other, but more distant from *curvirostra*. On the second hypothesis, there might be little genetic difference between any of the Crossbills. In fact this has now been investigated, and there is *no* genetic difference between the three 'species' (Piertney et al., 2001). Although they behave ecologically as three good species, even in Scotland where all three are present after Crossbill invasions (of small-billed *curvirostra*), they cannot be distinguished genetically. What appear to be morphologically good examples of each 'species' are totally confused in a phylogenetic tree (Figure 2.5), and although the Red Crossbills differ from the Two-barred Crossbill by about 44% in their mitochondrial DNA (suggesting an evolutionary split about 22 Ma, in the Miocene), the others show no difference. Genetically, they are not good species, not, at least, on the basis of the genes so far sampled. It is not surprising that ornithologists struggle to discern the taxonomic relationships of the various isolated populations or, indeed, to identify them in the field.

A further complication to this suggested history, that woodland birds survived in southern woodland refugia, is the evidence now available on the nature of those southern refuges. Older maps, for instance as reproduced by Moreau (1972), suggested that deciduous woodlands, of the sort now present in much of western Europe, survived at the maximum of the Last Glaciation in southern Spain, Italy, and Greece, so that the present-day woodland bird faunas would have been living there too. Increasingly, evidence from analysis of the pollen grains preserved in deposits of that age suggest that there was little or no woodland even in those southern refugia at 18,000 years ago. Trees such as oak, hazel, alder, and beech certainly did occur in these areas, but they were present only as scattered groves in sheltered locations, in what was generally a steppe or savanna environment (Adams & Faure, 1997). Birds might have retreated further south, into North Africa, an option not available to mammals and other terrestrial species, but the glacial maximum was not only cold, it was also a very dry time. The Sahara was at least as extensive as it is now, on modern estimations, and the Mahgreb was probably host only to a maquis-like Mediterranean scrub, not woodland. The conclusion seems to be that what we have supposed to be the refugia in southern Europe are likely to have been unsuitable for the modern European woodland fauna. In that case, where were the woodland birds? Molecular evidence from various small mammals strongly suggests that most of them had retreated much further east, to the region around the Black Sea, or even the Caucasus Mountains. Pollen analysis also suggests that woodland persisted in these south-eastern refuges– the only deciduous woodland for 'Europe' suggested by Adams & Faure (1997) is along the Black Sea coast of Turkey, running across to the Caucasus (Figure 2.6).

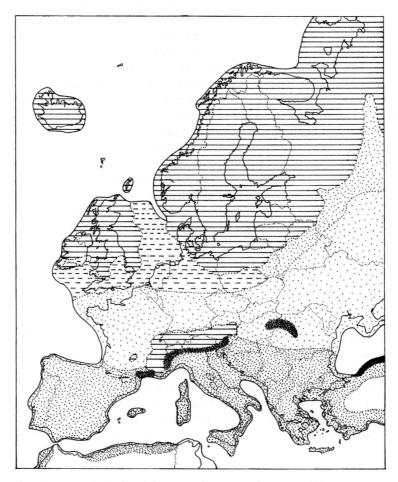

Fig. 2.6 Last Glacial vegetation in Europe, about 18,000 b.p., interpreting the evidence of pollen grains. Most of N Europe was covered in ice sheets (horizontal shading). The nearest deciduous woodland (black) was along the S shore of the Black Sea, though there was semi-arid scrubby temperate woodland (darkest stipple) S of the Alps and Carpathians. Most of Europe was covered in open, tundra or steppe-like, vegetation (stippling) (after Adams & Faure 1997).

So far, there is only a little direct evidence of bird faunas from these southern regions to support any of this theorizing. Two cave sites in southern Italy, Grotta Romanelli and Grotta del Santuario della Madonna a Praia a Mare, dated by ^{14}C to Late Glacial times, 12,000–9000 b.p., well after the Glacial maximum, have faunas dominated by Little Bustard, with Great Bustard also well represented, and with large numbers of geese, predominantly White-fronted, Brent and Bean Geese. Woodland birds are largely absent; other indicators of open conditions include both Choughs (Red-billed and Alpine) and two Sandgrouse (Pin-tailed *Pterocles alchata* and Black-bellied *Pt. orientalis*) (Tagliacozzo & Gala, 2002). The predominance of birds of open conditions, and of cold climates, is striking, for such southern sites. In Hungary, the extensive bird fauna of 68 species from the Glacial Maximum at Pilisszanto Rock Shelter (Janossy, 1986) is overwhelmingly dominated by Willow Grouse and Ptarmigan, with 2960 and 3112 bones respectively. There were much smaller numbers of Black Grouse (101 bones), and no other species contributed even 30 bones. Other indicators of open conditions include Golden Eagle and Snowy Owl, but a few bones of woodland species are also present – Great Spotted Woodpecker and Jay, for example – so there must have been small groves of trees in the area. In France, along the north side of the Pyrenees, Clot & Mourer-Chauviré (1986) report from Late Glacial (Würm 4) deposits Diver (Black-throated or Great Northern?), Sooty Shearwater, Mute and Whooper Swan, White-fronted and Greylag Goose, Mallard, Teal, Shoveller, Pochard, Ferruginous Duck, Long-tailed Duck, Velvet and Common Scoter, Smew, Goosander and Red-breasted Merganser, Griffon, Monk and Bearded Vulture, Golden Eagle, Common and Long-legged Buzzard, White-tailed Eagle, Sparrowhawk, Goshawk, Peregrine, Hobby, Kestrel, Merlin and Eleanora's Falcon, Willow Grouse, Ptarmigan, Black Grouse and Capercaillie, Grey Rock and Barbary Partridge, Quail, Crane, Water Rail, Corncrake, Little Auk, Wood Pigeon, Stock and Rock Dove, Snowy and Eagle Owl, Swift, and numerous passerines, including both Alpine and Red-billed Choughs, Raven, Snow Finch, Fieldfare, Ring Ouzel, Blackbird, and others. This certainly looks like a fauna of open rocky ground, in a cool climate, well to the south of its current range; the most abundant species, which include Snowy Owl, Willow Grouse, Ptarmigan, and Alpine Chough, certainly indicate this. The aquatic species, which were not confined to the Mediterranean end of the Pyrenees, also have a northern caste. Similarly, in Crimea, the commonest species in the Glacial fauna of Adzi-Koba are Grey Partridge, Alpine Chough and Song Thrush, with Willow and Black Grouse, Calandra and Crested Lark , Hawk Owl and Short-eared Owl, Red-footed Falcon, and Kestrel all suggesting rather open conditions, certainly not full woodland, though Hawfinch, Jay, Magpie, and Blackbird, like Black Grouse and Song Thrush, imply at least scrubby woodland in the area (Benecke, 1999). The southernmost faunas come from various caves on Gibraltar, of Late Glacial date: they include breeding Velvet and Common Scoter (juvenile bones of both), which suggest northern species breeding far to the south of their present range, but also such southern or woodland species, presumably exploiting the glacial refuge, as Griffon Vulture, Lesser Kestrel, Rock and Stock Dove, Wood Pigeon, ?Tawny Owl, Alpine Swift, Common/Pallid Swift and Azure-winged Magpie, as well as Eagle Owl, Jackdaw, and Red-billed Chough (Cooper, 2005). The Azure-winged Magpie is much the most interesting species here: not only does its presence this early, and in the expected glacial refugium (Cooper, 2000), confirm that it was indeed native (not a Portuguese Mediaeval introduction), but genetic evidence has since confirmed the distinction of Iberian from Asian populations (Fok et al., 2002), a good example of palaeontological and genetic evidence supporting each other.

If the glacial periods saw temperate faunas pushed south into refuges in the Mediterranean region, or south-east towards the Black Sea, the interglacial periods must similarly have seen the species that breed on the tundra pushed northwards into very small northern refuges. It is believed that both the Hoxnian and Ipswichian Interglacials were warmer than the present, Flandrian or Holocene, period. In that case, tundra must have been even more restricted than now, and broken into smaller patches (Kraaijeveld & Nieboer, 2000). It is suggested that the species and subspecies of various waders and geese reflect this fragmentation of range. For

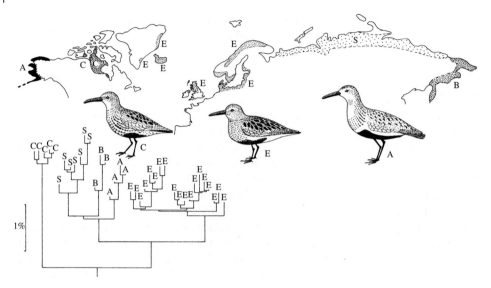

Fig. 2.7 Dunlin subspecies and phylogeny. The molecular phylogeny separates off the N Canadian (C) populations (*Calidris alpina hudsonia*) as the most distinct, then splits Siberian/Alaskan from European (with Greenland) populations. Within the former, Alaskan (A, *C. a. pacifica*), Beringian (B *C. a. sakhalina*) and N Siberian (S, *C. a. sibirica*) forms are distinct, but the various supposed European forms (E, *C. a. schinzii, C. a. arctica, C. a. arcticola*) are not separable from nominate *C. a. alpina* (after Wenink *et al.* 1996).

example, the three populations (now species) of Golden Plover now have a roughly continuous circumpolar distribution, but probably owe their separation to Eurasian Golden Plover *Pluvialis apricaria* being isolated on the tundra of central Europe during a glacial, Pacific Golden Plover *P. fulva* being isolated in the tundras of southern Siberia, and American Golden Plover *P. dominica* being isolated south of the Canadian ice sheet in the tundra of the American plains. During the interglacials, small areas of tundra in, respectively, northern Siberia, eastern Siberia, and northern Canada, would have hosted these diverging species, enforcing their evolving distinctions. Similar splits in range could have produced the subspecies of Bar-tailed Godwit, Brent Geese, and other northern species. The best documented species, for both morphological and mtDNA distinctions, is the Dunlin, *Calidris alpina* (Wenink *et al.*, 1996; Kraaijeveld & Nieboer, 2000). The Canadian form *C. a. hudsonia* differs by 3.3% in its DNA, suggesting that it became isolated from the Palaearctic forms about 225 ka, in the Pre-Ipswichian Interglacial (Figure 2.7). The European *C. a. alpina* differs by 1.73% from Siberian forms, suggesting a divergence about 120 ka, in the Ipswichian Interglacial. The three subspecies from central Siberia (*C. a. centralis*), eastern Siberia (*C. a. sakhalina*), and Alaska (*C. a. pacifica*) differ by 1.05–1.18%, suggesting divergences about 71–80 ka, around the end of the Ipswichian or beginning of the Wurm Glaciation. Interestingly, within the European *C. a. alpina* group there are three apparent subspecies (*C. a. alpina, C. a. schinzii, C. a. arctica*), which can be recognized by their measurements, but which show no genetic differentiation. These are presumed to be Postglacial divergences, associated with restriction to their respective present breeding grounds in northern Scandinavia/Russia, the Baltic and Spitzbergen/north-east Greenland areas.

Conclusions

The earliest bird, *Archaeopteryx*, was a contemporary of the pterosaurs, the flying reptiles, in southern Germany 150 Ma. It was very reptile-like in its skeleton, but its feathers, arranged much as in modern birds on its wings, leave no doubt that it could fly. During the subsequent Cretaceous period, birds from various parts of the world, including Spain, Mongolia, and China, show skeletons more like those of modern birds, shortening their tails, losing teeth and finger claws, and gaining the keeled sternum and tall coracoid that house the flight musculature needed for strong flight. By the end of the Cretaceous, however, there is little evidence for any of the modern orders of birds, which only appear in the early Tertiary, in the Eocene of Denmark, Britain, Germany, and the USA. A tropical bird fauna of early relatives of many modern orders occurs in the London Clay of this time, but there is then a gap in the fossil record of birds in Britain until the end of the Pliocene and the Pleistocene. By this time, modern genera and species are evident. The fluctuating climate of the Pleistocene period saw warm interglacial faunas alternating with cold glacial faunas, and must have caused substantial changes in range. Evidence of changes in DNA, and of modest evolutionary changes, produced subspecies and closely related species, in response to the ranges being split by climatic-induced change, but little evidence of the substantial evolutionary changes as shown by the contemporary mammals (where new genera and species of voles, lemmings, and elephants characterize different stages of the Pleistocene). It looks as though birds mostly responded to the changing climate by moving, rather than evolving (though the more sedentary owls and grouse may have changed size or diet). It is not clear whether migratory habits evolved during this time, as well, but it seems very probable, particularly for species that breed in the high arctic, such as geese and waders. Harrison (1980c) suggested that the appearance of Corncrake and Whimbrel remains in Bed 1 at Olduvai Gorge, Tanzania, dated back at about 1.9–1.7 Ma, imply that the long-distance Palaearctic–African migration system was already a feature of bird biology long before the more severe Pleistocene glaciations conspired to push temperate birds southwards or eastwards out of Europe. Longer northern feeding periods in summer might have attracted northwards movements in spring, and the shorter feeding periods of winter might have pushed them south in autumn, even if temperatures were not severe enough to enforce this. More fossil records from Africa are needed to strengthen these conjectures.

3
Coming in from the cold

The Last (Devensian, in Britain; Wurm, in the Alps) Glaciation ended in a period of oscillating climate that took some 5,000 years, from 15 thousand years ago (ka) to 10 ka, to settle into the post-glacial warm climate, and temperate habitats, that now characterize western Europe. At about 15 ka, as the ice caps of the glacial maximum melted, a fauna and flora typical of northern tundra moved into Britain: open vegetation with Reindeer, lemmings, and Woolly Mammoths feeding on it. Continued warming led to a period of birch scrub, at least in southern Britain, in which more southern species such as Red Deer and Aurochs made a short-lived appearance, though Wild Horse and Reindeer were probably still the most abundant large ungulates; this warmer period is termed the Windermere Interstadial (because it is well indicated in the muds of Lake Windermere). Then, about 11 ka, the climate reverted to being much colder; tundra vegetation returned, a small ice-cap reformed over the Scottish Mountains, and lemmings and Reindeer reappeared to dominate the fauna, along with some steppe species such as Steppe Pika *Ochotona pusilla*. Known as the Younger Dryas to archaeologists, as the Loch Lomond Readvance to geologists, and Pollen Zone III to the palynologists, this colder phase was probably responsible for the final extinction in the British Isles (Great Britain, Man, and Ireland) of the Giant Deer (Irish Elk) *Megaloceros giganteus*. At about 10 ka, however, a final warming, and an increase in mean July temperatures of about 8°C in 50 years, saw the end of the Last Glaciation and the beginning of the present Flandrian Interglacial, otherwise referred to as the Postglacial or Holocene period.

The Windermere Interstadial saw Human hunters of the Upper Palaeolithic culture established in many cave sites in southern and central Britain. So far as we know, they did not reach Ireland or Scotland. It seems likely that they hunted principally Wild Horse, Reindeer, and Mountain Hare (Campbell, 1977; Charles & Jacobi, 1994), but the cave sites they occupied have also given us some evidence of the bird life. The Younger Dryas was probably too cold for Humans to survive here, or they were so sparse that they left little evidence of their presence. As the climate warmed in the Postglacial, Humans quickly returned, but now of the Mesolithic culture, using a more refined armoury of stone tools that included tiny flints, microliths, that were probably used to make barbs on spears. They did spread into Scotland and Ireland; they too were hunter-gatherers, but the Reindeer and Wild Horse quickly died out, and the Mesolithic hunters pursued instead Red and Roe Deer, Elk, Wild Boar, Aurochs, and Beaver. Their hunting activities collected birds as a small part of their food supply, and the shoreline and lakeside sites which they occupied have left us a reasonably coherent account of the changing avifauna during these rapidly changing times. These Late Glacial and Mesolithic faunas mark the beginning of the present avifauna of Britain. Moreover, they suggest what species we ought to have in the absence of later human interference, so they merit examination in detail.

Late Glacial birds

The Late Glacial sites that have been excavated include some of the classic cave sites of British archaeology. At Creswell Crags, straddling the Derbyshire/Nottinghamshire border, the fauna from Robin Hood's Cave has ^{14}C dates on Mountain Hares, 12,600–12,290 b.p., that fit neatly in this period. The avifauna has been discussed by Campbell (1977) and Jenkinson (1984), relying on identifications by the late Don Bramwell. Most numerous are Willow/Red Grouse and Ptarmigan, contributing 16 of 41 birds identified. Also present were Tengmalm's Owl, Hawk Owl, two Short-eared Owls, three Kestrels, and two Goshawks as well as single ?Mallard, ?Goldeneye, Black Grouse, ?Grey Plover, ?Great Spotted Woodpecker, Jackdaw, ?Jay, ?Magpie, ?Ring Ouzel, ?Fieldfare, and some finches or buntings. This certainly looks like a northern fauna, perhaps with some woodland in the river valley but open moorland or tundra on the surrounding high ground. A similar fauna is described from the Mendip Hills of Somerset at about this time. From Unit 3 of Soldier's Hole, Cheddar Gorge, Willow/Red Grouse and Ptarmigan are present in most layers, and dominate the fauna (Bramwell, 1960a; Harrison, 1988). Also present are Mallard, Wigeon, Teal, White-tailed Eagle, Merlin, Black Grouse, Hazel Hen, Grey Partridge, Black-tailed Godwit, Rock Dove, Long- and Short-eared Owl, Hedge Sparrow, Blackbird/Ring Ouzel, Fieldfare, Snow Bunting, Raven, Jackdaw, and Magpie. Harrison (1988) commented on the uncertainty of dating of this fauna, but subsequently direct radiocarbon dates on a Grey Partridge femur (12,370 b.p.) and a Black Grouse tibiotarsus (12,110 b.p.) have been obtained, putting them at least firmly in the Windermere Interstadial (Jacobi, 2004). Gough's Old Cave, also in Cheddar Gorge, has Saiga antelope dated to 12,380 b.p. and Wild Horse dated 12,530–12,260 b.p., so well in this period; the similar avifauna with Greylag Goose, Mallard, Teal, Tufted Duck, Goosander, one of the few records of Golden Eagle from southern Britain, Hobby, Black Grouse, Willow/Red Grouse, Ptarmigan, Black Grouse, Grey Partridge, Lapwing, Rock Dove, Great Spotted Woodpecker, Blackbird/Ring Ouzel. Fieldfare, Song Thrush, Redwing, Stonechat/Whinchat, Snow Bunting, Red-billed Chough, and Jackdaw (Harrison, 1989b) is presumably contemporary. One additional species deserves particular notice – the Great Bustard is also present, one of its very few records from Britain. Again, this is essentially a fauna of open ground, though there must have been some woodland on the lower ground in nearby valleys or on the Somerset Levels, to support such species as the woodpecker.

The identifications of Willow/Red Grouse and Ptarmigan are interesting. The Willow Grouse *Lagopus lagopus*, Willow Ptarmigan in North America, is a circumpolar species characteristic of the scrubby birch and willow that constitute the transition zone between the boreal conifer forests (taiga) to the south and the open tundra to the north (Figure 3.1). The Ptarmigan *Lagopus muta*, the Rock Ptarmigan of North America, is the species that replaces it on full tundra, and higher up mountains. The Ptarmigan shows a classic 'glacial relict' distribution across Europe. It too has a circumpolar distribution on the northern tundra, but it also occurs at high altitude in the Alps, Carpathians, and Pyrenees, these being relicts of its former more widespread distribution in the Late Glacial Period, depicted by Tyrberg (1991a) (Figure 3.2). Ptarmigan bones mostly overlap in size with those of Willow Grouse, but while the wings are very similar, the hind legs are shorter. Thus the tibiotarsus and tarsometatarsus average significantly shorter and narrower, with little overlap, and can usually be identified safely. The Willow Grouse is represented now on the British Isles by a very distinctive race, the Red Grouse *Lagopus lagopus scotica*, which is adapted to live on

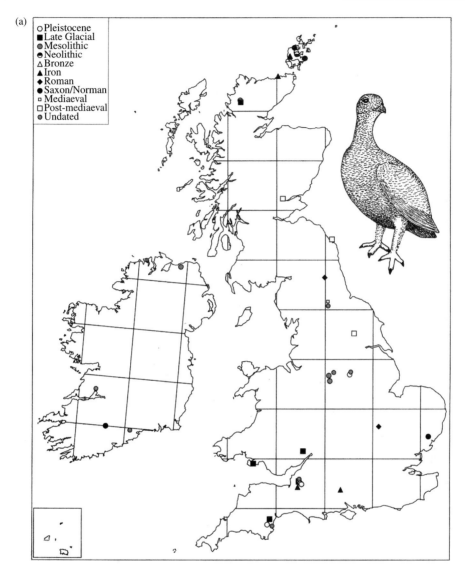

Fig 3.1 Distribution of Red/Willow Grouse. In the British Isles (a), mostly recorded in earlier periods, and in the upland, NW areas near or within its present range. In Europe (b), found in the Late Glacial as far S as the Pyrenees and Alps, where it no longer exists, but also extensively across the lowlands of central Europe, well S of its present (stipple) range (after Tyrberg 1991a).

heather moorland, and to feed largely on Heather *Calluna vulgaris* itself. For many years, it was regarded as a full species, the only British endemic bird, and as a consequence it appears on the front cover of the popular magazine *British Birds*. Unlike Willow Grouse, it does not have white wings, nor does it turn white in winter, which are presumably adaptations to the milder winters and less snow cover of the British Isles. However, as shown by

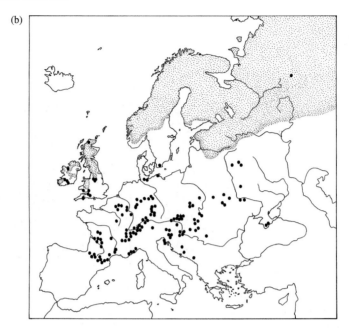

Fig. 3.1 *Continued*

Tyrberg (1995), their ranges were continuous in the Late Pleistocene, and it is only in the last 10,000 years, since the ranges of Willow Grouse and Red Grouse were separated by their retreat northwards and the opening of the North Sea between them, that these changes can have evolved. The fact that grouse, unlike many birds, do not migrate or even disperse very far, must have helped this microevolution. Studies of both ringed and radio-tagged Red Grouse confirm that they rarely move more than 20 km from where they hatched. It is an interesting point that several authors, describing the Late Glacial grouse from Britain, refer to them as having thicker shorter beaks than the modern Red Grouse with which they have been compared (Newton, 1924a; Bramwell, 1960a; Harrison, 1987b). While modern Red Grouse do not seem to differ in beak size from the continental Willow Grouse measured by Kraft (1972), it seems that their Late Glacial common ancestor did have a slightly different beak. In a more detailed analysis, in which, however, he does not consider beak sizes, Stewart (2007a) notes that the legs of these Late Glacial *Lagopus* are thicker than their modern British descendants, implying that they were heavier than modern Red Grouse and Ptarmigan.

Somewhat unexpectedly, evidence from mitochondrial DNA implies that Willow Grouse and Red Grouse have actually been distinct for much longer than 10,000 years. This topic has not been fully explored, but Lucchini *et al.* (2001) tentatively suggest that Scottish and Swedish grouse differ by 3.13% in their cytochrome *b* gene. There is some variability in the rate at which these genes change in the grouse subfamily, but an average tetraonine rate suggested by Drovetski (2003) is 7.23% change per million years. On that basis, the 3.13% difference between them should have required about 433,000, not 10,000, years to evolve;

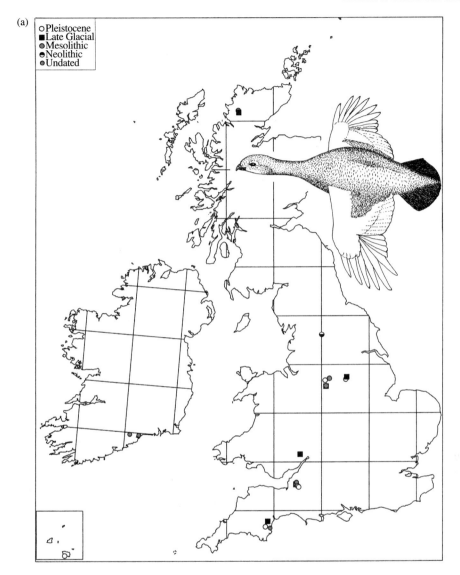

Fig. 3.2 Distribution of Ptarmigan. In the British Isles (a), once found much further S, along with Red Grouse. In Europe (b), had a more extensive range, but mostly in hilly areas, largely absent from the lowlands. Present range (stipple) mostly in the far north, but relics of its former range remain in the Pyrenees and Alps (after Tyrberg 1991a).

implicitly, they split sometime in the mid Pleistocene, and have remained distinct during at least four glacial/interglacial sequences since. It is possible that the difference identified between Scottish and Swedish grouse represents a difference between ancestral populations of each that survived later glacial periods in refuges in, respectively, Iberia and the Balkans.

Fig. 3.2 *Continued*

An alternative explanation is that these particular genes in *Lagopus* have changed much more quickly, so that applying an average tetraonine molecular clock is misleading us. It would be interesting to know if Norwegian Willow Grouse also differ so much, genetically, either from Scottish or Swedish grouse; Norwegian Willow Grouse on the outer islands are reported to be more like Scottish Red Grouse in their plumage.

The records of Hazel Hen are also worth comment. This is a species no longer found in Britain, but it remains widespread in central and northern Europe. It has a very extensive range through, mostly, the coniferous forest zones, though its diet is extensively the leaves, buds, and catkins of deciduous species, especially birch, alder, and hazel. Geographically its range overlaps broadly with both Willow Grouse and Black Grouse, though its habitat is rather different from both, as it prefers thick shrubby cover. There seem to be only five records of this species from Britain, all from the south-west (Table 3.1, Figure 3.3). One, much older than the others, comes from the Post-Cromerian levels of Westbury-sub-Mendip, probably contemporary with the Boxgrove site, about 500,000 b.p. The others are all from the Mendips, and probably all Late Glacial, though most were excavated some time ago, when dating was less assured than now. The Hazel Hen is much smaller than the other grouse, and the identifications at least seem secure.

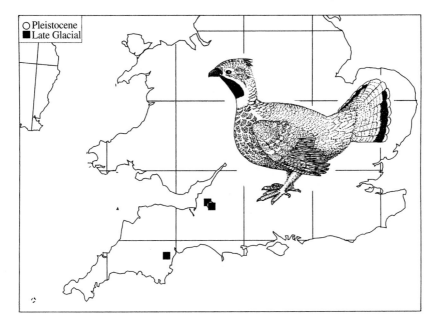

Fig. 3.3 Distribution of Hazel Hen. Known only from a few sites in SW Britain, in the Late Glacial, but possibly overlooked, and likely to have been more widespread in Postglacial woodlands (see Table 3.1).

Table 3.1 Records of Hazel Hen *Bonasia bonasus* from Britain.

Site	Grid Ref	Date	Citation
Westbury-sub-Mendip	ST 50 50	Post-Cromerian	Andrews (1990); Tyrberg (1998)
Bridged Pot Cave, Ebbor Gorge	ST 52 48	Late Glacial	Harrison (1987b)
Soldier's Hole, Cheddar	ST 46 54	Late Glacial	Harrison (1988)
Chudleigh Fissure, South Devon	SX 86 78	Late Palaeolithic	Bell (1922); Harrison (1980a); Harrison (1987b)
Chelms Combe Rock Shelter, Cheddar	ST 46 54	Late Glacial	Harrison (1989a)

Younger Dryas

The colder period of the Younger Dryas seems to have forced Human hunters to retreat south of Britain, and there is little evidence for them in cave sites of this period; consequently there are few well-dated avifaunas from this period either, as the best sources for subfossil birds are archaeological excavations of sites where Humans had been hunting them. A little further north from Gough's Cave, the birds from Chelm's Combe Shelter include again Willow/Red Grouse and Ptarmigan, along with Barnacle Goose, Hazel Hen, Little Auk, Eagle Owl, Blackbird/Ring Ouzel, and Song Thrush (Harrison, 1989a). This site has Reindeer dated to between 10,910 and 10,190 b.p., putting this well in this period. (Note though that Stewart (2007b) cautions that the Eagle Owl could in fact be a Snowy Owl, as the metacarpal is intermediate in size, and not morphologically distinct.) Bridged Pot Cave Shelter, also in the Mendips, is believed to be of a similar date, and has Smew, Hazel Hen, Willow/Red Grouse, Ptarmigan, Grey Partridge, and Skylark (Harrison, 1987b). Ossom's Cave, in the Manifold Valley of Staffordshire, is another site with dated Reindeer, of 10,780 and 10,600 b.p. (Scott, 1986); the birds from this cave include Ptarmigan, Black Grouse, Golden Plover, Eagle Owl, Jackdaw, and Raven (Bramwell, 1955, 1956). The birds seem not very different from those already recognized from the Late Glacial, and it may be that the paucity of sites has prevented finer discrimination. The boreal nature of the fauna is evident, and most are species of open ground. Merlin's Cave, in the Wye Valley, is another with Willow/Red Grouse and Ptarmigan, along with Mallard, Pine Grosbeak, Crossbill, Starling, Hawfinch, Jay, Jackdaw, and, rather improbably, House Sparrow (a contaminant, of later date, or a misidentification?). Harrison (1987b) lists this fauna as transitional between Late Glacial and Postglacial, and Price (2001), in a recent study of the small mammal fauna in the cave, gives four dates for small mammals that match this expectation, from Mountain Hare at 10,270 to Norway Lemming at 9685 b.p. Another small Late Glacial or transitional fauna has been described from Wetton Mill Rock Shelter, also in the Manifold Valley: it includes Red Grouse, Ptarmigan, Black Grouse, Capercaillie, and Grey Partridge, along with ?Greylag Goose, ?Mallard, and ?Jay (Bramwell, 1976a), suggesting open country on the limestone plateau above the cave, and scrubby woodland along the river below it.

Mesolithic birds

Early Mesolithic sites are also likely to be transitional between the open conditions of the Late Glacial and the fully wooded conditions of the later Mesolithic. Dowel Cave, near Buxton, has a Mesolithic fauna, identified as such by its associated small mammals (Yalden, 1999), that includes Water Rail, Red and Black Grouse, Capercaillie, Grey Partridge, Stock Dove, and Great Tit, as well as uncertainly identified finch, pipit/wagtail, chat, and thrush species (Bramwell, 1960b, 1971, 1978c). Demen's Dale, also in Derbyshire, though not well-dated, has a larger fauna, of ?Teal, ?Garganey, ?Gadwall, Wigeon, Shoveler, Pintail, Goldeneye, Goosander, Kestrel, ?Ptarmigan, ?Black Grouse, Grey Partridge, ?Grey Plover, Snipe, Dunlin, ?Knot, Eagle Owl, Tawny Owl, Mistle Thrush, ?Blackbird, Hawfinch, and Jay (Bramwell & Yalden, 1988). (This Eagle Owl at least is certain, an unmistakable tarsometatarsus, that may be the latest record of the species as a native in Britain.) The nearby River Wye must have been ponded at this time (evidence of Beaver in the fauna identifies one possible cause), and the mixture of woodland, wetland, and open-ground species may again reflect the juxtaposition of a sheltered wooded dale and open limestone plateau country above it. Alternatively, of course, this could be another transitional fauna, mixed in time.

One of the classic and best studied Mesolithic sites in Britain is Star Carr, in the Vale of Pickering about 8 km south of Scarborough (Clark, 1954; Legge & Rowley-Conwy, 1988). It is also one of the earliest sites, with dates of about 9,488 b.p., in other words within about 700 years of the abrupt warming at the end of the Younger Dryas. The hunters' camp was on the shore of a lake, with reed beds, sedges and birch scrub, and they primarily hunted large ungulates – Red and Roe Deer, Elk, Aurochs, and Wild Boar. A small bird fauna was also excavated, though, and the bones have been recently reviewed, re-identified in some cases, by Harrison (1987a). Not surprisingly, waterbirds dominate: Red-throated Diver, Great Crested Grebe, Dabchick, Brent Goose, Red-breasted Merganser, Common Scoter, and Common Crane. (Earlier identifications of White Stork, Common Buzzard, Pintail, and Lapwing have been discounted by Harrison, 1987a.) There has been some speculation that the absence of fish bones in this excavation indicates that the freshwater was still too isolated from continental source populations, following the Late Glacial, for a fish fauna to have colonized (Wheeler, 1978). A glance at the bird fauna is sufficient to indicate that, with four fish-eating specialists out of seven, the fish must have been present, but not preserved (or perhaps, just not recovered) (as Price, 1983, also pointed out). Another classic and well-dated Mesolithic site, Thatcham in the Thames valley west of Reading, has a small fauna of five species, Mallard, Teal/Garganey, Goldeneye, thrush sp., and, again, Common Crane (King, 1962). Dog Holes Fissure, in the Creswell Crags, also has a small avifauna of Robin, ?Song Thrush, ?Blackbird, ?Ring Ouzel, ?Blue Tit, and Jackdaw (Jenkinson, 1984; birds reidentified later by C. Walker (R. Jacobi pers. comm)). A woodland or woodland edge in present-day Derbyshire could produce the same species. However, the site contains an odd mixture of wild and domestic mammals, and the bird fauna might therefore also be confused across time and habitat.

A rather different Mesolithic fauna is provided by the coastal site of Morton in Fife (Coles, 1971). Marine species dominate – Fulmar, Cormorant, Shag, Gannet, Razorbill, Guillemot, Puffin, Great Black-backed Gull, and Kittiwake, with just thrush sp. and Crow to indicate a terrestrial element. A number of other coastal Mesolithic sites are on the small island of Oronsay, off Colonsay. Once thought to be Neolithic, the most recently dug, at least, has radiocarbon dates, 6,200–5,100 b.p., that show them ranging from late Mesolithic into early Neolithic (Mellars, 1987). As well as Cormorant, Shag, Gannet, Whooper Swan, goose sp., Shelduck, Red-breasted Merganser, Ringed Plover, gull sp., Common Tern, Razorbill, Guillemot, and Water Rail, Oronsay produced some of the earliest Great Auk bones ever excavated from an archaeological site (Grieve, 1882, Henderson-Bishop, 1913). Risga, another shell-midden site on a small island, is thought to be about the same age as Oronsay. Sited in the narrows at the mouth of Loch Sunart, it is less open to the sea, but has a very similar bird fauna of 11 species, all of them found also at Oronsay: Cormorant, Shag, Gannet, goose sp., Red-breasted Merganser, gull sp., Common Tern, Razorbill, Great Auk, Guillemot, and Water Rail (Lacaille, 1954). In Northern Ireland, the site of Mount Sandel, near the River Bann just south of Coleraine, produced a small bird fauna, including Capercaillie, Goshawk, Wood Pigeon, and Song Thrush, which suggest woodland, together with Red Grouse, Rock Dove, and Golden or White-tailed Eagle, which suggest open conditions. The Red-throated Diver, Mallard, Teal/Garganey, Wigeon, and Coot reflect the wetlands in the valley, while Snipe/Woodcock is of uncertain identification and therefore interpretation (Van Wijngaarden-Bakker, 1985).

The largest bird faunas of apparently Mesolithic age come from Port Eynon Cave in the Carboniferous Limestone of the Gower Peninsula, south Wales (Harrison, 1987b), and Wetton Mill Rockshelter in similar limestone in the Manifold Valley in the Staffordshire segment of the Peak District (Bramwell, 1976a). These record 43 and 22 species, respectively, though some degree of contamination seems to have taken place; the alleged appearance of Domestic Fowl at Port Eynon is hard to take seriously. Dating of the two sites is also uncertain – Port Eynon is given dates of 9,000 to 6,000 b.p., i.e. contemporary with the Mesolithic, which is the broad dating given for Wetton Mill. As a coastal site, Port Eynon has a fair number of (wintering or breeding?) marine species – Black-throated Diver, Manx Shearwater, Shag, Gannet, Long-tailed Duck, Common and Velvet Scoter, Little Auk, Puffin, Guillemot, and Razorbill. There are as well as a number of maritime species, which may well have frequented the coastal cliffs – Peregrine Falcon, White-tailed Eagle, Great Black-backed Gull, Raven, Red-billed Chough, Rock Pipit, and Black Redstart – or the mud-flats in the bays that characterize that coastline – Barnacle Goose, White-fronted Goose, Shelduck, Wigeon, Grey Plover, and Turnstone. However, there must also have been some woodland nearby, for the Sparrowhawk, Wood Pigeon, Song Thrush, Mistle Thrush, and Chaffinch/Brambling. There must also have been extensive open ground on the limestone hills, though, for the most spectacular bird was the Great Bustard, at another of its very few occurrences in the British record, along with Corncrake, Golden Plover, Wheatear, Crow/Rook, Blackbird/Ring Ouzel, Redwing, and Fieldfare, which might have shared its foraging grounds (Harrison, 1987b).

Conversely, as an inland site in a limestone dale only 3 or 4 km from the moors to its west, Wetton Mill has several gamebirds, including Capercaillie, Black and Red Grouse, Ptarmigan, and Grey Partridge. Their predators included Common Buzzard, while woodland species include Jay, Treecreeper, Great Tit, Chaffinch, Spotted Flycatcher, Blackbird, Song Thrush, Redstart, Robin, and Tawny Owl. As well as Black Grouse and Buzzard, species of the woodland/moorland fringe – those feeding in the open but sheltering or nesting in woodland – include Starling, Crow/Rook and perhaps the uncertainly identified bunting sp., which is most likely here to have been Reed Bunting, but could have been Yellowhammer instead. Greylag Goose and Mallard complete the fauna, probably present thanks to the river in the dale bottom, but perhaps associated with nesting sites on the neighbouring, more open, moorland (Bramwell, 1976a).

These Mesolithic faunas comply quite well with what we might expect at the present day in English wooded countryside, and on present-day coastlines. Species such as Mistle Thrush, Tawny Owl, Hawfinch, and Jay suggest the current fauna, not a more northern one, and the coastal species likewise match well the current fauna. More northern species, like Ptarmigan and Little Auk, are scarce. One or two species attract attention. Eagle Owl is not now present in Britain; rather, it is, just one or two pairs, as a consequence of recent escapes, and most birders regard this as a dangerous introduction of a non-native species that should be discouraged (e.g. Mead, 2000). All the evidence is to the contrary. As a large predator, it is never likely to have been especially common, but there is a trickle of records of Eagle Owls through the Pleistocene and into Mesolithic and perhaps Iron Age times (Table 3.2). The Iron Age specimen is a broken ulna, and its identification is uncertain, but the unmistakable tarsometatarsus from Mesolithic Demen's Dale leaves no doubt about identity (Stewart, 2007b). More speculatively, the Neolithic/Beaker barrows at Longstone Edge, Derbyshire, produce an abundance of small mammals and amphibians that have been interpreted as the contents of Short-eared and Eagle Owl pellets (Peter Andrews, unpublished), though no bones of the owls themselves have been found on the site. Most records are in southern England (Figure 3.4), though perhaps the Eagle Owl should have survived longest in Scotland. As a largely woodland predator, it has a wide range throughout Eurasia, and it would be surprising indeed if it had not been a native species to Britain in the Postglacial; it is also one most likely to have been exterminated by Humans, indirectly through habitat

Table 3.2 Claimed records of archaeological Eagle Owl *Bubo bubo* in the British Isles. In his review of these, Stewart (2007b) suggests that the Chelm's Combe specimen could be Snowy Owl, and that from Meare was an uncertain identification.

Site	Grid Ref	Date	Citation
East Runton	TQ 20 42	Pastonian	Harrison (1979, 1985)
Boxgrove	SU 92 08	Post-Cromerian	Stewart (2007)
Swanscombe	TQ 60 74	Hoxnian	Harrison (1979, 1985)
Tornewton Cave	SX 81 67	Wolstonian	Harrison (1980a, 1980b, 1987b)
Langwith Cave, Derby	SK 51 69	Devensian	Mullins (1913)
Chelms Combe Rock Shelter, Cheddar	ST 46 54	Devensian	Newton (1926); Harrison (1989a)
Merlin's Cave (Wye Valley Cave)	SO 55 15	Late Pleistocene	Newton (1924a); Tyrberg (1998)
Kent's Cavern	SX 93 64	Late Glacial/ Post Glacial	Bell, (1915, 1922); Bramwell (1960b); Tyrberg (1998)
Ossom's Cave	SK 09 55	Late Glacial	det. Cowles 1974 unpub; Bramwell, (1960a)
Demens Dale, Taddington	SK 16 71	Mesolithic	Bramwell, (1978c); Bramwell & Yalden (1988); det A Hazelwood unpub.
Meare Lake Village	ST 44 42	Iron Age	Bate (1966)

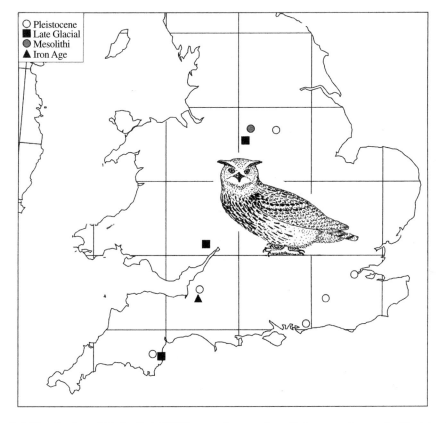

Fig. 3.4 Distribution of Eagle Owl. Well known from various Pleistocene sites, and with at least a marginal presence in the Postglacial period (Table 3.2).

change or directly by persecution. Modern Eagle Owls feed very extensively on Rabbits, which were not available to them in Mesolithic Britain. However, they are powerful enough to eat Hedgehogs, despite their prickles (17% of 784 prey items in Bavaria – Bezzel & Wildner (1970)), Water Voles would then have been much more plentiful (Yalden, 1999; Jefferies, 2003), and forest grouse such as Black Grouse and Hazel Hen would also have been acceptable prey. It has to be admitted, though, that the sparse British record is in sharp contrast with the archaeological record from Sweden, where the Eagle Owl is recorded from at least 20 sites of Viking age (Post-Roman, by British dating) burials, often associated with the remains of hawks or falcons (Ericson & Tyrberg, 2004). It is evident that it was used as a decoy in falconry, and of course it survives in the Swedish fauna to the present day, so it is less surprising that there are later archaeological records there than in the British Isles. By contrast, the absence of later records from Britain must surely be a reliable indication of its early extinction.

Reconstructing the Mesolithic bird fauna

One problem of comparing these archaeological bird faunas with modern faunas is that small passerines dominate modern faunas, and would surely have likewise dominated Mesolithic woodland faunas, but these are just the species most poorly represented in archaeological faunas, and the ones most difficult to identify with any certainty. A different approach to investigating the Mesolithic avifauna is to attempt to extrapolate, from the basis of what we know of the available habitats then, and analogous faunas now, what might have been expected. Tomiałojc (2000) provocatively asked 'Did White-backed Woodpeckers ever breed in Britain?'. He drew attention to the fact that many bird species characteristic of deciduous woodland have very patchy distributions in or are absent from western Europe; this has often been misinterpreted as some climatic or geographical limitation when it could well be the result of their extinction following the large-scale clearance of woodland from most of the continent (and woodland has, of course, been most drastically cleared in Britain, reduced to perhaps 4% coverage by 1895; Rackham, 2003). Woodpeckers are perhaps not the best species to investigate archaeologically, as they are always scarce in such sites, but the White-backed occurs at present very widely in eastern and northern European woodlands, as well as in the Pyrenees. It is then very likely that it once had a continuous range across the western part of the Continent. Hazel Hen, Black Stork, Three-toed and Black Woodpeckers, Ural and Tengmalm's Owls, and Nutcracker have similarly discontinuous ranges in western Europe, and might be predicted to have once had more extensive and continuous ranges. The presence of Eagle Owl, Hazel Hen, and Black Stork in the archaeological record from Britain indicates that this is more than a theoretical possibility.

The extent of habitats available in the Mesolithic is suggested by the woodland map provided by Bennett (1988), interpreting the pollen record. He shows deciduous woodland over most of southern and lowland Britain at about 7,000 b.p., with lime woodlands important in the south-east, oak in the west, probably ash woodland on the chalk and limestone, and alder woodland in fenlands. In mountainous regions, particularly Scotland, pine woodland would have persisted at lower altitudes with birch woodland at higher altitudes and further north. Only the highest hills and mountains would have had a mantle of open vegetation,

Table 3.3. Likely extent of habitats in Mesolithic mainland Great Britain (from Maroo & Yalden 2000). These figures are based on the averaged pollen record at about 7000 b.p., as described from 22 sites across Britain. The raw figures were modified by the various "correction factors" suggested by Faegri & Iversen (1975) to correct for the differential production of pollen by these plants.

Vegetation	% cover	Area (km^2)
Birch *Betula* woodland	9.28	20,426
Pine *Pinus* woodland	6.00	13,207
Deciduous *Quercus-Tilia-Corylus* woodland	43.23	95,154
Grassland Gramineae	19.25	42,371
Fenland Cyperaceae	8.11	17,851
Heathland/Moorland Ericaceae	8.49	18,687
Other herbs (herbs, ferns, sphagnum)	5.65	12,436
Total		220,111

something akin to modern moorland. However, in interpreting this map, it is important to remember that it is based on the pollen record. This is dominated by the abundant pollen produced by some of the wind-pollinated trees – especially oak, hazel, birch, pine, and alder. Insect-pollinated species (such as lime) and the shorter herbaceous species, even if wind-pollinated (as are grasses and sedges), are seriously under-represented. If the original pollen records are corrected for this bias (which pollen analysts themselves have suggested: Faegri & Iversen, 1975), rather more extensive grasslands, in among the trees, are predicted (Table 3.3). Any clearings might have been temporary, created by Mesolithic hunters or by the large mammals, particularly Aurochs, that they were hunting, or they might have been more permanent grasslands in river valleys, on chalk downland or in exposed coastal areas. It does seem that some expanses of open habitat were available for birds of open country, even in the most wooded of Mesolithic countrysides (Vera, 2000; Svenning, 2002).

The Białowieża National Park, on the Polish–Byelorussian border, has just such a mixture of habitats. There are open broad river valleys, filled with tall herbs, grasses, sedges, and rushes. Riversides are flanked by alder–ash woodland. On better soils, hornbeam-lime-oak woodlands dominate, while the poorest soils have pure spruce-pine woodland. On boggy ground, a strange (to more Western eyes) wet spruce–birch–alder woodland predominates (Jędrzejewska & Jędrzejewski, 1998). There are important differences from the woodlands of western Europe: oak is less dominant, spruce is naturally present and numerous, while beech is absent. Even so, it is probably the nearest we can get, ecologically and geographically, to the woodland countryside we might have had between 8,000 and 5,500 years ago. More importantly for present purposes, the bird faunas of each forest type have been very thoroughly and comprehensively surveyed (Tomiałojc *et al.*, 1984), enabling the sort of extrapolation that we wish to attempt. Throughout the various woodland types, the Chaffinch is the most abundant species there, averaging about 150 pairs/ha in deciduous woodlands, about 100 pairs/ha in conifers. Altogether, it contributes about 20% of all the bird territories recorded (Table 3.4). This matches well with the perceptions by Yapp & Simms that it is also the commonest species overall in present-day British woodlands (Yapp, 1962; Simms, 1971); they found it was the most numerous species in oakwoods (12–18% of all contacts), alderwoods

Table 3.4 The common breeding birds in the Mesolithic woodlands of Britain, about 7000 b.p., arranged in descending order of their estimated Mesolithic abundance. Numbers in deciduous woodlands estimated from the likely area of available woodland then (Table 3.3, above) and the average densities in the 7 oak-hornbeam census plots (W, WE, WI, CW, CE, MN, MS) sampled by Tomiałojc et al. (1984). Alder woodlands similarly estimated from their alder and alder-ash plots (L and H), and pine from their pine-bilberry plots (NW and NE). Birch woodland communities estimated from timed censuses in Scottish birch woods given by Yapp (1962), applied to plausible overall densities from Białowieża. Estimates of the current G.B. populations are from Gibbons et al. (1995), taking the higher figure when a range is suggested. The totals are of all woodland bird species present, not just the 23 most abundant species listed here.

Species	Deciduous Woodland	Coniferous Woodland	Alder Woodland	Birch Woodland	Total Population	Current GB Population
Area (km^2)	95,154	13,207	17,851	20,426		
Chaffinch	13,797,330	1,294,286	1,979,046	945,724	18,016,386	5,400,000
Robin	6,089,856	673,557	1,026,446	516,778	8,306,637	4,200,000
Wood Warbler	6,660,780	719,782	760,604	42,895	8,184,061	17,200
Pied Flycatcher	5,804,394	26,414	841,833	-	6,672,641	40,000
Hawfinch	3,996,468	26,414	25,884	-	4,048,766	6,500
Song Thrush	3,140,082	257,537	273,227	257,368	3,938,214	990,000
Willow Warbler	36,702	6,604	17,851	3,094,539	3,155,696	2,300,000
Great Tit	2,093,388	19,811	369,225	171,578	2,654,002	1,600,000
Wren	1,617,618	39,621	561,222	428,946	2,647,407	7,100,000
Blackcap	1,807,926	33,018	539,069	-	2,350,297	580,000
Blue Tit	1,712,772	6,604	339,687	171,578	2,224,037	3,300,000
Tree Creeper	1,427,310	191,502	413,552	-	2,032,344	200,000
Blackbird	1,522,464	92,449	383,994	-	1,998,907	4,400,000
Goldcrest	1,284,579	303,761	347,072	-	1,935,412	560,000
Chiffchaff	884,932	125,467	590,760	-	1,601,159	640,000
Nuthatch	1,237,002	6,603	339,687	-	1,576,689	130,000
Tree Pipit	566,846	33,018	26,777	774,145	1,400,786	120,000
Starling	1,237,002	-	8,123	-	1,237,002	1,100,000
Hedge Sparrow	489,963	92,449	428,301	-	758,635	2,000,000
Coal Tit	95,154	198,105	8,926	257,368	559,553	610,000
Golden Oriole	299,055	92,449	660,461	-	526,208	40
Sp. Flycatcher	312,649	6,604	30,347	85,789	435,389	120,000
Crested Tit	67,967	165,088	-	-	233,055	900
Total Birds	62,733,673	4,813,951	13,254,368	7,006,118		

(18%), beechwoods (17%), and most conifer woods (6–28%), as well as in some highland birch woods (displaced to second by Willow Warbler in others). Also abundant in Białowieża's deciduous woodlands are (in descending order) Wood Warbler, Robin, Collared Flycatcher, Hawfinch, Song Thrush, Starling, Blackcap, Great Tit, Blue Tit, Blackbird, Goldcrest, Wren, Tree Creeper, and Nuthatch. Coniferous woodlands there produce a similar list of commonest species – Chaffinch again the most abundant species, followed by Wood Warbler, Robin, Goldcrest, Song Thrush, Coal Tit, Tree Creeper, Crested Tit, Chiffchaff, Golden Oriole, Hedge Sparrow, and Blackbird. Greater abundance of Goldcrest and the replacement of Great and Blue Tit by Coal and Crested Tit is just what we might expect on ecological grounds and British experience. Some differences are harder to explain. Redstart is much less abundant in Białowieża woodlands than we might expect from British experience, languishing at 62nd place in the list of birds in deciduous woodlands, behind such more abundant species as Wryneck and Red-backed Shrike! Possibly competition from within the more abundant guild of hole-nesting passerines limits its abundance, but it also seems to have different habitat preferences in the east, favouring wet spruce woodlands (Fuller, 2002). Similarly the Willow Warbler, such a familiar species especially in northern Britain, is only 24th on the list of birds in the mixed birch–pine woodlands of Białowieża, well behind Wood Warbler (second) and Chiffchaff (10th). In this case, we can surmise that the rather open, scattered, birch scrub that suits it in Britain is not available in Poland, though it is also possible that the more oceanic climate of Britain suits the birch aphids and other insects on which it feeds better than the continental climate of Poland. Also notable is the fact that birds overall are much less abundant, per km^2, in Białowieża than in many British woods; for instance, densities of the nine commonest passerines combined are about 586 pairs per km^2, compared with 1946 per km^2 at Wytham Woods, Oxford, for the same species. Tomiałojc *et al.* (1984), pointing this out, discuss the reasons for it. There is no evidence that less food or more unoccupied space produces the lower densities in Poland, nor is there any evidence that the hole-nesting species suffer from a shortage of nest holes in the abundant old trees (though nest boxes at Wytham Woods certainly boost densities of tits there). They conclude that the much more numerous and complete range of predators, including species that can and do raid nests, such as Weasels, Stoats, Pine Martens, and various woodpeckers, produce a much lower density of tree- and canopy-nesting species. The greater density of trees, and of browsing ungulates, also means that ground- and shrub-nesters have less cover, so that such species as Wren and Blackbird have much larger territories than in Britain. An interesting argument in support of their thesis is that the New Forest, notable for its high density of grazing ungulates (albeit ponies and cattle as well as deer), has a combined density of the same species of about 744 pairs per km^2, much closer to the Polish figure.

Such differences make it more difficult to extrapolate from the Polish avifauna to a hypothetical British Mesolithic one, but it is still worth an attempt. Table 3.4 lists the likely numbers of the some of the most numerous species in British Mesolithic woodlands, as extrapolated from the nearest equivalent Polish plots. In a few cases, the species in Poland does not occur in Britain, and probably never has done. Collared and Pied Flycatchers both occur in Poland, even hybridize, and it is assumed that in Britain the equivalent biomass of *Ficedula* would all have been Pied Flycatcher. For British birchwoods, the rank order of the species present has been adopted from equivalent studies in Scotland (Yapp, 1962), and assigned the overall density of birds in the Polish mixed coniferous–deciduous woodlands (343 pairs/km^2). The commonest 10 or so species in each of the major habitat types are listed in Table 3.4, to allow some comparisons; it is evident that areas of conifer, and therefore absolute numbers of conifer specialists, were much smaller than areas of deciduous woodlands, and their specialists.

Tentative as these figures must be, they suggest a Chaffinch population some three times greater than now. Wood Warblers being so much more abundant than say Chiffchaffs may seem odd to present-day British ornithologists, but makes sense if the woodlands really were structured, as are Białowieża woodlands now, with a high canopy but sparse lower cover (because Aurochs and Red Deer would have eaten much of it). Modern beech woods, on the Cotswold and Chiltern Hills, for example, are much like this, and the Wood Warbler is indeed a characteristic bird, though not usually outnumbering the Chiffchaff (table 19 in Simms, 1971). There are other surprises, from a British (present-day) perspective. One is the already remarked absence of Common Redstart from a list of common birds present in deciduous woodlands, even more surprising when the Pied Flycatcher, so often its associate in present-day western oak woods, is so abundant. The Redstart was barely recorded in the censuses of deciduous woodland in Białowieża.

This rank order bears some discussion. If woodland was as extensive as all the pollen and other records indicate, then the much greater abundance in the Mesolithic of classic woodland species, not only Chaffinch but also Pied Flycatcher, Wood Warbler, Goldcrest, Tree Creeper, Nuthatch, Song Thrush, and Robin, matches expectations. But the greater abundance now of some of the other species also merits attention: Blackbird, Wren, and especially Hedge Sparrow were apparently much less common then, and their greater abundance now reflects the fact that they are essentially birds of woodland edge and hedgerow – habitats that are now much more abundant. Within Białowieża, Fuller (2000) found Hedge Sparrows, Chiffchaffs, and Blackcaps specifically associated with the small clearings created by treefalls. Further, although they didn't differ greatly in abundance between obvious gaps and closed-canopy woodland, Blackbirds and Wrens did prefer small gaps rather than large ones.

It is hard to know whether to take seriously the predicted greater abundance of Crested Tit, which seems to have now a wider niche in Europe than in Britain (there it is by no means confined to pine forests), and the Golden Oriole is another hard to take at face value. On the other hand, it was somewhat warmer in the Mesolithic than now, and the species has managed a modest colonization of East Anglian poplar plantations since the 1960s, with about 20–40 pairs each year. Perhaps it really was once much more abundant, but there is no archaeological evidence.

It is noteworthy that all the birds listed in Table 3.4 are passerines. There are, of course, many other species present in today's Polish woods, as there would have been in the Mesolithic woodlands of Britain, but they would not number in the top 15 or so. Wood Pigeon comes nearest, with densities ranging from 4.0 to 6.7 pairs/km^2, and an estimate of some 770,689 pairs for Mesolithic Britain. Collectively, though, woodpeckers, eight species in all, are among the most common non-passerines; in decreasing order of abundance, Middle Spotted, Great Spotted, Lesser Spotted, Three-toed, White-backed, Black, Grey-headed and Green, collectively about 26 pairs of woodpeckers per km^2, or the equivalent of about 3,350,000 pairs in Mesolithic Britain – assuming they or at least some of them, reached such densities in the mature, rot-infested, woodlands that then existed. The three species present today total only 50,000 pairs (Gibbons *et al.*, 1993). This returns us to Tomiałojc's perceptive question, of whether White-backed Woodpeckers ever bred in Britain. Unfortunately, woodpeckers do not have a good fossil record, though their bones are distinctive enough to be recognized when present. Tyrberg (1998) does list 184 records of Pleistocene woodpeckers in the Palaearctic, of 14 taxa (including Wryneck and some extinct forms), but unfortunately does not cover the post-Pleistocene, which would be most useful here. The Swedish Postglacial record includes only seven woodpeckers (three Green, three Great Spotted, one Black), emphasizing their scarcity as fossils (Ericson & Tyrberg, 2004). This matches the British record: there are only 14 records in our data base, nine of Greater Spotted Woodpeckers, two each of the other two species, and a Wryneck. Most of them are dated to the Late Glacial, and reflect the better faunas of smaller birds available from cave sites. Kear (2003) points out that hole-nesting ducks such as Goosander, Goldeneye, and Mandarin require the large holes excavated by Black Woodpeckers (or similar large species elsewhere in the world), and speculates that their rarity as nesting birds in Britain, until nest boxes were supplied in abundance, may well have been prompted by the absence of large woodpeckers here. Even this does not much help the present argument; Goldeneye (24 records) and Goosander (16 records) are present in sites all the way from Pleistocene to Medieval, including the Mesolithic records from Demen's Dale and Thatcham given above, but as they winter here as well as breed, this tells us nothing of their earlier status, nor does it imply anything about the former presence (or absence) of Black Woodpecker as a breeding species supplying them with nest holes. Tyrberg (1998) does include Late Glacial records of White-backed Woodpeckers from Austria and the Czech Republic, as well as eastern France, northern Italy, and Crete, suggesting their presence then in Alpine birch forests, while his few records of Black Woodpeckers include three from eastern France, one from Poland, and three from Georgia. Neither species occurs near enough to suggest that they might once have inhabited Britain, though one uncertain record of White-backed Woodpecker from Belgium might hint at this. While absence of evidence is not necessarily evidence of absence, at present we have no firm indication that other woodpeckers once occurred in Britain. Kear (2003), accepting this position, suggests that the scarce fauna of ants and longhorn beetles in Britain deprives the Black Woodpecker, in particular, of its important winter food supply, that this is due to the cooler more oceanic climate, which is therefore responsible indirectly for the absence of big woodpeckers, and scarcity of hole-nesting ducks. She emphasizes, though, the damaging effects of historical and modern woodland management, which respectively removed much woodland cover and removes dead and 'overmature' trees, the ones that have nest holes and wood-boring insects. Woodland clearance is evidently to blame for the absence now of woodpeckers from Ireland, where the records, albeit poorly dated, from the Newhall and Alice Caves of County Clare make it clear that Great-spotted Woodpeckers did once occur there (D'Arcy, 1999; Yalden & Carthy, 2004). It is worth noting that one of these 'poorly dated' Irish woodpeckers has in fact been dated now; a ^{14}C date of 3750 b.p. puts it in the Bronze Age (D'Arcy, 2006).

The group likely to be of greatest interest for historical reconstruction is the guild of predators; without serious human persecution or interference, their abundance and diversity in the pristine Mesolithic woodlands should tell us much about what bird life should be like in Britain. The commonest predatory birds in Białowieża now are Common Buzzard and Tawny Owl. Extrapolating to Mesolithic Britain, we might have had about 75,000 pairs of Buzzard and 160,000 pairs of Tawny Owl (Table 3.5). Given their present abundance, now that they have recovered from the pesticide era, it may seem surprising that Sparrowhawks are predicted to have been much less numerous than these two, though still common and widespread, at about 21,500 pairs. Goshawks were almost as common, at 14,000 pairs. There are two ecological points made here. One is that mammalian prey are much more abundant than avian prey, so predators of woodland mammals, Buzzard and Tawny Owl, are much more abundant than the two bird hawks. The other, of course, is that Sparrowhawks are themselves the prey of Goshawks: among 52 cases of other predators eaten by Goshawks

Table 3.5 How many predatory birds? Estimated numbers of the principle predatory birds in Mesolithic Britain. Arranged in decreasing order of their abundance then. Their present numbers and the number of Postglacial archaeological records are given for comparison. For woodland species*, derived from estimated densities in Białowieża (Jędrzejewska & Jędrzejewski 1998) and the areas of woodland suggested in Table 3.3. +Harriers from river valley densities in Białowieża, and areas of grassland in Table 3.3. For other species, modern densities in good British study areas and areas of grassland, heathland and herbaceous communities, or along river valleys, as appropriate (see text), were used. Current population estimates from Gibbons et al. (1995), updated from Ogilvie et al. (2003). Numbers of archaeological records in our data-base cover all periods from Mesolithic to Post-medieval.

Species	Density pairs/km²	Mesolithic Population	Current Population	Archaeological Records
*Tawny Owl (conifers)	0.55	(7,264)		
(deciduous)	1.6	(152,246)		
(total)		159,246	20,000	24
*Common Buzzard	0.585	75,404	17,000	107
*Sparrowhawk	0.167	21,507	32,000	44
*Honey Buzzard	0.136	17,515	70	0
*Goshawk	0.108	13,909	320	41
Kestrel	0.32	13,559	50,000	31
Osprey	0.06	3,832	158	5
+Montagu's Harrier	0.073	3,093	16	1
Hen Harrier	0.15	2,803	570	7
*Hobby	0.026	2,474	700	1
+Marsh Harrier	0.056	2,373	194	14
Peregrine	0.0026	2,257	1,200	22
White-tailed Eagle	0.002	1,858	23	50
Red Kite	0.4	1,128	440	70
Barn Owl	0.019	1,041	4,400	36
*Eagle Owl	0.0035	451	2	2
Golden Eagle	0.022	411	420	10
Merlin	0.02	374	1,300	6

in Białowieża were 22 Sparrowhawks, four Goshawks, and seven unidentified *Accipiter*, as well as 13 Tawny Owls (Jędrzejewska & Jędrzejewski, 1998).

It is not possible to extrapolate from Białowieża to the numbers of some of what we would now consider our more typical predators, because Kestrels do not breed there, Barn Owls are scarce or erratic, while Merlins and Hen Harriers occur only sporadically. To get some inkling of their putative population sizes, these have been estimated by taking the average densities in the main study sites used by Taylor (1998) for Barn Owl, Village (1990) for Kestrel, the Langholm study site of Redpath & Thirgood (1997) for Hen Harrier, and the better densities reported by Rebecca & Bainbridge (1998) for Merlin, applied to the relevant (grassland or heathland) areas in Table 3.3. Open habitats were then much less extensive than now, so it would be expected that these species were then less common. Red Kites, Peregrines, and Ospreys too are scarce or absent in Białowieża, making any estimate for them especially difficult, while White-tailed and Golden Eagles are limited to a pair or two each. Yet all of these must have been well established in Mesolithic Britain, to judge from their sub-fossil records (Table 3.5), and plausible (we hope) estimates for former times have, again, been derived from what seem like reasonable recent analogues for populations of former times (not reduced by habitat change or persecution). For Ospreys, the former territories are assumed to have been confined to lakes and larger rivers, at a spacing of 5.3 km, equivalent to the six pairs/100 km^2 quoted for Scandinavia and elsewhere by Poole (1989). Similarly, White-tailed Eagles are assumed to have been spaced at 16 km along freshwaters and the western coasts, reflecting the higher densities suggested by Love (1983). Both species can breed semi-colonially, especially on offshore predator-free islands, but it is assumed for present purposes that this did not happen in Britain. Golden Eagles were presumably confined to moorland, at the density of 0.2 pairs/km^2 suggested by Watson (1997) and Brown (1976). Brown (1976) doubted that White-tailed Eagles were ever more numerous than Golden Eagles in Britain. Contrarily, both these extrapolations and the archaeological record (Yalden, 2007) suggest that they were once at least three times, perhaps five times, more numerous. Red Kites were probably fairly scarce; as Lovegrove (1990) suggests, they are essentially birds of open country, perhaps then found mainly along river valleys. For Białowieża, where they are rare and irregular breeders, Jędrzejewska & Jędrzejewski (1998) certainly regard them as birds of river valleys. If they were so confined, at a density of about a pair to 2.5 km^2 (0.4 pairs/km^2), along riverine clearings 250 m wide, they would have had territories 10 km long, implying about 1,100 pairs. This is a very tentative estimation, as Red Kites are not really territorial, indeed can be semi-colonial, but implies a comparable abundance with the Barn Owl, another bird that would have been scarcer then through lack of habitat than in more recent times. The Peregrine is another species that is barely present in Białowieża, though it bred sparsely in the past, nesting in old crow's nests alongside the river valleys and clearings. In Britain, it must always have been more abundant, especially on coastal cliffs. It is harder to assess its former numbers inland, when moorland was less extensive and there were fewer grouse or pigeons. Jędrzejewska & Jędrzejewski (1998) suggest territories spaced 22 km apart, equivalent to a pair per 0.0026 km^2, while Brown (1976) suggests a spacing, on average, of 5.4 km on British sea cliffs. Applying the former figure to main river valleys and the latter to (the western) half of the British coastline, suggests a total of about 2,200 pairs. Ratcliffe (1980) detailed about 400 coastal eyries, including those on islands not accounted in the figure for the western coastline of mainland Britain, so perhaps 1,744 pairs of coastal Peregrines in former times is too generous an estimate. And as there is little evidence that British Peregrines ever nested in crow nests, perhaps the figure of 513 inland (riverine) pairs is also suspect. This is one species whose estimate must remain very tenuous. Ratcliffe (1980) thought that the combination of woodland cover and more numerous rivals (Ravens, eagles) for nesting cliffs might have limited it then to a few hundred pairs.

The fact that Białowieża is a more continental site makes some of the estimates contentious – it is a pleasant thought that Honey Buzzards might once have been nearly as common in Britain as Sparrowhawks, but it is very uncertain that wasps, their main food, were really sufficiently common in our more oceanic climate to have supported such an abundance. Worse, though Honey Buzzard bones are claimed by continental authors to be relatively easy to distinguish from other similar-sized raptors (*Buteo, Milvus, Accipiter gentilis, Circus aeruginosus* (Otto, 1981; Schmidt-Burger, 1982)), there are no records in our data base of their former presence in archaeological sites. Similarly, it seems unlikely that our impoverished reptile fauna was ever sufficiently abundant to support Short-toed Eagles, and we have no archaeological evidence of their presence. The same reasoning suggests omitting some of the boreal species that occur, albeit rarely, in Białowieża but were unlikely members of the Mesolithic bird community in Britain. Thus Pygmy, Tengmalm's and Great Grey Owl, which do breed in Białowieża, and Hawk Owl, a rare nomad there, probably never occurred in Postglacial Britain, though both Tengmalm's and Hawk Owl are recorded in Late Glacial Creswell caves (see above, p.38).

Overall, the suggested raptor community contained as expected far more woodland than open country species. Moreover, predators that rely on mammals (Tawny Owl, Buzzard) were more abundant than those reliant on birds (Sparrowhawk, Goshawk), which is appropriate as biomasses of available mammalian prey are likely always to have been an order of magnitude greater than those of avian prey (Harris *et al.*, 1995; Greenwood *et al.*, 1996; Maroo & Yalden, 2000). All the passerines listed in Table 3.4 would have contributed a biomass of about 3,300 tonnes, while the four common rodents (Field Vole, Bank Vole, Water Vole, and Wood Mouse) would have provided about 16,800 tons. Shrews and Moles would have been extra prey for those that eat them. Wood Pigeons, perhaps 807,682 tons, and gamebirds would have added substantially to the available biomass of avian prey, but then other, larger, mammals would also have been widely available, live or as carrion.

Birds of open country

As noted above, it has been customary, but possibly misleading, to regard Mesolithic Britain as completely wooded. The vegetation cover suggested by the 'corrected pollen rain' (Table 3.3) suggests instead substantial areas in total of grassland, sedges, and mixed herb communities. However, that is a calculation based on an averaged pollen rain for Britain, and whereas we have some ideas about the geographical distribution of different woodlands, we are very uncertain about the distribution of grasslands. Were these extensive plains on the chalk downlands or small glades in among the trees? We can suppose that larger river valleys and fenlands in low-lying areas were open country, on the basis of both comparisons with Białowieża and on the nature of the pollen rain – Godwin (1975) points out that even when woodland dominated the landscape, some sites had more grass than tree pollen. The fauna – Elk and Aurochs particularly – would have needed larger areas for grazing, and might have created or maintained them (cf. Vera, 2000), and Beavers, by causing waterlogging, would certainly have created openings in the landscape (Coles, 2006). The pollen record from chalk country is generally poor – wet acid bogs and pools preserve pollen well, dry base-rich sites do not. Although Bennett (1988) maps the chalk downs as Ash woodlands, he notes that the evidence for these is patchy, and they might well have been patchy themselves; it is even possible that the Ash was a secondary invader of cleared patches, not part of the original woodland cover at all. Patchy grasslands could well have supplied the grass pollen to the pollen rain, without being substantial enough to support such birds of really open country as Grey Partridge and Skylark, let alone Stone Curlew or Great Bustard. However, investigations of the chalk landscape along the Dorset/Hampshire border (Allen & Green, 1998) have found good evidence of woodland, not only Red and Roe Deer but remains of open Hazel and Ash woodland in the early Mesolithic, and more complete woodland cover somewhat later in the Mesolithic period. One category of open habitat seems clearly defined geographically – anything resembling moorland would have been confined to mountain tops in Wales, the Pennines, and, more generally, in Highland Scotland.

In evaluating the pollen rain (Table 3.3), it is assumed that the incidence of *Calluna* pollen represents this upland community.

So what were the populations of open-country birds likely to have been? Moorlands in the Highlands and elsewhere above the natural tree-line would have hosted populations of Red Grouse, Twite, Ring Ouzels, and Whinchat, with Ptarmigan, Dotterel, and perhaps Wheatear confined as now to the barer and stonier summits. Tree Pipits and Willow Warblers would, as now, have frequented the birch scrub at the edge of moorland, but will have been accounted in the previous estimates. Small patches of coastal and southern heathlands could similarly have harboured Stonechats, Wood Larks, and perhaps Dartford Warblers, but the evidence is that these were indeed small patches of open habitat in essentially a wooded countryside (e.g. Seagrief, 1960). In the Highlands and Islands, wind-swept coasts would surely have had heathland of some sort, with Twite, as now, one of the typical birds. Valley grasslands might have hosted Cranes, Corncrakes, Skylarks, and Meadow Pipits, as well as riparian species such as Sedge, Reed, and Marsh Warblers. Blanket bog, as dominated by either *Sphagnum* or *Eriophorum* at the present day, had not developed – that was a consequence of later tree clearance by Humans in combination with a wetter episode of the climate – so the disposition of its characteristic wader community – Golden Plover, Greenshank, Redshank, Dunlin, Curlew – is uncertain. Quite possibly, this community did not exist, and its species were spread between coastal salt marshes, river valleys, and wet forest clearings. Alternatively, the machair and moorland of the Outer Hebrides was never tree-covered, and might have carried a community somewhat like this before agriculture reached the islands.

To attempt some rough approximation of the numbers involved, let us suppose, somewhat arbitrarily, that valley grasslands occupied the 11,287 km of river (Order 3 or larger waterways in the classification of Smith & Lyle, 1979) with a width of 250 m on average; that would give about 2,822 km² of grassland, and account for only 7% of the grass pollen. The rest came from small patches of grassland within the woodlands. However, the suggested 17,851 km² of sedge community probably also occupied the river valleys, boosting the amount of open habitat available. We will work on the assumption that it amounted to 20,673 km² in the Mesolithic. Unmanaged grasslands might contain about 10 pairs/km² of Skylarks and 50 pairs/km² of Meadow Pipits, assuming enough large mammals grazing them to create the appropriate structure, suggesting Mesolithic populations of about 206,700 and 1,033,700 pairs, respectively. Red Grouse might number 20–25 pairs/km² on unmanaged moorland (HBWP), suggesting about 420,500 pairs formerly. With about 2,000,000 pairs of Skylarks and 1,900,000 of Meadow Pipits, they are now a rather more numerous, but Red Grouse seem scarcer, with 250,000 pairs currently estimated, despite the obvious expansion of moorland (Gibbons *et al.*, 1993). This perhaps surprising relative abundance of open country birds, at a time when closed woodland has been assumed to have been the predominant vegetation type, matches the abundance of Field and Root Voles suggested by Maroo & Yalden (2000) for this period, and might go some way to reconcile the evidence that woodland itself, and woodland animals in general (e.g. molluscs, beetles) was indeed predominant, yet grassland specialists, including downland flowers, for example, managed to survive through this wooded period (Svenning, 2002). Does bird archaeology have any direct evidence for this argument? Obviously, the record of passerines is weak, but the Skylark has a distinctive size and morphology among passerines, so offers a better prospect than most for addressing this

Table 3.6. The occurrence of open-ground birds in the archaeological record in the British Isles, showing the number of records of each species. The total for these open-ground species is compared with the total bird records for that time interval (% in brackets). Totals (right hand column) include a few extra records of uncertain age.

Spp.	Pleist	Late Gl.	Meso	Neo.	Bron. Age	Iron Age	Rom.	Ang-Norm	Med	Post-Med.	Total
Crane	-	-	2	3	11	17	34	35	34	11	155
Corncrake	-	1	2	1	-	4	5	4	4	-	24
Grey Partridge	11	7	3	1	2	3	14	10	50	17	126
Skylark	10	7	-	5	3	4	5	2	12	5	54
Crested Lark	2	2	-	-	-	-	-	-	1	-	6
Total open-ground	23	17	7	10	16	31	58	51	101	33	365
	(4.2)	(4.3)	(3.4)	(2.9)	(8.8)	(4.7)	(3.3)	(4.6)	(4.4)	(3.1)	(4.1)
Total Birds	539	398	203	344	181	664	1755	1108	2295	1075	8953

question. Other, larger, open ground birds might be more useful, and those with a reasonable record include Corncrake, Crane, and Grey Partridge. As Table 3.6 shows, these yield good numbers of Iron Age, Roman, and later records, which might seem to reflect the expansion of open ground as farming increased its impact on the landscape. However, when the records are expressed as a percentage of all bird records for each period, it looks as though there is in fact no trend with time: throughout, these species have contributed about 3–4% of the bird records, even in the apparently well-wooded Mesolithic. It seems as though the manipulations of the pollen record and estimation of attempted habitat types provide estimates of bird populations that are in fact confirmed by what we have of an archaeological record. The Mesolithic wooded landscape did indeed include enough grasslands, whether in river valleys, small clearings or the uplands, to support at least a selection of open-ground species. Some open-ground species do, though, seem to have been genuinely less common – the Red Kite, Kestrel, Lapwing, and Curlew for instance, seem genuinely scarcer then (see Appendix).

Conclusions

The archaeological record of birds in Late Glacial and early Postglacial, that is, Mesolithic, Britain provides a record that is good enough to document the expected change from a fauna dominated by species of open (tundra-like) habitats to one made up largely of woodland species. Despite the popular notion that most of the British Isles were covered in coast-to-coast woodland, both the pollen record and the archaeological bird fauna indicate that there were some patches of open habitat, grassland, sedge, or moorland. The archaeological record of small passerines, which would surely have been the most numerous birds then, as now, is too poor to give a direct indication of their relative numbers, and an extrapolation, based on a combination of likely habitat availability and densities of small birds present in Polish woodlands, is needed to complete an impression of the Mesolithic bird fauna in Britain. Numerically, the Chaffinch was probably the most numerous bird, as it still is in British woodlands, but there are hints, both archaeological and speculative, of a more exotic fauna than now. Extinct species such as Eagle Owl and Hazel Hen were certainly present, and woodland passerines might have included large numbers of Hawfinch, a species that does (because of its distinctive size?) get reported more frequently than expected from archaeological sites. It is interesting that the common woodland edge birds of the present day countryside (Wren and Blackbird, for example) were likely to be less numerous than Robins or Song Thrushes, which we tend not to appreciate as more specialist woodland birds. Among predators, Tawny Owls and Buzzards would have been much the most numerous, reflecting the abundance of Wood Mice and Bank Voles on the woodland floor. Sparrowhawks would, surprisingly, have been less common than now, held down by the predation of other predators (Goshawks and mammals such as the Pine Marten), and by the fact that small bird densities were themselves lowered by mammalian predation on their nests. And then there are some speculations that are so unexpected, and little supported, that they can only be put forward as interesting hypotheses, deserving further thought and investigation. Were Golden Orioles and Honey Buzzards really once as numerous as the extrapolations suggest? If so, why have we no archaeological evidence to confirm that? Is it that woodland animals generally are less likely to get into archaeological sites than coastal, cliff-dwelling, or open-ground species? Or are our reference collections too poor in such unexpected species that they have not been recognized – perhaps lost, respectively, among the thrushes or other buzzards? Wójcik (2002) describes how to tell oriole humeri from the very similar-sized thrushes, and the German doctoral theses are similarly helpful when it comes to discriminating *Pernis* from *Buteo* or *Accipiter gentilis* (Otto, 1981; Schmidt-Burger, 1982), but these were not available to earlier archaeologists.

4
Farmland and fenland

Culturally, the break from Mesolithic to Neolithic Britain, from hunter-gatherer to farmer, seems quite sharp. Archaeologists have been debating for many years whether immigrant farmers, bringing their crops and livestock, slowly displaced their Mesolithic forbears, or whether what they introduced was the idea of farming, to be taken up with enthusiasm by both existing inhabitants and newcomers. It is now possible to examine the ratios of two naturally occurring isotopes of carbon, ^{13}C to ^{12}C, to explore this point. Marine foods, which many Mesolithic communities exploited, are much higher in ^{13}C, while terrestrial foods, cereal crops, livestock, or game, give lower values. Some Mesolithic hunters exploited terrestrial sources, of course, so their $^{13}C/^{12}C$ ratios are not very different from those of later people. However, there is a very sharp change, at about 5,200 b.p., in the $^{13}C/^{12}C$ ratios of coastal peoples. Between 6,000 and 5,200 b.p., coastal Mesolithic people were still gathering seafood, but coastal Neolithic people very quickly and apparently completely abandoned this habit (Richards *et al.*, 2003).

This matches what archaeological records show. Sheep and goats, which have no wild ancestors in Europe, were brought in, already domesticated, by those first farmers, as were cereals. These first farmers must have been impressive sailors, for they very quickly reached Orkney in the north and Ireland in the west with their livestock. The exploitation of wild mammals, like the use of marine sources, also dropped very dramatically (Yalden, 1999). In creating fields for their crops and grassland for their livestock, they started to create the pattern of land use, and therefore the bird fauna, that we expect to see today in most of the countryside.

Neolithic birds

Some of the best Neolithic sites are, unexpectedly, not in south-east England, which we must suppose was the first to be settled by farmers, but in the far north, on Orkney, and in the west of Ireland. As coastal sites, their avifaunas are dominated by seabirds (Table 4.1).

Among the best (most diverse) avifaunas are those from Knap of Howar on Papa Westray (43 spp.; Bramwell, 1983c), Quanterness (40 spp; Bramwell, 1979a), and Isbister (22 spp.; Bramwell, 1983a) on Mainland, and Links of Notland (39 spp.; Armour-Chelu, 1988) and Point of Cott (27 spp.; Harman, 1997) on Westray. Collectively, four sites on Rousay also contribute 21 species (Davidson & Henshall, 1989), and Pierowall Quarry on Westray adds another three species (McCormick, 1984). While none of the species occurred on all sites, several were present in most, including Shag, Cormorant, Gannet, Oystercatcher, Guillemot, and Great Black-backed Gull. The occurrences of Fulmar, at three sites, are notable – it

Table 4.1 Neolithic non-passerine birds recorded in Orkney.

Site	Papa Westray	Isbister	Quanterness	Rousay	Links of Notland	Point of Cott	Others
Source	Bramwell 1983c	Bramwell 1983a	Bramwell 1979	Davidson & Henshall 1989	Armour-Chelu 1988	Harman 1997	
Gt N. Diver	+				+		
Bl-th Diver	+					(sp.?) +	
Mx Shearwater	+				+		
Shearwater sp.	+					+	
Leach's Petrel			+				
Fulmar	+				+	+	
Shag	+	+		+	+	+	
Cormorant	+		+	+	+	+	
Gannet	+		+	+	+	+	+
Bittern				+			
Whooper Swan	+			(sp.?) +	+		
Greylag Goose	+	+	+		+		
Pinkfoot Goose				+			
Barnacle Goose	+						
Shelduck	+						
Eider	+	+			+		
Velvet Scoter	+				+		
Goosander					+		
RB Merganser						+	
Mallard		+					
Teal						+	
C. Buzzard	+		+	+	(sp.?)+		
Wt Eagle		+		+	+	+	
Goshawk		+	+				
Kestrel		+			+		
Red Grouse		+	+				
Spotted Crake	+						
Water Rail					+		
Oystercatcher	+	+	+	+		+	
Grey Plover	+				+		
Lapwing						+	
Curlew	+	+		+			
Redshank	+				+		
Spd Redshank	+						
Greenshank					+		
Snipe	+	+	+		+	+	
Woodcock		+					
Turnstone	+						
Great B-b Gull	+	+	+		+	+	+
LBb/H Gull	+	+			+	+	
Common Gull		+					

Site	Papa Westray	Isbister	Quanterness	Rousay	Links of Notland	Point of Cott	Others
Source	Bramwell 1983c	Bramwell 1983a	Bramwell 1979	Davidson & Henshall 1989	Armour-Chelu 1988	Harman 1997	
Bl headed gull		+			+		
Kittiwake						+	
Great Skua	+						
Razorbill	+				+	+	
Great Auk	+			+	+		
Guillemot	+		+	+	+	+	
Puffin	+	+			+	+	
Black Guillemot	+						
Little Auk		+			+	+	
Short-eared Owl		+			+		

Table 4.2 The archaeological record of Fulmar *Fulmarus glacialis* in the British Isles.

Site	Grid Ref	Date	Citation
Morton, Fife	NO 72 57	Mesolithic	Coles 1971
Links of Noltland, Orkney	HY 42 49	Neolithic	Armour-Chelu 1988
Embo, Sutherland	NH 82 92	Neolithic	Clarke 1965; Henshall & Ritchie 1995
Westray – Point of Cott	HY 46 47	Neolithic	Harman 1997
Papa Westray, Orkney	HY 48 51	Neolithic	Bramwell 1983c
Howe, Orkney	HY 27 10	Iron Age	Bramwell 1994
Dun Bhuirg	NM 27 24	Iron Age	Bramwell 1981a
Old Scatness Broch, Shetland	HU 390111	Iron Age	Nicholson 2003
Crosskirk Broch	ND 02 70	Iron Age	MacCartney 1984
Skaill, Deerness, Orkney	HY 58 06	Iron Age	Allison 1997b
Niarbyl, Isle of Man	SC 21 77	Roman	Garrad 1978
Buckquoy	HY 36 27	Pictish	Bramwell 1977b
Buckquoy	HY 36 27	Norse	Bramwell 1977b
Lindisfarne	NU 13 41	Early Medieval	Rackham 1985
Hartlepool – Church Close	NZ 52 33	Early Medieval	Allison 1990
Rattray, Aberdeenshire	NO 17 45	Medieval	Murray & Murray 1993
Iona – Abbey	NM 28 24	Medieval	Coy & Hamilton-Dyer 1993
Hartlepool – Church Close	NZ 52 33	Medieval	Allison 1990
St Kilda – Hirta	NF 09 99	Post-Med	Harman 1996b
Guernsey – Le Dehus	WV358831	?	Kendrick 1928

has been pointed out that, prior to 1878, St Kilda was its only known British breeding station (Fisher & Lockley, 1954), and its spectacular modern increase in range only began in the late nineteenth century. These Neolithic occurrences indicate that it was formerly much more widespread, as do later records through to early Mediaeval times (Table 4.2). Thus either climatic changes or human hunting pressure reduced its range between prehistoric and modern times. Given its tendency to nest on more accessible slopes than other seabirds, and the low reproductive rate (single egg clutches, late maturation), human overhunting must be strongly suspected.

Nor is this the only species present in these early maritime faunas which has been severely hit by presumed human hunting. Two other notable species are the Great and Little Auk. The former is of course entirely extinct, but it bred on flat rocky islets, and would probably have been available to human hunters only in the breeding season; the latter is a northerly species recorded regularly in winter in British waters, particularly in the north, and perhaps indicative therefore of hunting by humans in winter. On the other hand, as Stewart (2002a) persuasively argues, the presence of Little Auks and other marine species in caves, often far inland, may simply indicate the contemporary 'wrecks' of marine birds that sometimes see seabirds occurring, often indeed dying, well inland in modern times, and does not necessarily provide evidence of human intervention. Other predators than humans could also have taken these as prey, but as these are genuine archaeological sites, human hunting is certainly the most likely explanation for their presence.

If this sounds to contradict the comments opening this chapter (that Neolithic peoples abandoned the earlier seafood predilections of their Mesolithic forebears), it should be emphasized that cattle, pigs, and sheep provided most of the meat they ate, and birds, including seabirds, were only a minor item of diet. At Isbister, for example, some 488 cattle bones and 206 sheep bones dominated the fauna, with only another 63 from Red Deer, Otter, Pig, Dog, and seal (Barker, 1983). At Knap of Howar, large cattle and small sheep dominate the fauna (Noddle, 1983), while at Quanterness, sheep were dominant (Clutton-Brock, 1979). Similarly, at Mount Pleasant, as a southern example, Harcourt (1979b) calculates that 60% of the meat eaten was cattle, 2% sheep and 16% pig, leaving only 21% to come from wild species, including deer and the very few birds.

A rather different species regularly recorded in these northern archaeological sites is the White-tailed Eagle, and its frequency in Isbister led to the soubriquet *Tomb of the Eagles* being applied to the site (Hedges, 1984). Among 745 identified bird bones there, no fewer than 641 were from this one species. At least 10 individuals were represented, scoring the bones directly, but the archaeologist interpreted them as coming from perhaps 14 individuals, when accounting for their distribution across the site and through the layers (Bramwell, 1983a; Hedges, 1984). They clearly had some symbolic purpose, being directly interred with human skeletons. The fact that they, and other carrion feeders (Raven, Great Black-backed Gull), were associated with human burial chambers was surely not an accident. Perhaps we see here some reflection of the 'sky-burials' performed by Parsees and others, in which human corpses are left exposed for vultures and crows to dismember, or perhaps the eagles conferred some status on the deceased. The fact that White-tailed Eagles may gather at carrion makes it easier to envisage the humans collecting some numbers of them, though exactly how they did this remains uncertain. One possible traditional method is illustrated by Love (1983); the eagles were drawn to bait placed near a sunken covered pit, in which the hunter hid himself. He could then grab the legs of any eagle that ventured close enough. Arrows or spears might also have been effective weapons at close quarters. An alternative was to put carrion in a trench, wide enough for the bird to walk in but too narrow for it to spread its wings. The frequency of White-tailed Eagle remains emphasizes that this species was much more widespread, and probably much more abundant, than the Golden Eagle in past times (Yalden, 2007), though their different ecologies mean that one was more likely occur in archaeological sites. The White-tailed Eagle's carrion-feeding propensity perhaps also made it more vulnerable to Humans when they decided to exterminate it, along with other raptors, in the nineteenth century.

Two mainland sites in Scotland have rather similar faunas to these Orcadian sites. Embo, on the east coast of Sutherland, is another site with Great Auk remains (Clarke, 1965; Henshall & Ritchie, 1995). Other sea birds there include Fulmar, Gannet, Shag, Razorbill, and Guillemot, with Red-necked Grebe and duck sp. suggesting freshwater. Capercaillie, Lapwing, Blackbird, and Starling represent the terrestrial fauna, and imply between them a mixture of open ground and woodland. At Carding Mill Bay, Oban, the small fauna similarly includes marine (Guillemot, Razorbill, Great Black-backed and Herring Gull) and terrestrial (finch sp., Crow/Rook, Swallow) species, but the site is most notably another with White-tailed Eagle (Hamilton-Dyer & McCormick, 1993).

In western Ireland, Carrowmore, in County Sligo, yielded few animal bones, and only one, an indeterminate goose *Anser* sp., of a bird (Burenhult, 1980). No more valuable is Newgrange, a famous Neolithic and Beaker Age site in County Meath, which produced remains of one Song Thrush, *Turdus philomelos*, but whose dating is uncertain – recent contaminants, including Rabbits, were also present (Van Wijngaarden-Bakker, 1982). Perhaps most interesting is the site at Ferriter's Cove, on the Dingle peninsula in County Kerry. This is thought to be a transitional site, from latest Mesolithic to earliest Neolithic. As testimony to the latter, there are a few bones of Sheep and Cattle, but the much more abundant remains of Wild Boar imply that hunting was still the major source of meat. A foraging, rather than farming, lifestyle is also suggested by the abundant fish remains, and the few seabirds (Gannet, Guillemot, Herring Gull) imply the same (McCarthy, 1999).

Neolithic sites in southern Britain tend to be associated with the chalk downlands. These may have been easier ground to clear of trees, to create either pasture for livestock or fields for crops, than the wetter soils of the valleys. Although calcareous soils should be good for preserving bones, few bird bones have in fact been recovered, suggesting that fowling was not a common practice at inland sites. It may be significant that sites such as Runnymede (Serjeantson, 1996), West Kennet enclosure (Edwards & Horne, 1997), Windmill Hill (Grigson, 1999), and Ascott-under-Wychwood (Mulville & Grigson, 2007), which do produce substantial numbers of animal bones, yield very few birds. Durrington Walls yielded duck, probably Mallard, and Cormorant, reflecting presumably its location by the River Avon, as well as Red Kite, Raven, and Woodcock, which presumably imply between them rather more woodland than now survives in the area (Harcourt, 1971a). Nearby Stonehenge, most emblematic of Neolithic sites, also produced evidence of Raven (Serjeantson, 1995). From Mount Pleasant, near Dorchester, a complex site ranging from Neolithic through Beaker to Bronze and Iron Age, the Neolithic levels produced Common Crane, and the Beaker age (Neolithic/Bronze Age transition) added Greylag/Bean Goose, Pintail, Song Thrush, and Mistle Thrush (Harcourt, 1971b). However, the best Neolithic faunas from England come from two neighbouring cave sites in the Peak District, 6 km south-east of Buxton. Dowel Cave lies in a small side-valley off the head of Dovedale, while the mouth of Fox Hole Cave overlooks it from the top of a nearby hill, High Wheeldon. Dowel Cave produced a large fauna, including numerous passerines, among which woodland species predominate. Great Tit is the most numerous, but Robin and Redstart, Hedge Sparrow, Bullfinch, Greenfinch, Hawfinch and Goldfinch, Song Thrush, Mistle Thrush, Blackbird and Redwing, Magpie and Wren are also represented.

Among non-passerines, Tawny Owl and Goshawk also imply woodland. However, some open-ground species are also present, including Skylark, Wheatear, Ring Ouzel, Jackdaw, Crow/Rook, Starling, and Linnet among the passerines, Grey Partridge, Kestrel, Barn Owl, and Stock Dove among the non-passerines. As several of the open ground species (Starling, Jackdaw, Crow/Rook, Stock Dove, Kestrel) nest in holes in trees, and several of the woodland species (Mistle Thrush, Blackbird, Magpie) forage out into grassland, it may well have been a rather open woodland, or, given its setting, wooded in the valley but more open on the limestone summits above (Bramwell, 1960b). One of the most unusual species in the collection is a shrike, which must, from its small size, be Red-backed Shrike. As shrikes generally prefer scrub, offering plenty of perches but also open areas to pounce on their prey, its presence is a strong indication of the mixed nature of the habitat at that time. Fox Hole also provided evidence of a more wooded environment than now, when it sits in limestone grassland on a quite treeless hill. Capercaillie suggests not just woodland, but perhaps coniferous woodland, though Black Grouse implies more open scrubby woodland. As one of the few English records of Golden Eagle also comes from this site, it may well be that the eagle nested here, perhaps hunting these grouse over the more acid moorlands about 6 km to the west. Nuthatch, Robin, Blackbird, Jay, and Great Spotted Woodpecker also indicate woodland, Fieldfare, Mistle Thrush, Magpie, and Crow/Rook imply at least some cover of trees, while the Skylark, like the Golden Eagle, suggests nearby open ground (Bramwell, 1978c).

The peat levels in the Cambridgeshire fens have also produced important bird faunas, in non-Human sites, but they probably cover some 4,000 years of Neolithic to Bronze Age times, and are discussed more fully below. Birds may well have been more abundant, and fowling may have been more important economically, there than in the drier uplands, even in the Neolithic.

The overall interpretation of these Neolithic bird faunas suggests that early farmers had made little impact on the landscape, and had little interest in bird resources, except at coastal sites. The birds present are much as might be expected currently in the same locations, though with a few surprises: in coastal sites, Little Auk and Great Auk, as well as currently more familiar seabirds, at inland sites, Golden Eagle and Capercaillie looking unfamiliar in an English setting.

Bronze Age

By the subsequent Bronze Age, starting in southern England about 4,000 b.p., farming had made a more severe change, and large areas of downland, in particular, were essentially treeless. Coneybury, near Stonehenge, produced Lapwing and White-tailed Eagle (Maltby, 1990), while various sites on the Marlborough Downs produced Rook/Crow, thrush sp., Grey Partridge, and Golden Plover, as well as Mallard, Kestrel, and Pigeon/Dove sp. (Maltby, 1992). Potterne, another Wiltshire site, on the Salisbury Plain, produced Teal, Mallard and Greylag Goose, Common Crane, White-tailed Eagle, Buzzard, Woodcock, Blackbird, (House?) Sparrow, Crow and Raven, plus, incongruously for such an inland site, Guillemot (Locker, 2000). Wigber Low, Derbyshire, yielded Skylark and thrush sp. as well as Raven and Woodcock (Maltby, 1983), while the coastal site of Brean Down in Somerset is another with Lapwing, Common Crane, and Starling, implying open country, as well as various thrushes (Robin, Redwing, Song Thrush, Mistle Thrush), Mute Swan, Greylag Goose, Mallard,

Woodcock, Snipe, and Guillemot reflecting other habitats in the area (Levitan, 1990). Sand Martin, Swallow, and Skylark at Wilsford Shaft, near Stonehenge, also imply open grassland, over which the two hirundines would have been hunting (Yalden & Yalden, 1989). Odd records of a few other species from various Bronze Age sites complete the record but add little to the story (Goose, Mallard? and Tawny Owl at Runnymede (Serjeantson, 1996), probable Wigeon at Anslow's Cottages, Burghfield (Coy, 1992), Crow/Rook on Cranborne Chase (Legge, 1991), Black Grouse and Long-eared Owl at Hindlow Cairn, Derbyshire (Bramwell, 1981b)). The occurrence of Lapwing and Golden Plover in these faunas might indicate the start of their habit of wintering on grasslands (well south of its breeding range for Golden Plover, at least), along perhaps with wintering thrushes.

Coastal Bronze Age sites do not differ markedly from the earlier Neolithic ones. Ardnave, Islay, only produced Common Crane and Curlew (Harman, 1983), but Dun Mor Vaul on Tiree (Shag, Cormorant, Gannet, Golden Plover, Puffin, Little Auk, Redstart, Song Thrush, Starling, and Crow; Bramwell, 1974) and Bu Farm on Westray, Orkney (Red Grouse, Teal, Kittiwake, Guillemot, Snipe, Gannet, and Greylag Goose; O'Sullivan, 1996) produced larger faunas. The Broch of Midhowe on Rousay, Orkney, contained Gannet, Shag, Heron, and possibly Oystercatcher but, improbably, Domestic Fowl (later contaminant, or misidentified, perhaps Black Grouse?) (Platt, 1933b). The biggest fauna is that from Jarlshof, Shetland, though this is a complex site (ranging from Bronze Age to Viking times), and identifications are sometimes uncertain, probably due to inadequate reference collections at this relatively early excavation. Species reported from Bronze Age levels there include Great Northern Diver, Storm Petrel, Gannet, Shag, Cormorant, Heron, Bittern, (White?) Stork, swan sp., goose sp., Eagle, Falcon, Lapwing, Turnstone, Herring and Great Black-backed Gull, Skua, and Raven (Platt, 1956). Somewhat oddly, Platt (1933a) earlier included Blue-eyed Shag, as well as Shag and Cormorant, in the list from Jarlshof. This essentially southern ocean species is never likely to have occurred in the North Atlantic, and the identification may simply indicate the scarcity of reference material available to bird bone specialists at that time. It would be valuable to get modern confirmation or greater precision on some of these identifications, but it is not known if the specimens have been kept, perhaps at the Royal Scottish Museum. In the Scilly Isles, at the other end of Britain, the site on Nornour has produced a very interesting and mixed fauna. The seabirds – Manx Shearwater, Cormorant, Gannet, Razorbill, Guillemot, Puffin – are unsurprising, except perhaps for the lack of gulls (Turk, 1971, 1978). Coastal occurrences of Redshank, Knot, godwit sp. and also Ruff seem plausible, though the latter implies more freshwater marsh than is now available, as do Mallard, White Stork, and Grey Heron. On what is now a much smaller island than it was then, the presence of Raven, Black Grouse, and Stone Curlew seems truly remarkable. At least the Blackbird is not surprising. A similar assemblage is reported from Caldicot in Gwent on the Severn marshes – Little Grebe, Grey Heron, Common Crane, ?Brent Goose, Mallard, ?Pintail, ?Wigeon, Tufted Duck, and Barn Owl (McCormick *et al.*, 1997).

Fenland

Mention of the Severn marshes leads reasonably to the most instructive, but in timing most uncertain, of these possible Bronze Age sites. The fenlands of Cambridgeshire and neighbouring counties were once part of a large area, over 3,000 km^2, of wet, bird-rich habitat. It was drained by a succession of Roman, Norman, and later engineers, to give the fertile peat-rich soils that are now prime arable land. Drains and ditches were mostly dug, even in the nineteenth century, by hand, giving many opportunities to encounter and excavate bones. Many of these are now in the University Museums of Geology and Zoology at Cambridge. The peats have, in general, been dated by pollen analysis – they show an earlier peat, mostly Neolithic in age (but with earlier, Mesolithic, channels in places) and a later peat, mostly Bronze Age. Occasionally bones can be dated by the pollen preserved in the peat retained inside them, but this is exceptional. A few of the mammal bones have been dated directly – Beaver to 3,079 and 2,677 b.p., Aurochs to 4,630 and 4,200 b.p. (details in Yalden, 1999) – but no bird bones have yet been so dated. The fauna is generally, but loosely, referred to as Neolithic in age – the radiocarbon dates for the mammals match this, as does such pollen as has been examined, but some of the bones could well come from Bronze Age or even later. As a whole, the avifauna accumulated from such localities as Burwell Fen, Burnt Fen, Feltwell Fen, Cambridge Fen, Swaffham Fen, and Lingey Fen, is an exciting one, and it is fortunate that Northcote (1980) re-examined all the available specimens, checking their identities and quantifying the fauna (Table 4.3).

Obviously the most striking occurrence here is the Dalmatian Pelican, a species which now nests no nearer than the delta of the R. Danube and the area of the Albanian/Montenegran border. However, according to Cramp *et al.* (1977), Pliny reported that it bred in the estuaries of the Rhine, Scheldt and Elbe, so its retreat from western Europe happened in historical times. As confirmation, it has also been reported from the archaeological sites of Vlaardingen, Netherlands, of early Neolithic age, and from Havno, Denmark (Andrews, 1917; Stewart, 2004). Drainage, and its sensitivity to Human disturbance on its breeding grounds, have conspired against it. The related White Pelican *Pelecanus onocrotalus* is similarly restricted in range, but reported more frequently in western Europe today, albeit probably as an escapee from zoos and bird gardens; this raises the question of identity. It averages somewhat smaller, though the two species overlap in size. Most of the sub-fossil bones are too large to belong to White Pelicans, but the best diagnostic character is

Table 4.3 The birds of the Cambridge fens, 7,000–3,000 years ago (from Northcote 1980)

Species	n. bones	n. individuals
Dalmatian Pelican *Pelecanus crispus*	3	3
Bittern *Botaurus stellaris*	130	21
Mute Swan *Cygnus olor*	306	32
Whooper Swan *Cygnus cygnus*	56	6
Greylag Goose *Anser anser*	26	6
Mallard *Anas platyrhynchos*	113	20
Smew *Mergellus albellus*	3	3
Red-breasted Merganser *Mergus serrator*	4	1
White-tailed Eagle *Haliaeetus albicilla*	1	1
Common Crane *Grus grus*	112	17
Moorhen *Gallinula chloropus*	5	2
Lapwing *Vanellus vanellus*	5	1
Woodcock *Scoloplax rusticola*	7	1
Razorbill *Alca torda*	8	1

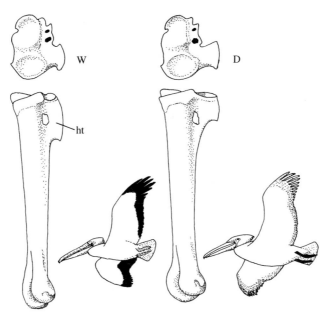

Fig. 4.1 Pelican bone identification. The Dalmatian Pelican (D) averages larger than the White Pelican (W), but with an extensive overlap that precludes specific identification of wing bones. However, the tarsometatarsus has a broader, deeper (antero-posteriorly) but less elongate (proximo-distally) hypotarsus (ht) that is distinctive (after Forbes *et al.* 1958).

the length of the hypotarsus on the tarsometatarsal bone. This is substantially shorter in *P. crispus* (Figure 4.1), and there is no doubt that the subfossil bones belong to this species (Forbes *et al.*, 1958; Joysey, 1963). One of the humeri in the Zoological Museum at Cambridge still had small amounts of peat adhering to it, and analysis of the pollen in this enabled it to be related to the peat deposits in the fenland, and to a radiocarbon dating on those peats. The most abundant tree pollen was from Oak, Alder, and Birch, with some Lime, Hornbeam, Ash, Pine, and Elm. Non-tree pollen came mostly from Hazel, grasses, and Bur-weed *Sparganium*, with a lot of fern spores. Hornbeam is a useful marker, because it is a tree that appeared late in Britain, and generally this is the pollen spectrum from Pollen Zone VIIb, known as the Sub-Boreal, covering the Neolithic and Bronze Ages, about 5,000 to 2,500 b.p. More precisely, the scrapings from the pelican bone match the lower part of the peat, before ferns became even more abundant, and those peats have been dated to about 4,000 b.p., when the fen was probably still somewhat brackish. Of course, this dating pertains to that particular bone; it is unlikely that all the bones listed in Table 4.2 have exactly the same date. However, the predominance of Mute Swan, Bittern, Crane, and Mallard in the fauna paint a coherent picture of a fenland with much open water, extensive shallows with reeds and grasses, and trees on higher islands. The other water birds mostly fit this picture. The Razorbill looks rather out of place, and Mergansers usually winter at sea, but the fens suffered several marine transgressions during the Postglacial period. Given that the birds accumulated over a long period, perhaps of 3,000 years or so, these species might have lived (rather, died) during one such period. Equally, they might have been the occasional 'wrecked' seabirds that sometimes turn up in odd places even today.

The abundance of Mute Swans in these Bronze Age fenlands is interesting for another reason, given past arguments over its status as a native bird. Their status as semi-domesticated birds in Mediaeval England is nicely summarized by Kear (1990). As is well known, Mute Swans on the River Thames belong to the Crown, if unmarked, or to the Dyer's or Vintner's companies, who indicate ownership with one notch or two notches, respectively, cut into each side of the bill in the annual July swan-upping. This is a remnant of a once widespread custom, in which Mute Swans everywhere (in England, at least) belonged to rich landowners, notably the Crown (Ticehurst, 1957). The large flock of Mute Swans at the Abbotsbury swannery in Dorset is another reminder that swans have been regarded as semi-domestic for a long period. This led to suggestions that it was not a native bird, but was imported variously by Romans or Normans as a source of food. One specific suggestion, apparently made by Yarrell in his *British Birds* of 1843 (Ticehurst, 1957) is that returning Crusaders brought them from Cyprus, though that rather dry island is hardly enhanced by an abundance of Mute Swan habitat. Even Professor Alfred Newton, in his *Dictionary of Birds*, considered that the degree of legal protection for Mute Swans in Medieval England pointed to a non-native status. In fact, Ticehurst (1957), reviewing the Medieval documentation, concluded that it was already well established in Britain by 1200, and that it must have been a native. This is made quite clear from the archaeological record: Mute Swans have been present in Britain for a long time. There is a continuous scatter of records all the way from Late Glacial, Mesolithic, and Neolithic times, through Roman to Medieval and Post-medieval. Although it is certainly true that most records (32 of 58) belong in these last two periods (Figure 4.2), this still leaves quite enough records to document the native status of the species (Table 4.4). Just three additional records of uncertain Mute/Whooper Swans raise the question of identification. In size, the two species overlap extensively, but Mute Swans walk less, so their tarsometatarsi are notably shorter and have narrower condyles for the toes (Figure 4.3). The skulls, of course, are very different, with a great boss for the knob in Mute Swans, and the sternum of Whooper (and the much smaller Bewick's) Swans is greatly excavated for a loop of the trachea, associated with their trumpeting calls (Figure 1.7). More subtle features, for instance muscle scars on the humeri, also allow the two to be distinguished. These early swans might have been eaten, but cut marks on the distal ulna and radius from Outgang Road, Market Deeping, which might be Mute or Whooper Swans, suggest removal of flight feathers (Albarella pers. obs.).

The presence of the Crane *Grus grus* in the Cambridge fenlands is less surprising. Though apparently lost as a British breeding bird in about 1600, its former status is well appreciated. An abundance of documentary evidence, including legal protection under Henry VIII, an abundance of placenames, and indeed an abundance of archaeological records (Boisseau & Yalden, 1999) all testify to its former widespread occurrence. Its bones are so much longer than those of other wetland birds as to be almost unmistakable. There has been some confusion in the past over its status because of the transfer of its name to the Grey Heron, an event likely to have followed its extinction. One other long-legged species does occur with it in similar habitats, the White Stork *Ciconia ciconia*, but that is far rarer. It also seems to have become extinct longer ago – the famed nesting in Edinburgh in 1416 being perhaps the last, or even only, recorded nesting. Of course, one or two pairs of Cranes have bred in Britain

FARMLAND AND FENLAND | **83**

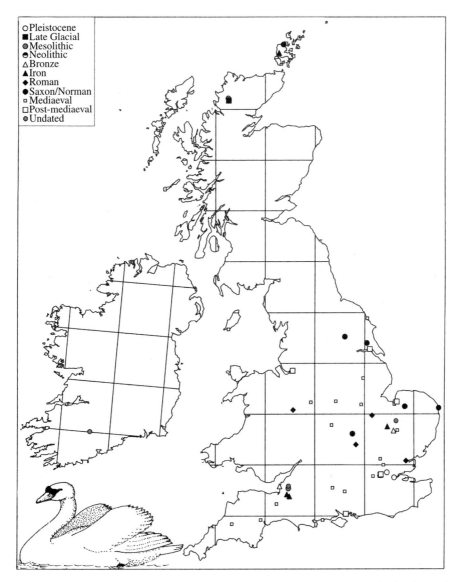

Fig. 4.2 Mute Swan distribution. Most records confined to England, but extending in time at least back to the Mesolithic, so clearly a long-standing native (see Table 4.4).

since about 1984, a welcome return of a lost native, and as this is written, a pair of Storks is reputedly nesting in Yorkshire.

Northcote (1980) makes the odd remark that the East Anglian fens are (were) more base-rich, so better for preserving bird bones, whereas those of Somerset are more acid, and so produce no comparable fauna. This comment ignores the substantial bird faunas excavated many years ago from Glastonbury Lake Village (Andrews, 1917) and from Meare Lake

Table 4.4 Archaeological records of Mute Swan *Cygnus olor* from the British Isles. The 3 records marked * are uncertain Mute/Whooper Swans.

Site	Grid Ref	Date	Citation
*Ilford, Essex	TQ 45 85	Ipswichian	Harrison & Walker 1977
*Grays, Essex	TQ 60 75	Ipswichian	Harrison & Walker 1977
Sutherland – Creag nan Uamh cave	NC 26 17	Late Glacial	Newton 1917; Tyrberg 1998
Castlepook Cave, Co. Cork	R 60 00	Late Pleist-Holocene	Bell 1915; Tyrberg 1998
Aveline's Hole, Somerset	ST 47 58	Late Pleist-Holocene	Davies 1921, Newton 1921b, 1922, 1924b, Tyrberg 1998
Gough's Cave, Somerset	ST 47 54	Mesolithic	Harrison 1980a; Harrison 1986
Inchnadamff, Sutherland	NC 25 21	Mesolithic	Newton 1917
Brean Down	ST 29 58	Bronze Age	Levitan 1990
Burwell Fen	TL 59 67	Bronze Age	Northcote 1980
Meare Lake Village	ST 44 42	Iron Age	Gray 1966
Haddenham	TL 46 75	Iron Age	Evans & Serjeantson 1988
Glastonbury Lake Village	ST 49 38	Iron Age	Harrison 1980a, 1987b
Howe, Orkney	HY 27 10	Iron Age	Bramwell 1994
Heybridge – Elms Farm	TL 84 08	Early Roman	Johnstone & Albarella 2002
Caister-on-Sea	TG 51 12	Roman	Harman 1993b
London – Billingsgate Buildings	TQ 32 80	Roman	Cowles 1980a; Parker 1988
Longthorpe	TL 15 97	Roman	King 1987
London – Lambeth	TQ 31 79	Roman	Locker 1988
York – General Accident Site	SE 60 52	Roman	O'Connor 1985b
Wroxeter	SJ 56 08	Roman	Meddens 1987
Bancroft Villa	SP 82 40	Roman	Levitan 1994b
Caister-on-Sea	TG 51 12	Anglo-Saxon	Harman 1993b
Northampton – Marefair	SP 75 61	Saxon	Bramwell 1979d
York – Coppergate	SE 60 52	Anglo-Scand	O'Connor 1989
Buckquoy	HY 36 27	Norse	Bramwell 1977b
Castle Acre Castle	TF 82 15	Norman	Lawrance 1982
Beverley – Lurk Lane	TA 04 40	11th – 13th C	Scott 1991
Scarborough Castle, Kitchen	TA 05 89	12–13th C	Weinstock 2002b
South Witham	SK 93 19	13th C	Harcourt 1969a
London – Baynard's Castle	TQ 32 80	1350	Bramwell 1975a
Portchester	SU 62 04	1350–1400	Eastham 1985
Taunton – Benham's Garage	ST 23 24	Medieval	Levitan 1984b
Hatch Warren, Brighton Hill South	SU 60 48	Medieval	Coy 1995
Writtle – King John's Hunting Lodge	TL 67 68	Medieval	Bramwell 1969
Ling's Lynn	TF 61 20	Medieval	Bramwell 1977a
Lincoln	SK 97 71	Medieval	Cowles (1973), Dobney *et al* (1996)
Launceston Castle	SX 33 84	Medieval	Albarella & Davies (1996)
London – Southwark	TQ 32 80	Medieval	Locker (1988)
Exeter	SX 91 92	Medieval	Maltby (1979a)
Castletown, Isle of Man	SC 26 67	Medieval	Fisher (1996)
Coventry – Town Wall	SP 33 78	Medieval	Bramwell (1986a)
Faccombe Netherton	SU 35 55	Medieval	Sadler (1990)
Hull – Scale Lane/Lowgate	TA 10 28	Medieval	Phillips (1980)

Table 4.4 (*Continued*)

Site	Grid Ref	Date	Citation
Christchurch – Dolphin Site	SZ 15 92	Medieval	Coy (1983a)
Brentford	TQ 17 78	Medieval	Cowles (1978)
Castle Rising Castle	TF 66 24	Medieval	Jones, Reilly & Pipe 1997
*Stafford Castle	SJ 92 23	15th C.	Sadler 2007
London – Baynard's Castle	TQ 32 80	1500	Bramwell 1975a
London – Baynard's Castle	TQ 32 80	1520	Bramwell 1975a
Hertford Castle	TL 32 12	High/Late Medieval	Jaques & Dobney 1996
Waltham Abbey	TL 38 00	Late Medieval	Huggins 1976
Donington Park	SK 42 25	Late Medieval	Bent 1978
Castle Rising Castle	TF 66 24	16th C	Jones, Reilly & Pipe 1997
Portchester	SU 62 04	16th-17th C	Eastham 1985
London – Aldgate	TQ 33 81	1670–1700	Armitage & West 1984
Norton Priory	SJ 55 85	Post-Med	Greene 1989
York – Aldwark	SE 60 52	Post-Med	O'Connor 1984a
London – Southwark	TQ 32 80	Post-Med	Locker 1988
Hull – Queen's Street	TA 10 28	Late Post-Med	Scott 1993
Guernsey – Le Dehus	WV358831		Kendrick 1928
Redmere, nr Littleport	TL 64 86		Harrison 1980a

Village about 5 km away (Gray, 1966). These are both slightly later, Iron Age sites, and the faunas reflect in part the more developed agriculture of those times, exemplified by early records of Domestic Fowl (Jungle Fowl, in Harrison (1987b); the original excavation reports seem not to include it in the bird lists), but their bird faunas largely support the conclusions drawn from East Anglian Fenland.

Glastonbury Lake Village, the earlier to be discovered and excavated, lies just 1.5 km north of the present town, which is perched on a small hill overlooking the flat fenland to the north. The Lake Village, first recognized in 1892 and excavated from then until 1907, was a classic crannog, a complex of 89 huts, indicated by gentle mounds in 1892, each built on a base of logs and brushwood. The basal layer of logs, mostly alder, were laid about 15–35 cm apart, and a second layer was placed at right angles to them, with a layer of brushwood on top. The floors of the living areas were made of clay, and as the huts gently sunk into the mire under their own weight, extra layers of clay were added. Raised hearths of clay, sometimes topped with limestone, were also a feature. Some, at least, of the huts were also joined by stone pathways. The whole village was surrounded by a palisade of vertical posts, again mostly alder but with some of birch and oak, often just a single row but in places two, three, or even four posts deep. Confirming that the site was set in water, to the east was a causeway, a complex and carefully crafted structure made of grooved oak planks with boards set in the grooves, and a landing stage, mostly of stone slabs with a wattle retaining wall one side and a side fence of morticed oak planks and rails the other. The inhabitants made use of their location by dumping most of their rubbish over the palisade into the shallow water, providing a very rich treasure trove for the archaeologists. Amber and glass beads, examples of iron sickle, saw and mattock with wooden handles still in place, bronze rings, antler combs, wooden ladles and bowls, even a wooden hand plough and a dugout canoe, were among the

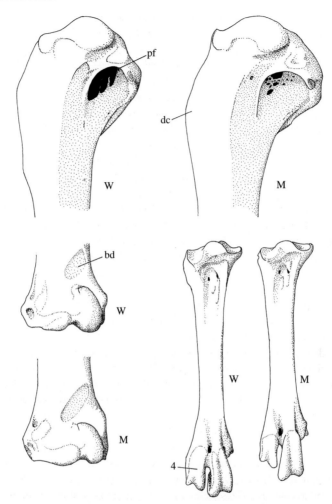

Fig. 4.3 Swan bone identification. Despite very similar sizes, most larger bones of Whooper (W) and Mute (M) Swans can be distinguished. For instance, the pneumatic fossa (pf) on the head of the humerus is deeper, but the deltoid crest (dc) less prominent but more elongate in Whooper Swans; distally, the brachial depression(bd) (for the attachment of the brachialis muscle) is more compact. Associated with their more regular grazing habits, the trochlea for the 4th toe is broader (4), and the whole tarsometatarsus is more elongate. (See also Fig 1.7).

many items discovered. The peat had also preserved large quantities of bones, notably of sheep (despite the wet site – presumably they lived on the nearby drier hills).

Meare, lying some 6 km north-west of Glastonbury, was actually discovered during the Glastonbury excavations, in 1895, and mapped in 1896, but digging was deferred until the former excavation was finished. The mapping showed in fact two settlements, Meare West and Meare East, about 200 m apart. Digging at Meare West began in 1910, and continued through to 1933. Only the eastern part of Meare West, involving 40 mounds (that is, hut sites), was excavated then; digging was confined to August-September, the driest time of the year, and even then had to be abandoned in three very wet summers. Compared with Glastonbury, the site seems to have been on a drier edge of the marshes, and while the more northern huts (those deeper into the marsh) had foundations of oak planks and brushwood, like Glastonbury, most had clay floors laid directly on the peat. There was no palisade, no landing stage or causeway (not, at least, in the parts excavated), and most of the archaeological specimens were found in the hut floors, sandwiched by the extra clay that was periodically added. A comparable wealth of archaeological material was reported, much of it closely matching that from Glastonbury. For instance, 155 beads, of amber, jet, and glass, 368 worked bones and 282 worked antler pieces, and 1469 pieces of flint, including flint arrowheads – despite the presence also of iron objects – were found. Writing up the excavations was delayed by the 1939–45 war, and then by the deaths of the two supervisors and excavators, first, of Arthur Bulleid in December 1951, and then of Harold St George Gray, so that the final reports did not appear until 1948, 1953, and 1966. Even then, there is a feeling that volume 3 was hurried out, to complete the series, and there is, for example, no report on the food bones.

Both sites produced large birds faunas: 58 species at Meare (Bate, 1966), 37 at Glastonbury (Andrews, 1917), and most of the identifications have been checked more recently (Harrison, 1980a, 1987b). At both sites, waterbirds dominate, and there must have been very extensive open water. Andrews (1917) thought that at least five individual Dalmatian Pelicans were represented, and gave measurements of 19 bones as evidence that they matched *P. crispus* rather than *P. onocrotalus*. Moreover, he emphasized that some of the bones were from juveniles (as indeed were some of the Cambridge fenland specimens), indicating that they bred locally. Other species include Dabchick, Bittern, Grey Heron, Cormorant, Common Crane, Moorhen, Coot, both Mute and Whooper Swans, and a long list of ducks – Mallard, Wigeon, Pintail, Scaup, Tufted Duck, Pochard, and Smew. Geese were surprisingly scarce – only one or two bones, species uncertain. Among predators and scavengers, White-tailed Eagle, Marsh Harrier, Red Kite, and Barn Owl were listed. (Records of Corncrake, Goshawk, Teal, Shoveler, Goldeneye, and Red-breasted Merganser given by Andrew (1917) were re-identified by Harrison (1980a) as other species in this list, but not all the bones reported by Andrews were available to Harrison.) Terrestrial birds were fewer, but possible Wheatear, Crow/Rook, and Song Thrush were reported. A single shearwater humerus (presumably Manx Shearwater, as it is the only breeding species in Britain) suggests perhaps that the locals had been exploiting seabirds of nearby islands in the Bristol Channel. However, the absence of most marine species – Puffins, for instance, which used to be common on Lundy – implies that they were exploiting the local waterbirds, not foraging further afield, and that the shearwater was therefore a stray. Mostly, the Glastonbury inhabitants fed on sheep meat, presumably reared on the Mendip Hills nearby, but cattle, horse, pig, and goat were also present, so the birds were presumably taken as a change in diet, and were certainly not a mainstay. At Meare similarly, an extensive fauna of waterbirds is preserved, though not pelicans. (More recent excavations at Glastonbury, at Wirral Park and the Mound, have added to the record of Dalmatian Pelicans in the area; Darvill & Coy, 1985.) Great Northern and Red-throated Diver, Dabchick and Great Crested Grebe, Cormorant, Grey Heron, Bittern, Moorhen, Coot, Water Rail, Common Crane, Whooper, Bewick's and Mute Swans, Barnacle, White-fronted and Greylag Geese, together with a good list of ducks – Shelduck, Teal, Garganey, Mallard, Wigeon, Gadwall, Pintail, Shoveler, Scaup, Tufted Duck, Pochard, Goldeneye, Smew and

Goosander. More predators were recorded than at Glastonbury, including Peregrine, Osprey, Marsh and Montague's Harriers, Red Kite, both White-tailed and Golden Eagle, ?Common Buzzard and perhaps the latest record of Eagle Owl from Britain (cf. Table 3.2) – only the shaft of an ulna, broken, and Bate (1966) was uncertain about its identity. (Owl ulnae have a characteristic triangular cross-section, but attempts to relocate this bone to check it have failed.) Rather more marine and terrestrial species were also listed – Gannet, perhaps Herring and Great Black-backed Gull, among the former, Grey Partridge, Black Grouse, Rook, and Song Thrush among the latter. Haddenham, in the Cambridgeshire fens, is another Iron Age site with a wetland fauna including Dalmatian Pelican (Fig 4.5), as well as Grey Heron, Crane, Mute Swan, Mallard, Coot, and Moorhen (Evans & Serjeantson, 1988). A pelican humerus from Haddenham bears cut marks, suggesting it was eaten, while the Crane was an almost fledged juvenile, proving, lest there was any doubt, that they bred locally.

Collectively, these Bronze and Iron Age fenlands tell us of a once much more extensive habitat, which must have covered some 8,427 km^2, to judge from the maps in Darby. Nearly half (about 3,164 km^2) of this would have been in the Fenlands of East Anglia, covering much of northern Cambridgeshire and southern Lincolnshire. The Somerset Levels, Romney Marshes and Pevensey Levels, the Humber marshlands, and the Vale of Pickering also had substantial fenlands in Roman times (Figure 4.4). The loss of this habitat began so long ago that its full

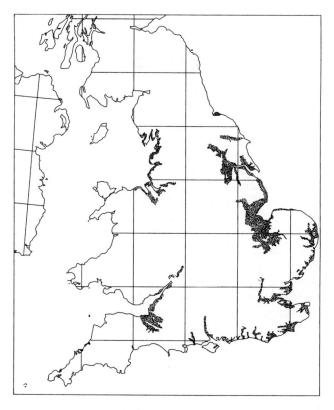

Fig. 4.4 The former extent of fenland in England (after Darby & Versey 1957).

extent is difficult to imagine. There have been at least three phases of drainage, starting with one in Roman times of which little direct evidence remains (Rackham, 1986). The pattern of contemporary settlements recorded in Domesday surveys of 1086 AD picks out some likely Roman embankments around fenland, particularly in southern Lincolnshire (Darby & Versey, 1975), but overall the pattern of settlements on their maps, which surround but rarely encroach on fenland, suggest that most fenland still remained in southern Britain into Norman times.

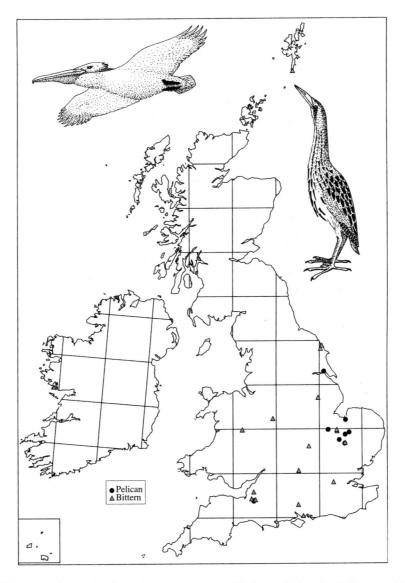

Fig. 4.5 Associated with the formerly greater extent of fenlands, Dalmatian Pelicans and Bitterns show a wide distribution in the archaeological record (see Table 4.5).

As late as 1540, when Glastonbury Abbey was dissolved by Henry VIII, there was a pool at Meare of some 200 ha, on which the Abbey had kept 41 pairs of swans (Bulleid & Gray, 1948). Bitterns and Marsh Harriers strongly suggest extensive reed beds, and likely large eel populations. Cormorants might be considered by present-day British observers to belong among the maritime species, but formerly they nested regularly in appropriate freshwaters; among other freshwater archaeological sites for them are Ulrome Lake, Yorkshire, also Iron Age (Harrison, 1980a), Lagore, County Meath, an early Christian crannog (Stelfox, 1938) and perhaps medieval Barnard's Castle, London (Bramwell, 1975a). They are now returning inland, for example, to such sites as the Lea Valley and Abberton reservoirs, to the annoyance of anglers. Stewart (2004) remarks that wetlands in Europe host the somewhat smaller, white-headed form, regarded as a subspecies *P. c. sinensis*, which has spread through the Netherlands to contribute extensively to this recent colonization of south-east England (Carss & Ekins, 2002). Ringing recoveries confirm the input of Dutch populations to these new inland colonies, but also indicate a contribution from native maritime populations (Newson *et al.*, 2007). These fenlands were also host to Beavers, as recorded at Haddenham, Welland Bank Quarry, Meare, and Glastonbury, and Otters, recorded at the last two. Both would have been hunted for their pelts, and many of the Beaver bones from Haddenham and Welland bear cut marks as testimony to this trade (Evans & Serjeantson, 1988; Albarella *et al.* in prep). Stewart (2004) concludes that persecution contributed to the loss of the Cormorant as a member of this inland fauna, just as Beavers were probably hunted to extinction. It is hard to quantify the relative impacts of hunting and drainage, but we have today a severely diminished fenland fauna.

Despite this conclusion, it is notable that a number of more southerly occurring members of the European fenland fauna have not been recorded as fossils in Britain, though of course their absence as fossils does not prove that they were necessarily absent from the faunas. At present, Little Bitterns, Purple Herons, and Little Egrets occur just across the North Sea in the Netherlands, but have not been recorded from archaeological sites in Britain. Nor have even more southerly species, such as Flamingos, Great White, and Cattle Egrets. If Bourne (2003) is correct, Little Egrets were once regular breeders in Fenland and given the numbers he reports being eaten, it is surprising that they have not been recognized from Mediaeval archaeological sites. It is possible that they have been confused with Bitterns (Fig 4.5), which have similar-sized bones: British archaeologists, at least, are unlikely to have reference skeletons of the Egret, and might not have been expecting to find it either. Little Egrets have of course started to nest (nest again?) in southern Britain and southern Ireland since 1996 (Mead, 2000).

There are three other, rarer, fenland species, Spoonbill, Night Heron, and Pygmy Cormorant, that are represented in the archaeological record (though not in the Iron Age sites just discussed) (Table 4.5). Two of them, Night Heron and Spoonbill, breed at present as near as the Netherlands, but the Pygmy Cormorant is a Balkan species. The Spoonbill is documented as having bred in a few sites in southern Britain up to the seventeenth century – in Pembrokeshire to 1602, and East Anglia to 1650. Its presence has been somewhat overlooked because early accounts called it the shoveller or shovellard, leading to confusion with Shoveler *Anas clypeata*. The Welsh account refers to the shovellards nesting in trees, appropriate for *Platalea leucorodia* but not *Anas clypeata*. Regular summering in East Anglia in recent years has been followed by at least two recent breeding attempts. Both Spoonbill records are Medieval. There are two archaeological records of Night Heron, a Roman record from London Wall (Harrison, 1980a) and an Elizabethan or early Stuart record from the Royal Navy Victualling Yard at Greenwich

Table 4.5 A summary of records of rare fenland birds in the archaeological record. (Ages abbreviated to Mes = Mesolithic; B.A. = Bronze Age; I.A. = Iron Age, Rom. = Roman: Rom-Br = Romano-British; Sax = Saxon; Norm = Norman; Med = Medieval).

Species	Site	NGR	Age	Source
Dalmatian Pelican	Kings Lynn	TF6120	?	Forbes *et al.* 1958)
	Burwell Fen	TL5967	B. A.	Northcote (1980)
	Burnt Fen	TL6087	?	Harmer (1897), Forbes *et al.* (1958)
	Feltwell Fen	TL6992	?	Forbes *et al.* (1958)
	Hull	TA1030	?	Newton (1928), Forbes *et al.* (1958)
	Glastonbury Wirral Park	ST4938	I.A.	Coy (1991)
	Glastonbury Lake Village	ST4938	I.A.	Andrews (1917), Harrison (1980a, 1987b)
	Haddenham	TL4675	I.A.	Evans & Serjeantson (1988)
Pelican sp.	Glastonbury – The Mound	ST4938	Med.?	Darvill & Coy (1985)
Bittern	Aveline's Hole, Somerset	ST4758	L. Pleis.	Davies (1921), Newton (1921b, 1922, 1924b), Jackson (1962); Tyrberg (1998)
	Rousay, Orkney	HY4030	PostGl.	Bramwell (1960a)
	Star Carr	TA0281	Mes.	Northcote (1980); Harrison (1980a)
	Rousay – Knowe of Ramsay	HY4028	Neo.	Davidson & Henshall (1989)
	Burwell Fen	TL5967	B. A.	Northcote (1980)
	Jarlshof	HU3909	B. A.	Platt (1933a, 1956)
	Glastonbury Lake Village	ST4938	I. A.	Andrews (1917); Harrison (1980a, 1987b)
	Meare Lake Village	ST4442	I. A.	Gray (1966); Harrison (1987b)
	Winnall Down	SU5029	Rom.-Br	Maltby (1985)
	Grandford	TL4195	Rom.	Maltby & Coy (1982); Parker (1988)
	Grandford, nr March	TL4098	M. Rom.	Stallibrass (1982)
	Portchester	SU6204	E-M.Sax	Eastham (1976)
	Oxford – Queen Street	SP5106	Sax.	Wilson, Allison & Jones (1983)
	Jarlshof	HU3909	9th C	Platt (1956)
	Hen Domen	SO2198	Sax-Norm	Browne (2000)
	Stafford Castle	SJ9223	12th C.	Sadler (2007)
	Scarborough Castle, Kitchen	TA0589	12–13thC	Weinstock (2002)
	Lincoln – Flaxengate	SK9771	Med.	O'Connor (1982)
	Scarborough Castle, Kitchen	TA0589	13–15th C.	Weinstock (2002)
	London – Baynard's Castle	TQ3280	1520	Bramwell (1975a)
Night Heron	London Wall	TQ2979	Rom.	Harrison (1980a)
	Greenwich	TQ3777	1560–1635	West (1995)

Table 4.5 (*Continued*)

Species	Site	NGR	Age	Source
Pygmy Cormorant	Abingdon	SU4947	15–16th C.	Bramwell & Wilson (1979), Cowles (1981)
Spoonbill	Southampton-Cuckoo Lane	SU4213	14th C.	Bramwell (1975c)
	Castle Rising Castle	TF6624	Med	Jones, Reilly & Pipe (1997)
Marsh Harrier	Lough Gur, Co Limerick	R 6441	?	D'Arcy (1999)
	Glastonbury Lake Village	ST4938	I. A.	Harrison (1980a, 1987b)
	Harston Mill	TL4150	I. A.	R. Jones, pers. comm.
	Meare Lake Village	ST4442	I. A.	Gray (1966); Harrison (1987b)
	Ballinderry Crannog	N 2239	E. Christian	Stelfox (1942)
	London – Westminster Abbey	TQ2979	Sax.	West (1991)
	Flixborough	SE8715	8th – 9th C	Dobney *et al.* (2007)
	Flixborough	SE8715	10th C	Dobney *et al.* (2007)
	Dublin – Woods Quay	O 1535	10–11th C	D'Arcy (1999)
	Dublin – Fishamble Street	O 1535	10–11th C	T O'Sullivan, in D'Arcy (1999)
	Beverley – Lurk Lane	TA0440	11th-13th C	Scott (1991)
	Portchester	SU6204	1100–1200	Eastham (1977)
	Faccombe Netherton	SU3555	Med.	Sadler (1990)
	Beverley – Dominican Priory	TA0440	Med.	Gilchrist (1986, 1996)
	Portchester	SU6204	16–17th C	Eastham (1985)

(West, 1995). Bourne (2003) makes a strong case that the Brewes or Brues mentioned in the accounts of Mediaeval and Tudor banquets, which have puzzled past historical ornithologists (Whimbrel? Godwit? Gurney, 1921) were in fact Night Herons.

A single individual of Pygmy Cormorant, represented by two unmistakable but unexpected metacarpals, was recovered from fourteenth century Abingdon, recognized by Don Bramwell, and passed to Graham Cowles for confirmation (Bramwell & Wilson, 1979; Cowles, 1981). This is, of course, a much later site than the Iron Age lake villages, and it is certain that much drainage had already been conducted by then. It is the only record of the species in an archaeological site anywhere in north-west Europe, and its significance is hard to evaluate. Does it indicate a limited and local colonization, perhaps in the Medieval warm period, which saw vines growing near Oxford, or does it represent an exotic animal brought home from the Balkans by some early traveller (Stewart, 2004)?

Conclusions

The archaeological record of coastal and fenland birds from Neolithic times onwards is a good one: these are mostly well-excavated sites, with good bone preservation, and the bones have in most cases been carefully examined or re-examined and conserved. They document some major losses from the bird faunas of the British Isles, including some unexpected species (Dalmatian Pelican, Pygmy Cormorant) and some more familiar ones (Great Auk, Crane). It also includes some unexplained twists and gaps in the historical record. We have become accustomed to believing that the Fulmar only spread around the British Isles from its St Kildan stronghold over the last 150 years. The archaeological record strongly indicates that this is actually another returnee from former persecution, like the Crane. Conversely, the historical record strongly hints that large numbers of Little Egrets and Night Herons once bred in eastern England, but if so, the archaeological record does little to confirm it. Clearly, there are some interesting hypotheses to be investigated here.

5
Veni, Vidi, Vici

Iron Age Britain

Julius Caesar is reputed to have remarked that he came, saw, and conquered. Although he came to and saw Britain, twice, in 54 and 53 BC, his short-lived expeditions hardly amounted to conquering even the south of England, and it was left to Claudius, in AD 43, to attempt a proper invasion and complete a conquest of England. However, Caesar's account, of sending his troops out to collect corn from the native's fields, their retreat, too, with their cattle, and his description of the farmed landscape through which he marched, emphasizes how extensively agriculture had spread. The Iron Age, Celtic, culture of southern Britain that he invaded had a well developed agricultural economy, with a network of hill forts established at regular intervals. Thus the former landscape, of extensive woodland with clearings (such as we inferred for the Mesolithic in Chapter 3), had been transformed into a farmed landscape with occasional woods. Claudius' invasion of England in AD 43, followed by Agricola's conquest of North Wales in AD 78, saw nearly 400 years of settled Roman rule in southern Britain, producing an abundance of archaeological sites that have yielded between them an extensive avifauna (Parker, 1988). Scotland was never conquered, though it was invaded at least as far north as the Antonine Wall, and Ireland seems barely to have been visited. It is convenient for us to refer to contemporary sites in Scotland and Ireland as of Roman date, even though this is clearly inaccurate historically.

The largest Iron Age faunas are those from Fenland sites, such as Glastonbury, discussed in Chapter 3. However, there are also good faunas from dryland sites in southern England, which confirm the impression of an agricultural landscape. The hill fort of Danebury, on the Downs south of Andover in Hampshire, yields one of the largest (Coy, 1984a; Serjeantson, 1991). Among farmland birds, represented by only a few bones each, there are Golden Plover, Lapwing, Quail, Skylark, Corncrake, Starling, and Wood Pigeon, though the last two imply nesting trees somewhere nearby. The presence of woodland is certainly indicated by Jay, perhaps also by Buzzard, Kestrel, Red Kite, Jackdaw, Rook, and Crow, which would have nested in trees somewhere near but foraged out into farmland. Blackbird, Song Thrush, Redwing, and shrike sp. imply scrubby woodland edges. More remarkable for southern England are the moorland and wetland birds reported – Red Grouse, Black Grouse, perhaps Long-eared Owl, among the former, Grey Heron, Bewick's Swan, Greylag and Barnacle Goose, Mallard, Teal, Wigeon, Gadwall, Goosander, Tufted Duck, and Kittiwake among the latter. The heathlands of the New Forest are only 20 km south-west, and could have been more extensive then, while the valley of the River Test is only 5 km east, and the Solent only 25 km south. All these species are thinly represented, one or two bones, as is a Peregrine, by a very distinctive skull. However, most remarkable is the abundance of

Raven bones, contributing 67% (533 of 798) bird bones (Coy, 1984a). Ravens are regularly reported from Iron Age sites in southern England – Wylye (Harrison, 1980a), Gussage All Saints (Harcourt, 1979a), Blunsdon St Andrews (Coy, 1982), Budbury (Bramwell, 1970), Maiden Castle (Armour-Chelu, 1991), Poundbury and Pennyland (Ashdown, 1993) are other examples. Given their size, Raven bones are unlikely to be overlooked, and their distinctive size, for a passerine, also makes them easily identified. Neither of these biases can explain their abundance at Danebury or their ubiquity at Iron Age sites, and a cultural explanation is implied. At Danebury and Winklebury they included complete skeletons, apparently buried deliberately in the bottom of pits. It is presumed that they represented some symbolic token, perhaps of the underworld. From an ornithological perspective, they are a reminder of how abundant and widespread Ravens were in England before nineteenth century persecution restricted them to the west.

The birds from other Iron Age sites in southern England also suggest a farmed landscape, albeit with some woodland, but the faunas are all small and not as informative as the wetland sites. Lark and thrush sp. at Winnall Down (Maltby, 1985), like Fieldfare at Maiden Castle (Armour-Chelu, 1991) suggest farmland. At Budbury (Bramwell, 1970), Stock Dove and Rook suggest wooded farmland, Jay suggests woodland, and Raven could belong anywhere – but Common Scoter seems very improbable at a site further inland from Bath. The presence of House Sparrow at Danebury, Abingdon (Ashville Trading Estate), Harston Mill, Slaughterford (Guy's Drift) and Old Scatness Broch certainly fit notions of cereal farming, and two earlier Bronze Age records of *Passer* sp., probably this species (Potterne, Poundbury) fit the suggestion by Ericson *et al.* (1997) that the species arrived in northern Europe with domestic horses, and somewhat earlier than Domestic Fowl. The presence of Common Gull, Wigeon, and Curlew at Poundbury, outside Dorchester, reflects the wet floodplain of the River Frome nearby (Buckland-Wright, 1987). The roughly contemporary site of Newgrange, in County Meath, also yields a small fauna of woodland (Woodcock, Goshawk, Blackbird, Dunnock, Greenfinch), wetland (Water Rail, Pied Wagtail), and farmland (Grey Partridge, Mistle Thrush, Song Thrush) birds (Van Wijngaarden-Bakker, 1974, 1986).

In the north of Britain, seabirds are inevitably better represented on the, mainly coastal, archaeological sites than land birds. The most important Iron Age sites are three in Orkney, at Bu (Bramwell, 1987), Skaill (Allison, 1997b), and Howe (Bramwell, 1994), with 44, 30, and an impressive 113 species recorded, respectively. The numbers of species testify to the excellent preservation produced by the shell sand that buried these sites. Across on the north coast of Caithness, Crosskirk Broch provides records of 26 species (MacCartney, 1984). Naturally seabirds dominate. Gannet, Cormorant, Shag, Guillemot, Black Guillemot, Little Auk, Razorbill, and Puffin are present at all three Orkney sites, and five of them (not Black Guillemot, Little Auk, Puffin) also at Crosskirk. Both Fulmar and Great Auk, the latter in some abundance, are present at Crosskirk, Howe and Skaill, while the Great Northern Diver is present at Crosskirk, Bu, and Howe, and Manx Shearwater is present at Crosskirk. A range of waders (Lapwing, Grey and Golden Plover, Curlew, Whimbrel, Oystercatcher, Greenshank, Redshank, Dunlin, Green Sandpiper, Snipe, Woodcock), ducks (including Eider, Common and Velvet Scoter, Teal, Wigeon, Smew, Goosander, and Merganser), and other seabirds (various gulls, Great Skua, Sandwich Tern) is also present. Raptors are fewer, but include White-tailed Eagle at both Skaill and Howe, Golden Eagle, Rough-legged Buzzard, Red Kite, Kestrel and Peregrine at Howe, Merlin at Bu and Howe, and somewhat surprisingly, given the tree-less nature of Orkney, Goshawks at both Howe and Skaill. Common Buzzard is the only raptor listed for Crosskirk. Cranes are represented at Howe by juvenile bones, undoubted evidence of breeding there. Despite the dominance of seabirds, some terrestrial species are also recorded, notably Red Grouse at all four sites, Black Grouse only at Crosskirk. The Howe list includes such unusual (for archaeological sites) identifications as Corn, Reed and Snow Bunting, Waxwing, Great Grey Shrike, and Wren. The Swallow, Skylark, Starling, various thrushes (Blackbird, Ring Ouzel, Song Thrush, Redwing, Mistle Thrush), and Raven are more regular members of such assemblages. A Tawny Owl, perhaps a wind-blown stray, seems as unlikely on Orkney as the Goshawk; the Short-eared Owl, still a regular breeder there and reliant on the Orkney Vole (which had been introduced in Neolithic times), is a more expected record. Bu, too, has the fairly predictable Skylark, Redwing, and Raven, but the Chough identified there is one of only 15 archaeological records of the species in the British Isles, and the Quail is a reminder of how widespread that little migrant gamebird can be.

Early domestication

Caesar mentions one other important detail for a faunal history, when he says of the British Celts that they had hens, geese, and hares, though they did not eat them. This introduces an important aspect of our bird population, the extent to which the avifauna of Britain has been transformed by introduced alien species. Ask any ornithologist what is the commonest bird in Britain, and he (or she, but usually he) will probably answer Wren, estimated by Gibbons *et al.* (1993) at 7.1 million pairs. He might, alternatively (and particularly after a hard winter, to which Wrens are susceptible), suggest Chaffinch (5.4 million) or Blackbird (4.4 million). The correct answer of course is Domestic Fowl with some 155 million adults in June – though far fewer pairs! About 117 million are table birds, 29 million are egg-laying hens, and 11 million are the breeding stock (http//statistics.defra.gov.uk/esg/publications). As a measure of their rate of production, about 877 million are killed each year for meat. Geese, ducks, and doves were also very common in the past, less so now (about 10 million turkeys, ducks, and geese combined), and it is interesting to speculate also on their domestication, about which far less has been written.

Domestic Fowl

It is odd that we have no satisfactory specific name for our most common bird. Frequently called Chickens, but that strictly refers to young females in their first year, or Hens and Cocks, but they could be female and male of any bird, the Domestic Fowl (and Fowl strictly is Anglo-Saxon for any bird, as in Fowlmere – bird lake), formally *Gallus domesticus*, is a native of South-east Asia. There are four wild species of *Gallus*, the Grey Jungle Fowl *G. sonnerati* of south-west India, Green Jungle Fowl *G. lafayettei* of Sri Lanka, Black Jungle Fowl *G. varius* of Java, and the Red Jungle Fowl *Gallus gallus* of India, Burma, and South-east Asia (Figure 5.1). The latter is certainly the main ancestor of the Domestic Fowl, and it is usual to apply the name *Gallus gallus* to the domestic form as well. However, it has now been

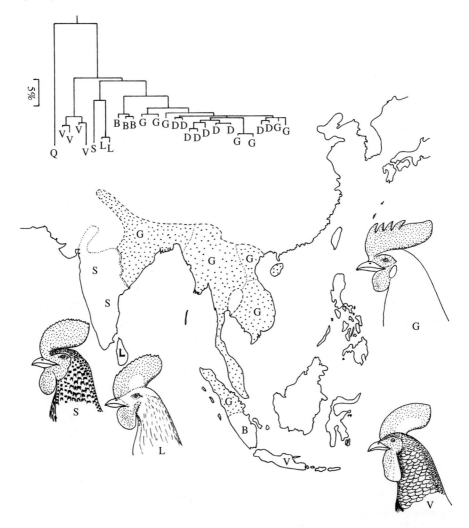

Fig. 5.1 Distribution and phylogeny of wild *Gallus* species. The serrated comb and golden neck cape strongly suggest that the Red Jungle Fowl *Gallus gallus* (G) of SE Asia (stipple) is the ancestor of the Domestic Fowl (D). The Javan (Green) Jungle Fowl *Gallus varius* (V), Grey Jungle Fowl *G. sonnerati* (S) and Ceylon Jungle Fowl *G. lafayetti* (L) have simpler combs and scalloped, marbled or streaked capes. The molecular phylogeny (top) confirms, relative to Quail (Q), that the three southern species are well distinct from Red Jungle Fowl, puts Domestic Fowl clearly within the Red Jungle Fowl lineage, but suggests that the southern form on Sumatra, usually regarded as a subspecies *G. g. bankiva*, is a full species (B) (after West & Zhou 1988 (map) and Fumihito *et al.* 1994, 1996 (phylogeny)).

ruled by the International Commission on Zoological Nomenclature (ICZN, 2003, Opinion 2027) that wild mammal and insect species should, for clarity and lack of confusion, retain separate specific names from their domestic descendants where available (so correctly *Canis familiaris* for the Dog, and *Canis lupus* for the Wolf, even though we know that biologically and historically they are the same species). Particular confusion has been caused when the domestic form has a name which takes taxonomic priority over that of the wild form (Gentry *et al.*, 2003) – it is not helpful to refer to the Grey Wolf as *Canis familiaris*. Though not formally covered by that Opinion, the same reasons make it sensible to use *G. domesticus* for the domestic bird. This adds particular clarity to the discussion of whether *G. gallus* is the only ancestor for *G. domesticus*, or if the other species have been involved, and whether different populations have been involved – whether, that is, the Domestic Fowl was domesticated only once, or several times, perhaps in different places and from different races or species. Good archaeological and genetic evidence is now available to settle this matter.

Zeuner (1963) considered that the earliest record of Domestic Fowl was from the site of Mohenjo-Daro, in the Indus Valley (now Pakistan, but India when excavated by Sir Mortimer Wheeler), dated to about 2,000 BC. At the time, this was certainly the earliest record of the species in an archaeological site, and as it is outside the natural range of *G. gallus*, it is reasonable to regard it as domesticated. Indeed, the occurrence of any species in an archaeological context outside its natural range is generally regarded as a sure indication of domestication. (Compare, for example, the appearance of domestic goats at Beidda, in the Jordan Valley, or indeed of sheep and goats in Europe.) The realization that Chinese archaeologists had already discovered much earlier evidence was not really appreciated until West & Zhou (1988) reviewed the subject. They report the earliest sites as Peiligan, dated to 5,935 BC, and Cishan, dated to 5,405 BC, both in the Hwang-Ho valley in northern China. Moreover, 16 of 18 Chinese archaeological records that they list are earlier than Mohenjo-Daro, and all are well north of the natural range of *G. gallus* (Figure 5.2). This implies that domestication took

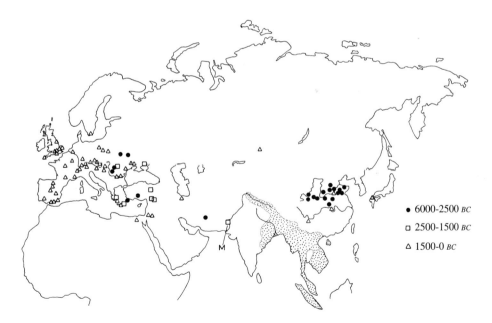

Fig. 5.2 Archaeological sites for *Gallus*. Mohenjo-daro (M) was long thought to be the earliest site for Domestic Fowl, but the more extensive archaeological record now shows many much earlier sites in China, well north of the natural range of *Gallus gallus* (stipple), as well as broadly contemporaneous sites further west. It spread into W Europe by Iron Age times (after West & Zhou 1988).

place somewhere in southern China, Thailand or perhaps Viet Nam (where, however, there are as yet no archaeological sites), and that hens were then taken northwards into China.

Genetic evidence fortunately confirms the impression that Domestic Fowl were first domesticated in South-east Asia. Fumihito *et al.* (1994, 1996) examined both the relationships of the different species of *Gallus* to each other and to the Domestic Fowl, and then the relationships of Domestic Fowl with three of the five supposed subspecies of *Gallus gallus*. They fully confirmed that *Gallus domesticus* is derived from *Gallus gallus*, and moreover that the form found in Thailand, *Gallus gallus gallus*, is apparently its sole ancestor. Even domestic birds on Sumatra, where two subspecies, *G. g. gallus* and *G. g. bankiva*, occur in the wild, are more closely related to Thai *G. g. gallus* than to Sumatran *G. g. gallus*.

The spread of Domestic Fowl out of South-east Asia and into Europe remains to be discussed. Evidently, they spread quite rapidly northwards into China. How did they spread westwards? When Mohenjo-Daro was supposed to be the earliest site for them, it was natural to suppose that they spread first into the Middle East, already by then with a long history of farming, and then around the Mediterranean lands. However, West & Zhou (1988) also point out that we now know numerous sites much further west that have chicken bones earlier than Mohenjo-Daro, including sites in Iran, Turkey, Ukraine, and Romania, but no early records from Iraq, Israel, or Jordan, the classic area for early domesticates. It looks as though Domestic Fowl may have spread from northern China along the route of what later became the silk road, through southern Russia, to enter south-east Europe and the Middle East from the north-east, though at present there are virtually no archaeological data from along that route to confirm this particular piece of the thesis. MacDonald (1992) remarks that the earliest archaeological record close to Egypt was from Sweyhat, Syria in about 2,400 BC, and from Egypt itself not until the XVIIIth dynasty, about 1567–1320 BC. Although wild and domestic waterfowl were beautifully illustrated in ancient Egypt, the Domestic Fowl was not depicted until the time of Rameses, about 1,200 BC, and then as an exotic import. It only became common in the Nile Valley about 300 BC (Houlihan, 1996), so it arrived late, and perhaps from the north, not the east. For it is clear that Domestic Fowl were well established in south-east Europe (Greece, Crete) by Bronze Age times, and spread through most of Europe in Iron Age times, reaching Italy by 500 BC, Holland and Poland by 700 BC, and France by 100 BC (West & Zhou, 1988). The earliest records from England are Late Iron Age sites including Uley (Cowles, 1993), Colchester (Bate, 1934), Gussage All Saints (Harcourt, 1979a), Aylesbury, Ashville Estate (Bramwell, 1978a), Brean Down (Levitan, 1990), Nornour (Turk, 1971), and Thorpe Thewles, Cleveland (Rackham, 1987). All of these are likely to date to the period 200 BC to A.D 50, matching the assumption, interpreting Julius Caesar's remark, that it had probably not been present long, and was too highly regarded to be eaten as a common item of diet. In all of these sites, there are just a few bones, again matching this status as a recent immigrant. Collectively, just under 6% of the Iron Age bird records are of this species (Table 5.1). Notable is the presence of Domestic Fowl at both Howe and Skaill, the Orkney sites discussed above, and at Crosskirk, an indication that domestic species spread very rapidly even to the far north of Britain.

It has already been noted in previous chapters, and is made explicit by Table 5.1, that there is a scattering of even earlier archaeological records. There are two explanations for these. One is the notion, elaborated by Harrison (1978), that there was a species

Table 5.1. The reported occurrence of domestic birds (and Pheasant) in the archaeological record in the British Isles, showing the number of records of each species, and their % of the total bird records for that time interval. Totals (right hand column) include a few extra records of uncertain age. Wild relatives are included for comparison: all wild geese (Greylag, White-fronted, Bean, Barnacle, Brent); all wild *Anas* (Mallard, Gadwall, Garganey, Teal, Wigeon, Pintail, Shoveler); all wild *Columba* (Stock Dove, Rock Dove, Wood Pigeon).

Spp.	Pleist	Late Gl.	Meso.	Neo.	Bron. Age	Iron Age	Rom.	Ang-Norm.	Med.	Post-Med.	Total
Domestic Fowl	2 (0.4)	1 (0.3)	3 (1.5)	2 (0.6)	3 (0.2)	39 (5.9)	259 (14.8)	128 (7.3)	305 (13.3)	143 (13.3)	900 (10.1)
Domestic Goose			1 (0.5)	1 (0.3)	1 (0.6)	7 (1.0)	75 (4.3)	82 (7.4)	137 (6.3)	62 (5.8)	369 (4.1)
Wild Geese	25 (4.6)	11 (2.8)	7 (3.4)	5 (1.5)	9 (5.0)	15 (2.3)	27 (1.5)	38 (3.4)	26 (1.1)	7 (0.7)	179 (2.0)
Domestic Duck	1 (0.2)					8 (1.2)	44 (2.5)	31 (2.8)	65 (2.8)	34 (3.2)	187 (2.1)
Wild Ducks	31 (5.6)	20 (5.0)	21 (10.3)	7 (2.0)	13 (7.2)	48 (7.2)	142 (8.1)	63 (5.7)	132 (5.8)	56 (5.2)	547 (6.1)
Domestic Dove	2 (0.4)	1 (0.3)				1 (0.2)	35 (2.0)	29 (2.6)	50 (2.2)	21 (2.0)	138 (1.5)
Wild Pigeons	3 (0.6)	11 (2.8)	5 (2.5)	2 (0.6)	1 (0.6)	10 (1.5)	49 (2.8)	30 (2.7)	55 (2.4)	12 (1.1)	184 (2.1)
Peacock							2 (0.1)	5 (0.5)	21 (0.9)	7 (0.7)	35 (0.4)
Turkey							2 (0.1)		16 (0.7)	26 (2.4)	47 (0.5)
Pheasant	2 (0.4)	3 (0.8)				1 (0.2)	8 (0.5)	7 (0.6)	27 (1.2)	5 (0.5)	58 (0.6)
Total Birds	539	398	203	344	181	664	1755	1108	2295	1075	8953

of wild *Gallus* present in Europe at least during the Ipswichian Interglacial, which he named *G. europaeus* (see p. 34), and that it might have recurred in Postglacial times. Tyrberg (1998) is sceptical about all early *Gallus* reported from western Europe, though the undoubted presence of *Gallus* in Late Pleistocene Levant and Transcaucasia allows the possibility that a species of *Gallus* might have occurred in southern Europe in earlier interglacials. Given the poor dating assigned to most of these records, and the absence of any comparable species elsewhere in Europe, a far more likely explanation is that these are bones of much later domestic birds that have got into archaeological layers, either through the agency of burrowing mammals, such as Wolf or Fox (which routinely take large prey back to their dens) or Badger (which, however, does not), or by bones simply infiltrating loose cave floor deposits. In most cases, the excavators have dismissed the bones as intrusive, and they are surely right to have done so. Of the sites listed above, Brean Down is assigned to Bronze Age, but the dating of the site, and individual records within it, is uncertain – it also contains House Mouse, which is surely an Iron Age import (Levitan, 1990; Yalden, 1999). James Fisher (1966) waxes eloquently on the problems of the Hoodwink *Dissimulatrix spuria* in regard to such occurrences. It is pertinent here to remember that early Domestic Fowl were about the same size as Pheasant, and later, larger, breeds overlap Black Grouse as well, with large cocks of some modern breeds even rivalling Capercaillie in size. Although there are morphological features that allow most bones to be differentiated (Erbersdobler, 1968, and see Chapter 1), the possibility of genuine error in identification (particularly by earlier excavators, who lacked the full range of comparative material) must also be born in mind.

The subsequent history of Domestic Fowl here, as everywhere else in the world, is of an increasingly common animal, exploited for eggs and meat as well as for sport (cock-fighting) and for religious purposes. Parker (1988), reviewing the occurrences of bird remains on 86 Roman sites, concentrated on wild species, but observed that Domestic Fowl are reported on virtually all of those sites. From the now larger number of about 253 Roman sites in our data base, Domestic Fowl contribute 259 records (some sites have several layers), so they are essentially ubiquitous. They contribute 14.8% of the 1755 bird species records for this period, a sharp increase over their prevalence in Iron Age times, becoming and remaining, as today, the most abundant and ubiquitous British bird species (Table 5.1).

Domestic geese

Two wild species have contributed to the stock of domestic geese: the Greylag *Anser anser*, a common wild species in much of the temperate Palaearctic (from Iceland and the British Isles to Manchuria), and the Swan Goose *Anser cygnoides*, an eastern Palaearctic species common in China. Their relative contributions to domestic stock, the timing and place of domestication, and the genetic make-up of modern domestic geese, have received much less attention than have Domestic Fowl. Kear reviewed what is known. Most waterfowl, including all the geese, can interbreed with their nearest relatives, even in the wild; in captivity hybrids between species of geese are frequent. Obviously, modern genetic techniques would do much to assist in the interpretation of their history, but have barely been applied yet to this problem. However, archaeological data are some help.

The domestic goose in Europe is considered to be descended from the Eastern Greylag *Anser anser rubirostris*, if only because all European domestic geese share pink beaks with this race – western race Greylags, *Anser anser anser*, have an orange beak (Kear 1990). This is also in accord with an early depiction of what Kear (1990) considers to be a domestic goose in ancient Egypt: this has the features of an Eastern Greylag, and, because it is evidently in lay (further south than its natural range), is presumed to be domestic. However, there are also superb tomb paintings of Red-breasted, Bean and White-fronted Geese, which are likely to have been wild geese wintering in the Nile Delta (Houlihan 1996). Also depicted are Egyptian Geese *Alopochen aegyptiacus*, which had already been domesticated in ancient Egypt but were displaced by domestic Greylag Geese. Geese are depicted in cages, penned and in a variety of colours, indicating domestication, and by the eighteenth Dynasty, 1450–1341 BC, such evidence is so common as to leave no doubt (Albarella, 2005). It seems likely that either Domestic Geese, or the idea of domesticating them, passed from ancient Egypt into Europe via Greece and Rome. Geese were known as domestic animals by the Ancient Greeks, because Penelope, Odysseus' wife, is described as having a small flock in Homer's *Odyssey*. They passed from them to Roman culture (Toynbee, 1973), to whom they were both food, and sacred to some gods. The geese in Juno's temple in Rome cackled to warn the Romans of an attack by the Gauls in 390 BC, when the dogs failed to bark. The concept of fattening geese for pate de foie gras was known to Pliny, as was the value of white geese, for food and down. That white geese could be preferred indicates clearly that they were indeed domesticated.

The Greylag is the largest of the European geese, though the Bean Goose rivals it. One obvious consequence of domestication is selection for larger, heavier birds that are less able to fly. Their leg bones therefore get somewhat longer and especially stouter, to carry the greater weight, but wing bones are less affected. Because domestic geese are protected, they have less need of flight, or may be prevented from flying, for instance by clipping the wing feathers (as a temporary measure, until the next moult) or, more permanently, by pinioning them (surgically removing the distal wing bones and their feathers). Thus there is no pressure for wing bones to lengthen as the other bones of the skeleton get larger, and they may stay the same (absolute) size, so becoming relatively smaller. Occurrence of robust goose leg bones in an archaeological site is therefore a good indicator that domestication has occurred, but it requires an adequate (statistically) large sample to prove this point. At Viking-age Haithabu, in Denmark, goose metatarsal bones average about 3 mm wider, at any given length, than wild Greylag bones, though the wing bones were thinner; clearly, these were Domestic Geese (Reichstein & Pieper, 1986). Obviously, though, bones of the earliest Domestic Geese would be indistinguishable from Greylag. Wild Greylag, as well as other species, would also have been hunted, both by Mesolithic and Neolithic people prior to domestication, and in subsequent times. The various species of wild geese overlap in size, so are aggregated in Table 5.1 to provide a comparison with occurrences of Domestic Geese. These do indeed suggest a modest presence of Iron Age Domestic Geese, and a more pronounced increase in Roman times that persists to modern times. Indeed, the overall pattern matches quite well that of Domestic Fowl, adding archaeological strength to the interpretation of Julius Caesar's remark. Iron Age records of Domestic Geese in Britain are claimed from Gussage All Saints (Harcourt, 1979a), Budbury (Bramwell, 1970), West Stow (Crabtree, 1989b), and Harston Mill (Jones pers. comm.) in England, Howe, Orkney (Bramwell, 1994), Skaill (Allison, 1997b), and Crosskirk Broch (MacCartney, 1984) in Scotland. Earlier records from Mesolithic Gough's Cave (Harrison, 1980a, 1986), Neolithic Point of Cott, Westray (Harman, 1997), and Bronze Age Dowell Cave (Bramwell, 1960b) are perhaps correctly identified but time-transgressive, like the spurious Domestic Fowl records discussed above, but they might instead be large wild Greylag Geese. In later times, Domestic Geese are abundantly recorded in Anglo-Saxon sites, and even more abundant in Medieval times (Albarella, 2005), only being displaced as the choice of Christmas dinner when Turkeys became more common, from the seventeenth century.

We are not aware of any archaeological records from Britain of Swan Geese/Chinese Geese, but suspect they would get hidden among the other species anyway. The historical record suggests that they were not imported to western Europe until the end of the eighteenth century – Bewick illustrated one in 1797 (Kear 1990). Given their knowledge of Domestic Fowl, it is likely that the Chinese also domesticated geese, and ducks, at least as early as they were domesticated in the Mediterranean area, but the Chinese archaeological literature is obscure, to us, and we have no detailed record of this.

Domestic Duck

There is general agreement that the Mallard *Anas platyrhynchos* is the ancestor of all domestic ducks (with the obvious proviso that the entirely different South American Muscovy Duck *Cairina moschata* has also been domesticated). The drake has the curled tail feathers that also characterize all breeds of domestic drake. However, the Mallard is a very widespread species across the Palaearctic, breeding from Iceland, Ireland, and Spain in the west to northern Japan and Kamchatka in the east. It also breeds right across the Nearctic from Alaska to Labrador. Although it does not breed in Egypt, Iraq, or China, it does winter in all three. Thus, potentially, it could have been domesticated in any of the classical centres of early agriculture in the Old World, or in none of them. There is a little more genetic evidence to help unravel its history than for the domestic goose, but documentary and archaeological evidence has to support most of the debate.

Kear (1990) reports that the duck has been domesticated for only half as long as the goose, and that it was not known to Egyptian, Assyrian, or Babylonian civilizations, nor to the Jews or Ancient Greeks. She suggests that the Romans in the west and Malays in the east were responsible for starting to domesticate it. She further comments that domestic ducks are not listed in the poultry trade in London until 1363 (when listed as 'tame Mallard'), though Teal were listed as early as 1274, implying that domestic ducks might have been a late addition to the English farmyard. Perhaps contradicting this, Houlihan (1996) says that Egyptians in the eighteenth Dynasty (about 1550–1307 BC) were rearing domestic ducks and geese on agricultural estates to supplement those taken from the wild. Toynbee (1973) reports on the recommendations of Varro and Columella for keeping ducks, but it is not clear that these were being farmed commercially. Their descriptions sound more like those for the husbandry of pets or cage birds than for food animals.

The interpretation of the archaeological record of ducks, especially Domestic Ducks, is even more bedevilled by problems of identification than that of geese. Mallard are larger than most other ducks, certainly larger than all the other dabbling ducks *Anas* sp., and approaching Eider and Shelduck (which can, however, be distinguished anatomically on most bones; Woelfle, 1967). Further increase in size makes domestic ducks more distinctive, albeit still problematic on partial remains, and many of the records grouped in Table 5.1 as Domestic Duck are in fact given by the archaeologist concerned as 'Domestic Duck/Mallard'. Examples confidently identified as Domestic Duck come from Iron Age Glastonbury Lake Village (Andrews, 1917; Harrison, 1980a, 1987b), Gussage All Saints (Harcourt, 1979a), Howe, Orkney (Bramwell, 1994), and Ashville Trading Estate, Abingdon (Bramwell, 1978a). Less confident Iron Age records come from Dragonby (Harman, 1996a) and Micheldever Wood (Coy, 1987a). For the Roman period, there are abundant records of Domestic Duck (at least 17) as well as Domestic Duck/Mallard (another 19). So many archaeologists are involved, with so many records, that it seems certain that Domestic Ducks were common in Britain by Roman times, and Albarella (2005) notes that duck bones (wild or domestic uncertain) were more common than goose bones in Roman sites. The statistical evidence of Table 5.1 also suggests a genuine increase in representation of ducks during the Iron Age–Roman transition, just as with Fowl and Goose. Moreover, the identifications of Iron Age Domestic Ducks strongly support the notion that, like Domestic Fowl and Domestic Geese, Domestic Ducks were already part, albeit a much smaller part, of the farmyard stock before the Romans arrived. It seems as though the notion of having a range of domestic poultry had spread through the Celtic world to Britain. Notice, in Table 5.1, that Domestic Fowl, Goose and Duck all increase markedly in representation from Iron Age to Roman times, but none shows any sharp increase in Norman or Medieval times. Obviously, this contradicts Kear's (1990) opinion that Mallard were not domesticated until medieval times, but Albarella (2005) finds duck bones much less abundant than goose bones in Anglo-Saxon sites, and still less abundant than goose bones in Medieval sites. Perhaps the notion of keeping Domestic Ducks died out between Roman and Anglo-Saxon times, and ducks were, as it were, re-domesticated later on. There is an alternative view, that these are in fact wild ducks of one species or another (mostly Mallard, undoubtedly), and that what we are describing is in fact an increasing interest in the harvesting of wild birds (Albarella & Thomas, 2002), perhaps associated with the expansion of hawking, and with high-status sites. This would apply equally to the record of geese. There are now sufficient records from sufficient different sites for a review of the evidence to be undertaken, considering both reliability of the identifications as indeed Domestic Ducks, and their abundance relative to other species.

So what help do we get from genetic evidence on the ancestry of Domestic Ducks? So far, it is not conclusive, but Hitosugi *et al.* (2007) indicate a major split between Domestic Ducks from south-east Asia and north-east Asia. They argue that ducks were domesticated in China about 3000 bp, and certainly demonstrate that breeds such as Indian Runners and Khaki Campbells are related to ducks in south-east Asia. Breeds from Taiwan and Japan belong to the different lineage. Data on old European breeds, such as Aylesburies and Call Ducks, are urgently needed.

Domestic Dove

The term 'dove', as used by both archaeologists and ornithologists, is a deceptive one. Domestic Dove or Domestic Pigeon usually implies Domestic Rock Dove *Columba livia*. The Romans certainly had domesticated *Columba livia*, using them as food, as pets and as carrier pigeons (Toynbee, 1973). Varro and Columella apparently give instructions for feeding, housing, and fattening pigeons, and report that one house could contain as many as 5,000 pigeons. That certainly sounds like commercial production. However, Barbary Doves *Streptopelia risoria*, looking remarkably like our now familiar Collared Dove *S. decaocto*, have also been domesticated for many centuries, and also appear in Roman mosaics. Archaeologists do not always specify exactly what they mean by 'Domestic Dove'. Some records are actually more helpful for being uncertain – 'Domestic/Rock Dove' and 'Domestic/Stock Dove' are certainly *Columba* of the size of *C. livia* (or *C. oenas*), not *Streptopelia*. Some 38 records in the literature are identified only as 'Dove sp.' and another 29 are designated 'Dove/Pigeon'. However, our records include only three specifically attributed to *Streptopelia*, two of them the long-standing native Turtle Dove *S. turtur*, though Don Bramwell (Bramwell, 1985b) identified a possible Barbary Dove *S. risoria* bone from Roman Barnsley Park, Gloucester. In the absence of contrary evidence, we have assumed for the purposes of compiling Table 5.1 that all the uncertain Doves and Pigeons reported by archaeologists are of *Columba livia*-sized animals. (Even considerably enlarged

Domestic Pigeons are significantly smaller than Wood Pigeon *C. palumbus*, but the overlap with Stock Doves is complete, nor can Stock and Rock Doves be easily separated on archaeological specimens; they are the same size, and the morphological differences are small (Fick, 1974)).

Assuming that these doubts are not overwhelming the evidence available, what can we learn from the archaeological record presented in Table 5.1? For all periods with a reasonable sample, wild pigeons contribute 2–3% of the bird records; of these 184 records, 86 are Wood Pigeon and 98 are Rock or Stock Doves. However, from Roman times onwards, there is an additional component of another 2–3% that appear to be Domestic Doves. There are only four earlier records included among these; a Devensian (Pleistocene) identification of Domestic Dove (Langwith Cave, Derbyshire; Mullins, 1913) may well be correctly identified but a time-transgressive 'Hoodwink', but the Ipswichian (Pleistocene) record from Kirkdale Cave and the Late Glacial record from Merlin's Cave (Harrison, 1980a) are actually cautious 'Pigeon sp.' and the Iron Age record is similarly a cautious 'Dove sp.' from Skaill, Orkney (Allison, 1997b). In other words, these get incorporated in Table 5.1 as 'Domestic Doves' only by our cautious (or incautious) collation of the records. As with the other domestic birds, these data suggest that the Domestic Dove was indeed an early introduction to Britain, albeit slightly later, Roman rather than Iron Age.

Other Roman introductions

Three other gamebirds are represented in Roman mosaics, and were certainly familiar birds in Roman times: Pheasant, Peacock, and (Helmeted) Guinea-fowl.

The status of the Pheasant as a wild bird in Britain, and in western Europe generally, has been and remains controversial. It has never been a farmyard bird, but it is now well established as a wild bird, and moreover is bred on a commercial scale for sporting estates. Toynbee (1973) remarks that Pheasants were known to the Greeks of fifth to fourth century BC, that Ptolemy VIII (145–116 BC) apparently had them in captivity in Egypt, and that various Roman authors repeat the point that the species originated in Colchis, near the River Phasis, which provide both its vernacular and scientific names. Now in Georgia, the River Phasis (now Rioni) runs into the eastern end of the Black Sea, while Colchis was the coastal region, now extending from Turkey into Georgia, around it. Toynbee (1973) reproduces a superb Roman mosaic from the Justinian's church, Sabratha, in Libya, showing an obvious cock Pheasant and what is surely a hen, though, perhaps misled by the long tail, Toynbee identifies it as a parrot (long-tailed parakeets often also appear in Roman mosaics). It seems likely that the native range of the Pheasant stretched eastwards from Transcaucasia across the southern USSR to China, Korea, and northern Japan (Cramp *et al.*, 1977–1994; Tyrberg, 1998). Though Cramp *et al.* (1980) add Turkey, Thrace, and south-east Bulgaria to the native range, it seems unlikely that Greeks and Romans would have regarded it as native to Colchis had it occurred so much nearer. Certainly, it was not native to western Europe, let alone the British Isles, so when did it arrive here? Conventional opinion recently has been that the Normans imported it (Fitter, 1959; Lever, 1977; Cramp *et al.*, 1980). This opinion follows Lowe (1933), who refuted the earlier consensus of Roman introduction. He demonstrated that the supposed Romano-British Pheasant bones from Silchester were in fact Domestic

Fowl, checked also supposed Roman Pheasants from York, Verulamium and the neighbourhood of Shrewsbury, which were also Domestic Fowl, and asserted confidently that he knew of no certain Roman remains. How does the evidence look 70 years later?

A quick resumé of the apparent record for Pheasant shows only eight that might be Roman in age, but these include the refuted Silchester bones, while the bones from Studland (King, 1965) were among a group that included Turkey, and were attributed to the efforts of a modern, not Roman, Fox. Possibly valid records include Barnsley Park (Bramwell, 1985b), Quinton (Field, 1999), Hardingstone (Gilmore, 1969), Latimer (Hamilton, 1971), Colchester (Luff, 1982, 1993) and perhaps Barrow Hills, Radley (Roman/Saxon: Barclay & Halpin, 1998). Even these are very few records, given the abundance of Roman sites. Barnsley Park is interesting, in that both Pheasant and Domestic Fowl were numerous, with respectively eight and 12 individuals represented. The Pheasants were thought to have been reared there for the table. Given that the Romans did know, and eat, Pheasant, a few records from this time seem likely, but do not establish that the bird itself became established in the Roman countryside. In the succeeding Saxon period, the Pheasant is equally scarce or absent, and only claimed from York–Fishergate (O'Connor, 1991), Lincoln (Cowles, 1973; Dobney *et al.*, 1996), Lewes (Bedwin, 1975), and the uncertainly dated Barrow Hills, Radley specimens already mentioned. The contrast with later periods is striking; there are at least 27 records of Norman or later Medieval date. Particularly interesting is a small series of records that seem to be very late Anglo-Saxon/Anglo-Scandinavian or early Norman–Hen Domen (Browne, 1988, 2000), York–Coppergate (O'Connor, 1989), Jarlshof (Platt, 1956), and Flixborough (Dobney & Jaques, 2002). Taken together, the increase in records in Norman–Medieval times, and these late Saxo-Norman records, certainly support the idea that the species was not established as a British bird by the Romans, but by later introductions around the ninth to tenth century, possibly just pre-Norman.

The Peacock, a native of India and South-east Asia, was also certainly well known to the Romans, and illustrated on mosaics, coins, and cooking vessels (Toynbee, 1973). Originally, it was kept for pleasure, but later became also a prized food. It was not known to the ancient Egyptians, but seems to have been introduced there, too, in the Graeco-Roman period, under Ptolemy II (285–246 BC) (Houlihan, 1996). The only Roman records for Britain are from Portchester (Eastham, 1975) and Great Staughton (Bramwell, 1967). As Table 5.1 shows, it is another introduction that became more frequent in Medieval times. The Guineafowl is a native of Africa, though very rare now in North Africa and extinct in Egypt; it is also illustrated, along with the Pheasant, in the mosaics at Justinian's church in Sabratha, Libya. Roman accounts of its husbandry make clear its familiarity to them (Toynbee, 1973), but there is no archaeological record of its presence in Britain, then or later. MacDonald (1992) does not include any early records of archaeological Guineafowl in Africa or Europe, and notes that in West Africa introduced Domestic Fowl seem to have preceded the use of the native Guineafowl as a domestic bird. Luff (1982) does mention an archaeological specimen from the Roman camp at Saalburg in West Germany. It was reputedly brought to Europe from West Africa by the Portuguese in the Middle Ages, but the date of its first introduction to Britain seems uncertain, perhaps not until the seventeenth century.

The Turkey was domesticated in Mexico, perhaps as early as 4,000 b.p., and was imported into Europe during the sixteenth century. Its importation to England is closely dated to between 1525 and 1532, and Shakespeare mentions it twice (though not the Guineafowl).

Remarkably, Turkeys were taken back already domesticated to New England, where wild Turkeys of a different subspecies occur. Of 46 records in our data base, nearly all are dated, as expected, to late Mediaeval or Post-mediaeval times. The exceptions are four cave records with no precise date ('Unknown' or 'Late-Pleistocene-Holocene') from Keshcorran Cave, County Sligo, Castlepook Cave, County Cork, Aveline's Hole, Somerset and Catacomb Cave, County Clare, and two more troublesome records, a 'Romano-British' record from Ossom's Cave, Staffordshire (Bramwell, 1954) and 'Roman' Keston, Kent (Locker, 1991). Clearly, these are likely to be either misidentified or time-transgressive 'hoodwinks', bones that have inserted themselves in inappropriate archaeological layers somehow, perhaps with help from Foxes, or excavators, or porous rubble deposits.

Wild birds in Roman Britain

Of the 1755 records of bird species from Roman sites in our data base, 413 (23%) are of the four domestic species, fowl, duck, goose, and dove, just discussed. That leaves a wide range of records of wild species. Dissecting these to establish their utility to the Romans, and to derive some indication of the avifauna of Roman Britain, was initiated by Parker (1988). He accumulated records of 94 species from 86 sites, plus a few extra records of species aggregates (plover, small wader, thrush, tit, small passerines, finch/bunting), which may well be the same as species identified more precisely at other sites. Our more substantial list records 136 species (not counting aggregates such as 'Crow/Rook') from about 244 sites (Figure 5.3), though this total includes as 'Roman' some Pictish and Irish sites, notably Balinderry, which we presume to be contemporaneous. The largest fauna comes from Ossom's Eyrie Cave, a rural site in which a mixture of Barn Owls feeding on small prey, a breeding Golden Eagle taking much larger prey, and a brief Romano-British occupation contributed a diversity amounting to 63 species, though this site probably extends from Romano-British into Anglo-Saxon times; we have discussed the whole fauna as though it were Roman in age. Of more conventional Roman sites, a range of villas, forts and towns has been investigated. Many contain just a few bird bones, often imprecisely identified ('thrush', 'finch', 'duck'). Those with the largest faunas include the following (Table 5.2).

Parker (1988) found Raven the most frequent species, and this is certainly still true. With 95 records, it is approached only by Mallard (75 occurrences) in frequency. It is tempting to assume that these were scavengers around the outskirts of towns and villages. The numerous records of other obvious scavengers, notably White-tailed Eagles (19), Red Kites (14), and Common Buzzards (19), support this interpretation. The numerous records of other crows, including nine Carrion Crow, 18 Rook and 71 'Crow/Rook' or 'crow sp.' (the two are very difficult to distinguish, though Rooks average smaller, and have, of course, longer thinner bills), might also support this notion. On the other hand, Ravens had a symbolic status (cf. the Iron Age sites discussed previously), and may well have been kept as pets, while Rooks were possibly eaten – Rook pie was after all a frequent dish in recent times, perhaps still is.

Some of the other species were certainly taken as food. The frequency of Woodcock, identified at 68 sites, is particularly striking. Collectively, plovers, with 63 records (including 10 Lapwing, 30 Golden Plover and 7 Grey Plover, as well as 15 uncertain 'plover sp.') also were

Table 5.2 Some Roman and Roman-age sites with diverse wild bird faunas, see also Parker (1988).

Site	Type	Number of Species, and examples	Ref.
Wroxeter	Fortress	30, inc. Mute Swan, Raven, Water Rail	Meddens (1987)
Uley Shrines	Temple	24, inc. W-t Eagle, Mute Swan, Raven	Cowles (1993)
Stonea	Rural camp	19, inc. W-t Eagle, Kite, Raven	Stallibrass (1996)
Silchester	Town	22, inc. Wh Swan, Crane, Stork, Raven	Parker (1988)
Ossom's Eyrie	Rural	63, inc. G Eagle, Bl Grouse, Raven	Bramwell *et al.* (1990)
Portchester	Fort	16, inc. Gt N Diver, Raven	Eastham (1975)
London Wall	Town	18, inc. Crane, Little Egret, Night Heron	Harrison (1980a)
Ilchester	Town	17, inc. Kite, Peregrine, Goshawk	Levitan (1994a)
Heybridge, Elms Farm	Town	31, inc. Mute Swan, Peregrine, Raven	Johnstone & Albarella (2002)
Frocester	Villa	26, inc. Kite, Buzzard, Quail	Bramwell (1979b)
Filey	Fort	21, inc. Razorbill, Guillemot, Puffin	Dobney *et al.* (2000)
Exeter	Town	19, inc. Crane, Cuckoo, Raven	Maltby (1979a)
Dorchester	Town	23, inc. Crane, Corncrake, Kite, Buzzard.	Maltby (1993)
Colchester	Town	32, inc. Crane, Raven, Grey Shrike	Luff (1982, 1993)
Castle Copse, Bedwyn	Villa	19, inc. Goshawk, Golden Plover	Allison (1997a)
Annetwell St., Carlisle	Town	26, inc. Kite, Crane, Bl Grouse	Allison (1991)
Caister-on-Sea	Fort	17, inc. Mute Swan, Crane, Raven	Harman (1993b)
Caerwent	Fortress	18, inc. Kite, Raven	Bramwell (1983d)
Caerleon	Fort	15, inc. W-t Eagle, Avocet, Crane, Raven	O'Connor (1986)
Buckquoy	Pictish	19, inc. G N Diver, Osprey, L Auk	Bramwell (1977b)
Barnsley Park, Gloucs.	Villa	27, inc. Crane, Bl-t Godwit	Bramwell (1985b)

a regular item of diet. Wild ducks and geese (as summarized in Table 5.1) were also important, particularly Mallard, as mentioned, and Teal (49 records), but most species that might be expected have been reported from at least a few Roman sites, including not only the more numerous Wigeon (16 sites), Gadwall (five), Garganey (five), Pintail (four), Pochard (four), and Shoveler (four), but rarer species such as Goosander and Red-breasted Merganser. One notable absentee is the Eider, which perhaps only bred north of Roman Britannia, but might have been expected at Pictish sites of this age. Barnacle Goose is quite frequent, not only at the northern sites of Carlisle and Papcastle, close to its modern wintering grounds, but also at York and Gloucester. Did they then winter on the Severn Marshes and the Humber floodlands, or did the Romans trade wildfowl extensively? Whooper Swans occurred at Carlisle, Lincoln, Doncaster, and Piercebridge, perhaps reflecting their wintering on the marshes and floodplains of the Solway, Humber, and Tees, but also at Silchester and Over Purbeck, southern sites that would be more usually the haunts of Mute Swans. Mute Swans were indeed recorded at 8 sites, all in the southern half of Britain – York (O'Connor, 1985) is the northernmost, with Wroxeter, Heybridge, Caister-on-Sea, Longthorpe, and three in London the others. At Annetwell Street, Carlisle, small wild geese, probably mostly Barnacle Geese, which of course still winter on Solway, were more numerous as food remains than Domestic Fowl, unusually for a Roman site (Allison, 1991).

Some of the other wetland birds, probably also food items, are notable. There are 35 records of Crane, of which 31 are specifically identified as Common Crane and three as

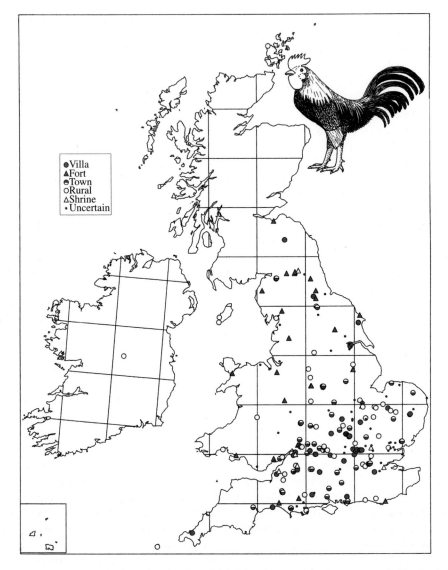

Fig. 5.3 Roman archaeological sites in the British Isles. Unsurprisingly, most are in England, but there are a few contemporaneous sites elsewhere, including some genuine Roman sites in Wales and S Scotland.

'Crane sp.' – but it is hard to imagine what other species they would be. They are widely spread across England, from Carlisle, Newstead, Housesteads, Papcastle, and Piercebridge in the north to Exeter in the south-west and London and Silchester in the south-east. There are a couple of Welsh records (Caerleon; Pentre Farm, Flint) and one from Ireland – Balinderry crannog – but no apparently contemporaneous sites from Scotland. Cranes were certainly eaten – a tibiotarsus from Carlisle bore cut-marks (Allison, 1991), and the skull from Caerleon had the back of the braincase removed as recommended by Roman writers on cookery (Hamilton-Dyer, 1993). By contrast, there is only one record of White Stork from Roman Britain, at Silchester (Newton, 1905; Maltby, 1984). Perhaps as a more southern species, it has never been common here. Interestingly, there is also one record of another southern wetland species, Night Heron, at London Wall (Harrison 1980a), to accompany three records of Bittern (twice from Grandford, in the fens, and Winnall Down, near both Winchester and the Itchen valley) and eight of Grey Heron, seven from the southern half of England, and one from Balinderry, again. Smaller waders, presumably taken from mudflats, estuaries, and other wetlands, also turn up frequently, though identifications are sometimes uncertain. As well as seven records of 'Wader sp.', 15 Curlew and 10 Snipe, identifications of rarer waders include three each of Bar-tailed Godwit (Colchester, Ilchester, and London Wall) and Black-tailed Godwit (Colchester twice, Barnsley Park), two of Dunlin (Caerwent, Colchester) and Greenshank (Over Purbeck, Ower), and one each of Avocet (Caerleon), Redshank (Caister-on-Sea), Green Sandpiper (Thenford), Knot (Camulodunum), and Turnstone (Buckquoy). As another bird characteristic of wet grasslands, the five occurrences of Corncrake, at Camulodunum and Colchester, Dorchester, Farmoor, and Rudston, reflect both habitat and Roman food interests.

Clearly, wetlands were still extensive in southern Britain. However, farmland was surely the main habitat, and reflected in the avifauna. Grey Partridges, reported from 13 sites, were presumably another of the food species, and two other quintessential farmland birds, the House Sparrow and Starling, were reported from 10 and 24 sites respectively, despite that most small passerines are severely under-recorded. However, there are only five records of Skylark, which we associate with rough grassland and cereal fields. By contrast, another striking indication of the open nature of the countryside is the frequency of Barn Owls, reported from 14 sites, mostly in southern England but including Catterick and Piercebridge. Conversely, there are no records of Tawny Owl, despite good records from both earlier (Neolithic Runnymede Bridge, Iron Age Howe and Slaughterford) and many later sites. Was Roman Britain so bereft of woodland, or were Tawny Owls just too secretive? It seems unlikely that they were rarer then than now, and interesting that the other two native owls, which are now much rarer, do at least have a Roman record. Short-eared Owl is recorded from Ossom's Eyrie Cave, not far from the North Staffordshire moorlands (Bramwell *et al.*, 1990), and Long-eared Owl from Wroxeter (Meddens, 1987). Another indication of extensive open farmland might be the single record of Roman Great Bustard, from Fishbourne (Eastham, 1971), though such an important table bird might just have been imported, like the Red-legged Partridge reported from the same site (and a recent rumour, not yet published, suggests that the relevant bone is in fact from Crane).

If Grey Partridge are a good indicator of the extent of farmland, the equivalent indicators of moorland and scrubby moorland-edge woodland are Black Grouse. These, too, have quite a strong Roman record, from 12 sites. Black Grouse is numerous at Ossom's Eyrie Cave, where it was the main prey of the contemporary Golden Eagle, but this is its southernmost site. Elsewhere there are four records from Roman Carlisle (Allison, 1991, 2000; Stallibrass, 1993), two from York (O'Connor, 1985; Parker, 1988), and records from Doncaster (Carrott *et al.*, 1997), Ribchester (Stallibrass & Nicholson, 2000), Birdoswald (Izard, 1997), Piercebridge (Parker, 1988), and Corbridge (Bell, 1922). The association with

Roman military sites in the north, particularly along Hadrian's Wall, is noteworthy, as is the fact that most of these are from recent excavations, when the possibility of errors in identification have been well appreciated. At Carlisle, Black Grouse were numerous, as well. Less certain, in dating at least, is the single record of Capercaillie, from Wookey Hole, Somerset (Balch & Troup, 1910), which might have been Iron Age (Parker, 1988, therefore omits this record), but at least the identification, based in part on an unmistakable beak, is certain. By contrast with the good record of Black Grouse, there are only three records of Red Grouse (Corbridge: Bell, 1922; Great Staughton: Parker, 1988; Ossom's Eyrie Cave: Bramwell *et al.*, 1990) – and two of 'Grouse sp.' (Thornborough Farm near Catterick: Stallibrass, 2002; Victoria Cave near Settle: Geikie, 1881), which must surely also be this species. At Ossom's Eyrie, the 39 Black Grouse greatly outnumbered the three Red Grouse, matching this disparity in representation by sites, and it seems that heather moorlands, prime Red Grouse habitat, were indeed much more limited in extent then than now.

The culinary interests of the Romans in small birds are well documented. As small birds are usually under-recorded, the identification of Blackbirds at 14 sites is strong confirmation of this interest. To these must be added eight records of Song Thrush, seven Redwing, six Mistle Thrush, three Fieldfare, and 25 'thrush sp.'. Most of these are from villas (e.g. Frocester, Bedwyn) and military camps (Housesteads, Birdoswald), and surely indicate food remains, although a few, including an additional species, Ring Ouzel, from Ossum's Eyrie, come from other sites. The abundance of House Sparrows and Starlings may be another dietary reflection.

Less certain is the use by Romans of seabirds. Parker (1988) contrasted their slight representation on Romano-British sites with their much better representation in Medieval excavations. In particular, he noted the absence of two now numerous species, Black-headed Gull and Puffin, from his lists of Roman birds. There are now two records of this gull in our list, one from the Pictish site of Buckquoy (Bramwell, 1977b), well outside the Roman province, but the other from Filey Signalling Station, where Dobney *et al.* (2000) thought that the locals were indeed exploiting seabirds as food. The same two sites also yielded Puffin remains, as did the apparently contemporary site of Perwick Cave on the Isle of Man (Garrad, 1972). Razorbills and Guillemots were also recorded at both Filey and Perwick Cave. Additional records of Guillemot came from Over Purbeck, Ower, Perwick Bay, and Rope Hole Lake. The one Roman-dated example of Great Auk also came from Perwick Cave, while the one contemporary Little Auk came from Buckquoy. The final auk species that might be expected, Black Guillemot, was also recorded from Buckquoy and Filey. Other seabirds recorded include Shag (Birsay – Saevar, Dorchester, Filey, Iona, Perwick Bay, Stonea), Cormorant (Birsay – Saevar, Buckquoy, Chester (Northgate Brewery), Filey, Iona, Perwick Bay, Stonea), Gannet (Howe, Buckquoy), Fulmar (Buckquoy, Niarbyl on the Isle of Man), Glaucous/Great Black-backed Gull (Buckquoy), Herring/Lesser Black-backed Gull (Birsay – Saevar, Howe, Caerleon, Kenchester, Dragonby, Rope Lake Hole, Segontium) and Common Gull (Ballinderry, Chelmsford, Pevensey, Poundbury). Clearly, several of these are also well outside the Roman Province, and tell nothing about the dietary or other interests of the Romans themselves. However, Dobney *et al.* (2000) remark that the assemblage from Filey is unusual for a Roman site, and moreover a Guillemot humerus has definite cut marks while one of the Cormorant tibiotarsi seems to bear chop marks. The locals at Filey were clearly eating seabirds, though this was evidently not a frequent Roman habit.

Conclusions

An abundance of Roman archaeological sites, many with well-preserved bird bones, give a good record of the bird life of those times. It is a very familiar avifauna, with few surprises. It seems as though the Romans ate a very wide range of wild birds, from a range of habitats, albeit in small absolute numbers. Birds of farmland were certainly present, and the symbolic importance of Raven, a feature of Iron Age Britain, persisted into Roman times. The Roman sites document the increasing importance of domestic birds, which begins a little before the Romans arrived, with at least Domestic Fowl and Domestic Goose present in the late Iron Age. These two were the most abundant and ubiquitous bird species through Roman times, and subsequently, with Domestic Duck and probably Pigeon added to the farmyard. There is a little evidence that the Romans had, at least ate, Pheasants, in Britain, and knew Peacocks too.

6
Monks, monarchs, and mysteries

It used to be thought that the end of Roman Britain, about AD 410, meant the end of any organized civil life, that farms and towns were abandoned, and that much farmland reverted to woodland. More recent knowledge does confirm that most towns fell into disuse, and the Anglo-Saxon newcomers seem to have preferred to settle in farms and small villages, but farming certainly continued, and the countryside remained largely agricultural. With the adoption of the eight-ox plough, Saxon farmers were able to break the deeper clay soils that had been largely intractable previously, so they extended farmland into previously wooded valleys and lowlands. However, the most remarkable consequence of their arrival is the complete change of social order and language, at least in most of England. The Celtic tribal society of Roman Britain disappeared. In its place appeared the modern counties that we still recognize (or did until 1974), many of the estates and parishes, and most of the placenames. Rivers sometimes retained their Celtic names – Avon, Derwent, Ouse – but very few of the settlements, and even fewer of the landscape features, retained either Celtic or Roman names. This has interesting consequences for the historical naturalist. Often landscape features, sometimes settlements too, were bestowed with names that referred to animals. This is clearly true for mammals (Yalden, 1999): places named for Wolf, Badger, Fox, and Red and Roe Deer are frequent; Beaver names are rare, and there seem to be no Bear names. Domestic mammals appear even more frequently than wild ones, emphasizing the fact that the early Saxons were essentially farmers. A glance at placenames is just as useful for the ornithologist, and reveals our earliest intimation of named British birds.

Birds in placenames

Scholars of the English Place-Name Society (EPNS) have been garnering the early spellings of modern placenames, to divine their original meanings, since 1924. Their interpretations have been published in a remarkable series of county volumes, which have become more detailed as the series has progressed; the earliest, those from 1920s and 1930s, concentrated on main settlements – towns and villages, manors and farms – but the later ones cover minor features too, even field names. So far, the series includes 80 volumes, but does not cover all counties; in particular, several of the eastern counties (Kent, Suffolk, Norfolk, Lincolnshire), which the newcomers might have settled first, are only partially covered, or not at all. For those counties with coverage, extracting the placenames that make some animal reference provides the most complete lists, but these are necessarily biased to the better known counties. Such lists are also biased to English placenames; there are books on Irish, Manx, Scottish, and Welsh placenames, but they are mostly not so complete in their coverage as the EPNS volumes. However, there are some sources that, while they concentrate on major settlements, offer more even coverage across either the whole of England (e.g. Ekwall, 1960) or the whole of the British Isles (e.g. Mills, 2003). Many places are named after people, and of little interest here, except that sometimes people bore the names of animals, a serious source of confusion. More interesting are places named after adjacent landscape features, a topic reviewed for England by Gelling & Cole (2000); their volume makes a good starting point to appreciate the variety and balance of bird placenames (Table 6.1).

Table 6.1 Bird-derived place-names in England.
This is a sample drawn evenly from across England of bird names associated with landscape features, extracted from Gelling & Cole (2000). For several species, much more comprehensive lists have been extracted from the available literature (e.g. Boisseau & Yalden 1999, Gelling 1987, Moore 2002), but they are necessarily uneven in geographical coverage. Some names which might derive either from birds, or from personal names of people bearing the same/similar names, are indicated by "(or pn?)". (ME = Middle English; OE Old English = Anglo-Saxon; ON = Old Norse.)

Place	Co.	NGR	old name	meaning
Algrave	DRB	SK4545	OE *ule, graef*	owl grove
Amberden	ESX	TL5530	OE *amer, denu*	bunting valley
Ambrosden	OXF	SP6109	OE *amer, dun*	bunting hill
Anmer	NFK	TF7429	OE *ened, mere*	duck lake
Andwell	HMP	SU6952	OE *ened, well*	duck spring
Areley Kings	WOR	SO8070	OE *earn, leah*	eagle clearing
Arley	WAR	SP2890	OE *earn, leah*	eagle clearing
Arley	WOR	SO7680	OE *earn, leah*	eagle clearing
Arley	CHE	SJ6780	OE *earn, leah*	eagle clearing
Arley	LNC	SD5327	OE *earn, leah*	eagle clearing
Arley	LNC	SD6707	OE *earn, leah*	eagle clearing
Arncliff	YON	SD9371	OE *earn, clif*	eagle cliff
Arncliff	YOW	SD9356	OE *earn, clif*	eagle cliff
Arnecliffe	YON	SD9371	OE *earn, clif*	eagle cliff
Arnewas	HNT	TL0997	OE *earn, waesse*	eagle wash (wetland)
Arnewood	HMP	SZ2895	OE *earn, wudu*	eagle wood
Arnold	NTT	SK5945	OE *earn, halh*	eagle nook
Arnold	YOE	TA1241	OE *earn, halh*	eagle nook
Birdbrook	ESX	TL7041	OE *bridd, broc*	bird brook
?Birdshall	YOE	SE8165	OE *bridd, halh*	bird nook (or pn?)
?Bonsall	DRB	SK2758	ME *bunting*, OE *halh*	bunting nook (or pn?)
Bridgemere	CHE	SJ7145	OE *bridd, mere*	bird lake
Buntingford	HRT	TL3629	ME *bunting, ford*	bunting ford
?Buntsgrove (now Birchgrove)	SSX	TQ4029	ME *bunting*, OE *graef*	bunting grove
Caber	CMB	NY5646	ON *ca, berg*	jackdaw hill
Cabourne	LIN	TA1301	OE *ca, burna*	jackdaw stream
Carnforth	LNC	SD4970	OE *cran, ford*	crane ford
Cavill	YOE	SE7730	OE *ca, feld*	jackdaw field
Cawood	LNC	TF2230	OE *ca, wudu*	jackdaw wood
Cawood	YOW	SE5737	OE *ca, wudu*	jackdaw wood
Chickney	ESX	TL5728	OE *cicen, eg*	chicken island
?Chignall	ESX	TL6709	OE *cicen, halh*	chicken nook (or pn?)
?Coggeshall	ESX	TL8522	OE *cocc, halh*	cock, or mound, nook

Place	Co.	NGR	old name	meaning
Cookridge	YOW	SE2540	OE *cucu, ric*	cuckoo strip
Cople	BDF	TL1048	OE *cocc, pol*	cock pool
Cornbrook	LNC	SJ8295	OE *corn, broc*	crane brook
Corney	CMB	SD1191	OE *corn, eg*	crane island
Corney	HRT	TL3530	OE *corn, eg*	crane island
Cornforth	DRH	NZ3034	OE *corn, ford*	crane ford
Cornsay	DUR	NZ1443	OE *corn, hoh*	crane height
Cornwell	OXF	SP2727	OE *corn, well*	crane spring
Cornwood	DEV	SX6059	OE *corn, wudu*	crane wood
Coxwold	YON	SE5377	OE *cucu, wald*	?cuckoo forest
Crakemarsh	STF	SK0936	ON *craka*, OE *mersc*	crow marsh
Crakehall	YON	SE2490	ON *craka*, OE *halh*	crow nook
Crakehill	YON	SE4273	ON *craka*, OE *halh*	crow nook
Cranage	CHE	SJ7568	OE *crawena, laecc*	crow's bog
Cranborne	DOR	SU0513	OE *cran, burna*	crane stream
Cranbourne	HMP	SU9272	OE *cran, burna*	crane stream
Cranbrook	KNT	TQ7735	OE *cran, broc*	crane brook
Cranfield	BDF	SP9542	OE *cran, feld*	crane field
Cranford	NTP	SP9277	OE *cran, ford*	crane ford
Cranford	GTL	TQ1077	OE *cran, ford*	crane ford
Cranoe	LEI	SP7695	OE *crawena, hoh*	crows' heel
Cransford	SFK	TM3164	OE *cran, ford*	crane ford
Cranshaw	LNC	SJ4885	OE *cran, sceaga*	crane wood
Cranmere	SHR	SO7597	OE *cran, mere*	crane lake
Cranmore	SOM	ST6843	OE *cran, mere*	crane lake
Cranwell	LIN	TF0349	OE *cran, well*	crane spring
Cranwich	NFK	TL7795	OE *cran, wisc*	crane marshy meadow
Crawley	BUC	SP7011	OE *crawe, leah*	crow clearing
Crawley	ESX	TL4440	OE *crawe, leah*	crow clearing
Crawley	HMP	SU4234	OE *crawe, leah*	crow clearing
Crawley	OXF	SP3312	OE *crawe, leah*	crow clearing
Crawley	SSX	TQ2636	OE *crawe, leah*	crow clearing
Crawshaw	LNC	SD6951	OE *crawe, sceaga*	crow wood
Creacombe	DEV	SS8119	OE *craw, cumb*	crow valley
Cromer	NFK	TQ2142	OE *crawe, mere*	crow lake
Cronkshaw	LNC	SD8133	OE *cranuc, sceaga*	crane wood
Cronkston	DRB	SK1165	OE *cranuc, dun*	crane hill
Crowell	OXF	SU7499	OE *craw, well*	crow spring
Crowborough	SSX	TQ5130	OE *craw, beorg*	crow hill
Crowcombe	SOM	ST1336	OE *craw, cumb*	crow valley
Crowholt	CHE	SJ9067	OE *crawe, holt*	crow wood
Crowhurst	SSX	TQ7512	OE *crawe, hyrst*	crow wooded hill
Crowhurst	SUR	TQ3947	OE *crawe, hyrst*	crow wooded hill
Crowmarsh	OXF	SU6189	OE *craw, mersc*	crow marsh
Croydon	CAM	TL3149	OE *crawe, denu*	crows valley
Croydon	SOM	SS9740	OE *crawe, dun*	crows hill
Cucket Nook	YON	NZ8413	OE *cucu, wald*	?cuckoo forest
Cuckfield	SSX	TQ3024	OE *cucu, feld*	?cuckoo field
Cuxwold	LIN	TA1701	OE *cucu, wald*	?cuckoo forest
Duffield	DRB	SK3443	OE *dufe, feld*	dove field
Duffield	YOE	SE6733	OE *dufe, feld*	dove field

Table 6.1 *(Continued)*

Place	Co.	NGR	old name	meaning
Dukinfield	CHE	SJ9497	OE *ducena, feld*	ducks' field
Dunkenshaw	LNC	SD5755	OE *dunnoc, sceaga*	dunnock wood
Dunnockshaw	LNC	SD8127	OE *dunnoc, sceaga*	dunnock wood
Earley	BRK	SU7571	OE *earn, leah*	eagle clearing
Earnley	SSX	SZ8096	OE *earn, leah*	eagle clearing
Earnwood	SHR	SO7478	OE *earn, wudu*	eagle wood
Eldmire	YON	SE4274	OE *elfitu, mere*	swan lake
Elveden	SFK	TL8279	OE *elfitu, denu*	swan valley
Enborne	BRK	SU4365	OE *ened, burna*	duck stream
Enford	WLT	SU1351	OE *ened, ford*	duck ford
Enmore	SOM	ST2335	OE *ened, mere*	duck lake
Eridge	SSX	TQ5535	OE *earn, hrycg*	eagle ridge
Exbourne	DEV	SS6002	OE *geac, burna*	cuckoo stream
Finborough	SFK	TM0157	OE *fina, beorg*	spotted woodpecker hill
Finburgh	WAR	SP3372	OE *fina, beorg*	spotted woodpecker hill
Finchfield	STF	SO8897	OE *finc, feld*	finch field
Finchhale	DRH	NZ2947	OE *finc, halh*	finch nook
Finchley	GTL	TQ2890	OE *finc, leah*	finch clearing
Finkley	HMP	SU3848	OE *finc, leah*	finch clearing
Finmere	OXF	SP6333	OE *fina, mere*	spotted woodpecker lake
Foulden	NFK	TL7699	OE *fugol, dun*	fowl hill
Foulness	ESX	TR0494	OE *fugol, naess*	fowl spit
Fowlmere	CAM	TL4245	OE *fugol, mere*	fowl lake
Fulbourn	CAM	TL5256	OE *fugol, burna*	bird stream
Fulmer	BUC	SU9985	OE *fugol, mere*	fowl lake
Gaisgill	YOW	NY6405	ON *gas, gil*	goose ravine
Gazegill	YOW	SD8246	ON *gas, gil*	goose ravine
Gledholt	YOW	SE1416	OE *gleoda, holt*	kite wood
Gledholt	YOW	SE0910	OE *gleoda, holt*	kite wood
Glydwish	SSX	TQ6923	OE *gleoda, wisc*	kite marshy meadow
Goosewell	DEV	SS5547	OE *gos, wella*	goose spring
Goosey	BRK	SU3591	OE *gos, eg*	goose island
Gosfield	ESX	TL7829	OE *gos, feld*	goose field
Gosford	DEV	SY0997	OE *gos, ford*	goose ford
Gosford	OXF	SP4913	OE *gos, ford*	goose ford
Gosford	WAR	SP3478	OE *gos, ford*	goose ford
Gosforth	CMB	NY0603	OE *gos, ford*	goose ford
Gosforth	NTB	NZ2467	OE *gos, ford*	goose ford
Hampole	YOW	SE5010	OE *hana, pol*	cock pool
Handforth	CHE	SJ8883	OE *han, ford*	cock ford
Hanford	STF	SJ8642	OE *han, ford*	cock ford
Hannah	LIN	TF5079	OE *hana, eg*	cock's island
Hanney	BRK	SU4193	OE *hana, eg*	cock's island
Hanwell	GTL	SP4343	OE *hana, wella*	cock's spring
?Hanwood	SHR	SJ4409	OE *hana,* or *han, wudu*	cock's, or stone, wood?
?Hauxwell	YON	SE1595	OE *hafoc, wella*	hawk's well (or pn?)
Hawkhill	NTB	NU2212	OE *hafoc, hyll*	hawk hill
Hawkhurst	KNT	TQ7630	OE *hafoc, hyrst*	hawk wooded hill
Hawkridge	BRK	SU5472	OE *hafoc, hrycg*	hawk ridge
Hawkridge	SOM	SS8630	OE *hafoc, hrycg*	hawk ridge

Place	Co.	NGR	old name	meaning
Hawkeridge	WLT	ST8653	OE *hafoc, hrycg*	hawk ridge
Hawkedon	SFK	TL7952	OE *hafoc, dun*	hawk hill
Hawkwell	NTB	NZ0771	OE *hafoc, wella*	hawk well
Hawkwell	SOM	SS8725	OE *hafoc, wella*	hawk well
Hawridge	BUC	SP9405	OE *hafoc, hrycg*	hawk ridge
Haycrust	SHR		OE *hafoc, hyrst*	hawk wooded hill
Hendred	BRK	SU4688	OE *henn, rith*	wild birds' stream
Henhurst	KNT	TQ6669	OE *henn, hyrst*	bird wooded hill
Henhull	CHE	SJ6453	OE *henn, hyll*	hen hill
Henmarsh	GLO	SP2035	OE *henn, mersc*	wild bird marsh
Hinnegar	GLO	ST8086	OE *henn, hangra*	bird hanger
Howler's Heath	GLO	SO7435	OE *ule, hlid*	owl slope
Iltney	ESX	TL8804	OE *elfitu, eg*	swan island
Kaber	WML	NY7911	ON *ca, berg*	jackdaw hill
Kidbrooke	KNT	TQ4076	OE *cyta, broc*	kite brook
Kidbrooke	SSX	TQ4134	OE *cyta, broc*	kite brook
Kigbeare	DEV	SX5496	OE *ca, bearu*	jackdaw wood
Kitnor (=Culbone)	SOM	SS8348	OE *cyta, ora*	kite bank
Larkbeare	DEV	SX9291	OE *lawerce, bearu*	lark wood
Larkbeare	DEV	SY0697	OE *lawerce, bearu*	lark wood
Ockeridge	WOR	SO7762	OE *hafoc, hrycg*	hawk ridge
Oldberrow	WAR	SP1165	OE *ule, beorg*	owl hill
Ousden	SFK	TL7359	OE *ule, denu*	owl valley
Peamore	DEV	SX9188	OE *pawa, mere*	peacock lake
Pinchbeak	LIN	TF2425	OE *finc, baec*	finch ridge?
Pitshanger	GTL	TQ1687	OE *pyttel, hangra*	kestrel hanger
Poundon	BUC	SP6425	OE *pawan, dun*	peacock's hill
Pudlestone	HFE	SO5659	OE *pyttel, dun*	kestrel hill
Purleigh	ESX	TL8301	OE *pur, leah*	?dunlin clearing
Purley	BRK	SU6676	OE *pur, leah*	?dunlin clearing
Putney	GTL	TQ2274	OE *puttoc, hyth*	kite's landing place
Rainow	CHE	SJ9575	OE *hrafn, hoh*	ravens' height
Raincliff	YOE	TA1475	OE *hrafn, clif*	raven cliff
Raincliffe	YON	TA0182	OE *hrafn, clif*	raven cliff
Ravendale	LIN	TA2300	ON *hrafn, dalr*	raven valley
Ravenfield	YOW	SK4895	OE *hraefn, feld*	raven field
Ravenscliffe	DRB	SK1950	OE *hrafn, clif*	raven cliff
Ravensdale	DRB	SK1773	ON *hrafn, dalr*	raven's valley
Ravensden	BDF	TL0754	OE *hraefn, denu*	raven's valley
Raven's Hall	CAM	TL6554	OE *hraefn, holt*	raven wood
Ravensty	LNC	SD3190	ON *hrafn, stigr*	raven path
?Ravensworth	YON	NZ1407	ON *hrafn, vath*	raven ford, (or pn?)
Rawerholt	HNT	TL2596	OE *hragra, holt*	heron wood
Rawreth	ESX	TQ7793	OE *hragra, rith*	heron stream
Renscombe	DOR	SY9677	OE *hraefn, cumb*	raven valley
Rockbeare	DEV	SY0294	OE *hroca, bearu*	rook wood
Rockbourne	HMP	SU1118	OE *hroc, burna*	rook stream
Rockford	HMP	SU1508	OE *hroc, ford*	rook ford
Rockwell	BUC	SU7988	OE *hroca, holt*	rook wood
Roockabear	DEV	SS5230	OE *hroca, bearu*	rook wood
Roockbear	DEV	SS6041	OE *hroca, bearu*	rook wood

Table 6.1 (Continued)

Place	Co.	NGR	old name	meaning
Rookhope	DRH	NY9342	OE *hroca, hop*	rook valley
Rookwith	YON	SE2086	ON *hrokr, vithr*	rook wood
Roxhill	BDF	SP9743	OE *wrocc, hyll*	buzzard? hill
Roxton	BDF	TL1554	OE *hroca, dun*	rook hill
Roxwell	ESX	TL6408	OE *hroc, well*	rook spring
Ruckholt Farm	ESX	TQ3886	OE *hroca, holt*	rook wood
Ruckler's Green	HRT	TL0604	OE *hroca, holt*	rook wood
Saniger	GLO	SO6701	OE *swan, hangra*	swan hanger
Scargill	YON	SD9771	ON *skraki, gil*	merganser ravine
Snitterfield	WAR	SP2159	OE *snite, feld*	snipe field
Snydale	YOW	SE4020	OE *snite, halh*	snipe nook
Spexhall	SFK	TM3780	OE *speot, halh*	green woodpecker's nook
Stinchcombe	GLO	ST7298	OE *stint, cumb*	sandpiper valley
Stinsford	DOR	SY7191	OE *stint, ford*	sandpiper ford
Sudbroooke	LIN	TF0276	OE *sucga, broc*	sparrow brook
Sugnall	STF	SJ7930	OE *sucga, hyll*	sparrow hill
Sugwas	HRE	SO4541	OE *sucga, waesse*	sparrow wash (wetland)
Swalcliff	OXF	SP3738	OE *swealwe, clif*	swallow cliff
Swalecliffe	KNT	TR1367	OE *swealwe, clif*	swallow cliff
Swallowcliffe	WLT	ST9626	OE *swealwe, clif*	swallow cliff
Swalwell	DRH	NZ2062	OE *swealwe, well*	swallow spring
Swanbourne	BUC	SP8027	OE *swan, burna*	swan stream
Swanmore	HMP	SU5816	OE *swan, mere*	swan lake
Tarnacre	LNC	SD4742	ON *trani, akr*	crane ploughland
Tivetshall	NFK	TM1787	OE *tewhit, halh*	lapwing nook
Tranwell	NTB	NZ1883	ON *trani*, OE *wella*	crane spring
Trenholme	YON	NZ4502	ON *trani, holmr*	crane island
Ulcombe	KNT	TQ8449	OE *ule, cumb*	owl valley
Ullenhall	WAR	SP1267	OE *ule, halh*	owl nook
?Warmfield	YOW	SE3720	OE *wraenna, feld*	wren's, or stallion's, field
Wraxhall	DOR	ST5601	OE *wrocc, halh*	buzzard? nook
Wraxhall	SOM	ST5936	OE *wrocc, halh*	buzzard? nook
Wraxhall	WLT	ST8174	OE *wrocc, halh*	buzzard? nook
Wraxhall	WLT	ST8364	OE *wrocc, halh*	buzzard? nook
Wroxhall	IOW	SZ5579	OE *wrocc, halh*	buzzard? nook
Wroxhall	WAR	SP2271	OE *wrocc, halh*	buzzard? nook
Yagdon	SHR	SJ4619	OE *geac, dun*	cuckoo hill
Yarnscombe	DEV	SS5523	OE *earn, cumb*	eagle valley
Yarner	DEV	SX7778	OE *earn, ofer*	eagle ridge
Yarnfield	STF	SJ8632	OE *earn, feld*	eagle field

Most placenames combine a landscape feature with a qualifier, sometimes an adjective but more often a noun, either genitival or in apposition; Table 6.1 shows a range of these. An obvious example is Cranwell, Lincolnshire, but less obvious is Tranwell, Northumberland, the same name but derived from the Old Norse (ON) *trani*, crane, instead of the Old English (OE) *cran*. Presumably, at some time, Cranes were seen, perhaps regularly but possibly only on one memorable occasion, near the springs in question. Note that names generally make ecological sense – Cranes are combined with springs or marshes, whereas Eagles and Ravens appear with cliffs and dales. A systematic listing of all the bird placenames in Gelling & Cole (2000) produces 201 names that refer to wild birds (Table 6.1), compared with 164 referring to wild mammals. There are a further 33 names that refer to domestic birds, or are ambiguous; these include 16 goose and six duck names, which might be wild or domestic, as well as four 'hen' and six 'cock' names, which probably refer to Domestic Fowl, but could refer to females and males of any bird.

Of the wild birds, Cranes appear most frequently. The usual roots are *cran* or its Scandinavian equivalent *trani*, but are sometimes less obvious: Corney, Cumbria, is OE *corn*, eg crane isle – the Germanic root *cran* getting transposed into *corn*, and Cronkshaw, Lancashire shows a change from something more like the modern German *kranich*. A more extensive listing by Boisseau & Yalden (1999) gives details for 225 crane placenames (Figure 6.1), along with some discussion of their significance. Many placename accounts, including Mills (2003), refer assiduously to 'crane, or heron' places. These imply that our Anglo-Saxon forbears did not know the difference between the two species. But they clearly did, equating their *cran* with the Latin *grus*, and their *hragra* with the Latin *ardea*. In the much later illustrated manuscripts, again, the difference between the two was also very clear. This is particularly evident in the Sherborne Missal (Yapp, 1982b; Backhouse, 2001), probably illustrated in Dorset around 1400, which identifies a *heyrun*, a long-legged grey bird with a whispy crest, and twice illustrates (though does not name) the Crane with its characteristic 'bustle' of secondary feathers and a red nape patch. The confusion arises because of the tradition in recent times of referring to Herons as Cranes in many counties, notably in East Anglia. As Greenoak (1979) points out, John Clare was describing the Heron when he wrote, in *The Shepherd's Calendar*,

> While far above the solitary crane
> Swings lonely to unfrozen dykes again
> Cranking a jarring melancholy cry
> Thro' the wild journey of the cheerless sky

Cranes had by then long been extinct as breeding birds in England, and it seems a frequent occurrence for the name of a lost species to be transferred to another loosely similar one when they are no longer both around to be distinguished. Yalden (1999) pointed to a similar case, of Latin *castor* and *fiber* in a tenth century dictionary being correctly identified with *befer* Beaver, but being equated with Badger and Otter, respectively, by the fifteenth century when the correct owner of the name was no longer a familiar British animal. There seems no reason to doubt that when the Anglo-Saxons named the place Cranfield, they knew that it was Cranes, not Herons, that they were remarking. They did notice Herons occasionally: Rawreth, Essex (OE *hragra, rith*=heron stream) and Rawerholt, Huntingdonshire (OE *hragra, holt*=heron wood) are the two examples in Table 6.1, and the holt of Rawerholt could well have contained the local heronry. However, these were evidently much less striking birds than Cranes. Another case of name transference was discussed by Yapp (1981a). He drew attention to the eighth century glossary rendering of Latin *fasianus* by the Anglo-Saxon *wórhana*, repeated in an eleventh century example, which had been (mis)used as evidence that Pheasants were known in Britain well before the Norman conquest. Presumably the Latin *tetrao* was unknown to the glossator, and he attempted an equivalent. As Yapp points out, *wórhana* is the same bird as the German *Auerhuhn*, that is, Capercaillie; Ekwall (1936) had

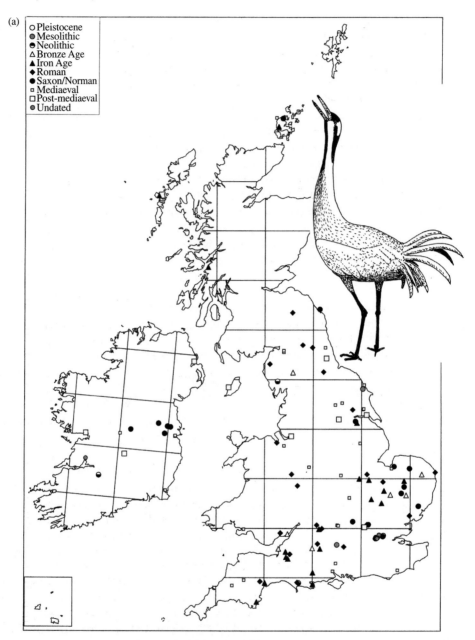

Fig. 6.1 Maps of Crane archaeological sites (a) and place-names (b) (after Boisseau & Yalden 1999).

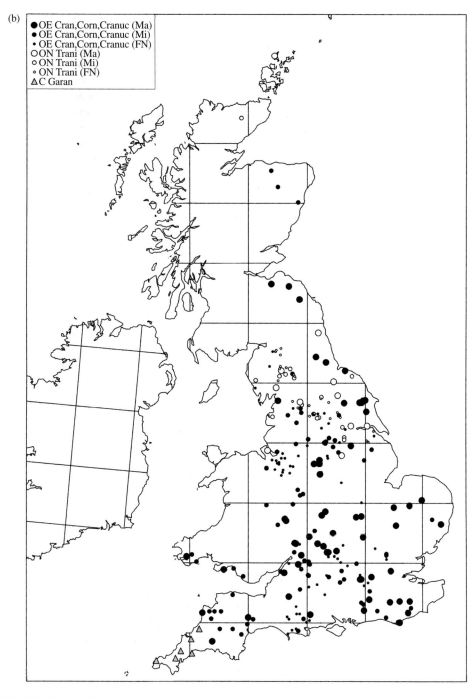

Fig. 6.1 *Continued*

already made essentially the same identification with regard to the placenames Woodspring (*Worsprinc* in 1086) and Worle (= *wor-leah*, capercaillie wood), Somerset. However, as the Capercaillie was already extinct in southern England by the time Pheasants appeared, this became another example of a name transferred to an ecologically equivalent animal.

The Erne and Raven, next most frequent after Crane in Table 6.1, are also striking birds. Erne is now rarely used as a vernacular name, but if at all, it implies White-tailed Eagle *Haliaeetus albicilla*. It is uncertain how rigorously the Anglo-Saxons differentiated White-tailed Eagles from Golden Eagles, and Greenoak (1979) reports the recent use of erne for both species in modern dialect usage. In Anglo-Saxon England, it seems certain that most or all of the places named after earns were in fact referring to White-tailed Eagles, which are far more likely to occur in lowlands and near freshwater, nesting sometimes on cliffs but often in large trees (Yalden, 2007). The Anglo-Saxon Chronicle for AD 937 ends its description of the Battle of Brunnanburgh by relating that the dead of the opposing Welsh/Scots/Irish army were left on the battlefield to be scavenged by Ravens, White-tailed Eagles, and Wolves. While this was probably a conventional end to the retelling of a saga, it certainly implies that White-tailed Eagles – the *'earn aeftan hwit'* eagle white behind, of the saga – were familiar to those who heard the story. The Golden Eagle is everywhere, in North America and in Europe as in Scotland, a species of the uplands, and less devoted to scavenging. Gelling (1987) summarized 33 major placenames derived from OE *earn*, which often became Arn- or Yarn- (Yarner, Devon = OE *earn*, *ofer*, eagle ridge; Yarnscombe, Devon = OE *earn*, *cumb*, eagle valley). A fuller list, including minor places, contains 53 placenames (Figure 6.2). As Gelling (1987) pointed out, many of these are in broad river valleys, appropriate sites for a largely fish-eating bird.

Raven placenames are slightly more problematic than those derived from *cran* or *earn*, for OE Hraefn and ON Hrafn were well-attested personal names. Mills (2003) cites Ravenstone, Leicester and Bedford, as *Hraefn's tun*, farmstead of a man called Hraefn – birds do not generally own farmsteads, villages, or towns (though they might conceivably frequent them), and the earliest, Domesday Book (1086), spellings, respectively *Ravenestun*, *Raveneston*, show these to be settlements (*'tun'* being the precursor of the modern 'town'). Conversely, people are unlikely to have inhabited cliffs, so that names such as Ravencliffe (Derbyshire), Ramscliff (Wiltshire), and Raincliff in Yorkshire surely refer to the birds. There are 16 Raven names in the sample from Gelling & Cole (2000), but a fuller listing by P.G. Moore (2002) includes over 400 names (Figure 6.3). Perhaps more surprising is the fact that Crows figure slightly more frequently than Ravens, and Rooks are as numerous, in Table 6.1, as these are much less striking birds than Ravens. It seems possible that large black birds, in general, had a symbolic importance, as birds associated with the gods, with the afterlife, or as omens. Some of the Crow placenames look as though they should be referring to Cranes (Cranage, Cheshire = OE *crawena*, *laec*, crows' bog; Cranoe, Leicestershire = OE *crawena*, *hoh*, crows' heel, in the sense of a heel-shaped hill), emphasizing the need to examine the original spellings to get at the basic meaning of placenames and not rush to a hasty interpretation.

As a group, birds of prey also figure largely in these placenames, though their ornithological interpretation is sometimes difficult. Hawkridge seems easy enough (OE *hafoc*, *hrycg*, hawk ridge), as hawks frequently soar over ridges, but are these specifically Goshawks or Sparrowhawks, or is this a more general use of 'hawk' for any large bird of prey? The

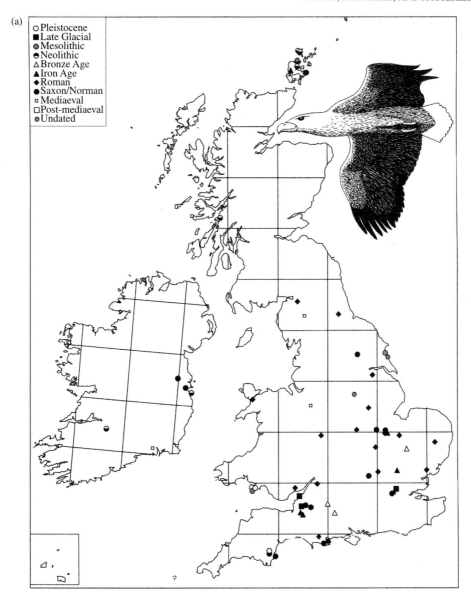

Fig. 6.2 Maps of White-tailed Eagle archaeological sites (a) and Eagle place-names (b) (cf. Yalden 2007).

OE *cyta* gives us the modern Kite, but the Anglo-Saxon glossaries give OE *cyta*=Latin *buteo*, buzzard; Kitnor, the old name for Cudmore, Somerset (*cyta ora*=kite bank) might have been a buzzard bank. Conversely, the two places called Gledholt in West Yorkshire and Gleadless, Sheffield, which include the root of the modern 'glider' and the dialect name glead, used for kites and harriers in different counties (Greenoak, 1979), seem most properly

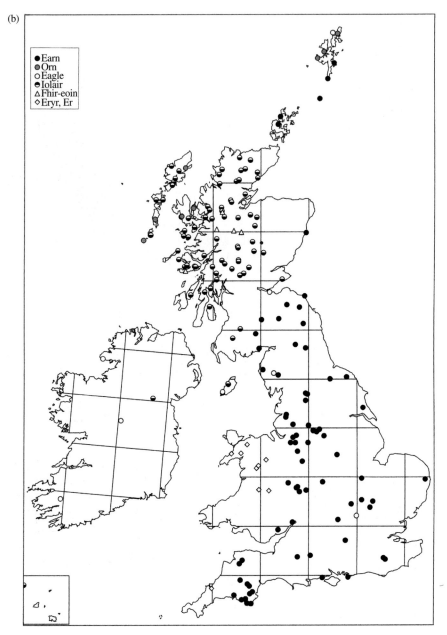

Fig. 6.2 *Continued*

interpreted as kite placenames, as harriers are not woodland birds (*gleoda*, *holt*=kite wood, *gleoda*, *leah*=kite clearing). Whether Glead Hill along the Pennine Way in Derbyshire referred originally to kites or harriers is a moot point, but the Anglo-Saxon glossaries do give OE *gleada*=Latin *milvus*. The use of *gleoda* seems more frequent in northern England, and *cyta* in southern England, so some dialectic difference seems to have been established

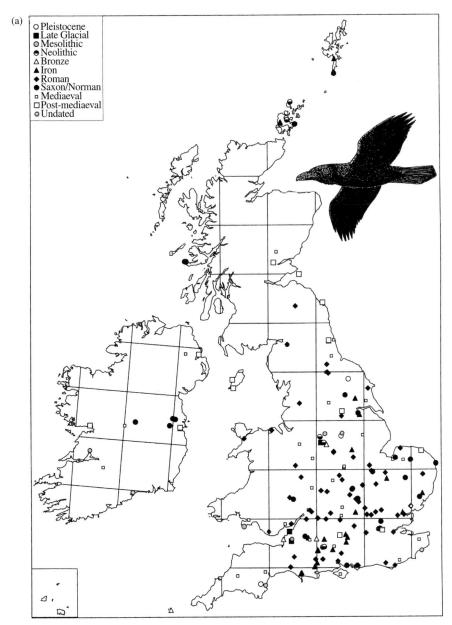

Fig. 6.3 Maps of Raven archaeological sites (a) and place-names (b) (based in part on Boisseau 1995, Moore 2002).

early. The noun *puttoc* also appears as a modern dialect name puttock, used variously for Kite and Buzzard; in placenames, it appears in Putney, London (*puttoc, hyth*=hawk's/ kite's landing-place) – did kites scavenge regularly here? Two Anglo-Saxon raptor names have caused some discussion and uncertainty, *pyttel* and *wrocc*. In Anglo-Saxon glossaries, *bleripyttel* is equated with Latin *soricarius*, and *mushafoc* is also given as *siricarius*,

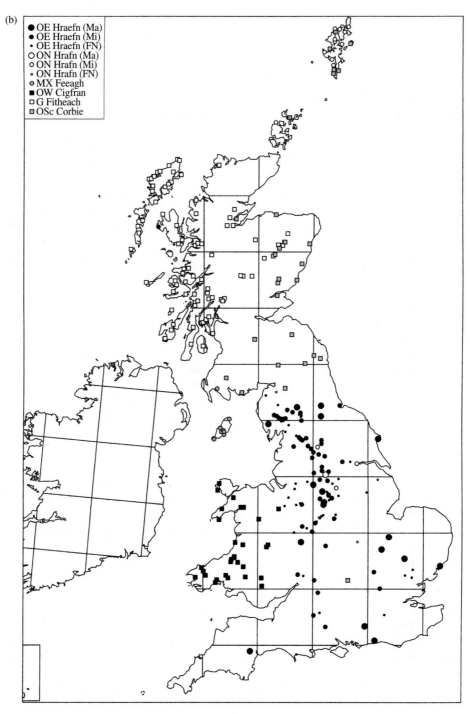

Fig. 6.3 *Continued*

implying shrew-hawk in both cases. The dialect Mousehawk is used in some northern counties for Kestrel (Greenoak, 1979) and, although other raptors, notably Buzzards, could be described as 'mouse-hawks', it seems likely that *pyttel* and *mushafoc* were terms for Kestrel. Kitson (1998) has another, perhaps better, suggestion for *bleripyttel*, remarking that *bleri* indicates a white blaze, and pointing out that a white rump patch characterizes Hen Harriers, being especially conspicuous on the brown females and immatures. The term *wrocc* does not appear in the glossaries, but has been invoked to explain a series of placenames (Roxhill, Bedfordshire, four cases of Wraxhall and two of Wroxhall) that seemingly referred to remote places, not human settlements (OE *halh*=nook, hollow). Ekwall (1936) surmised that it might refer to a bird of prey, perhaps a Buzzard, and his opinion has been widely followed, though without any additional confirmation. A Norwegian zoologist friend tells me that in parts of western Norway, the dialect term *våk*, sometimes spelt *vråk*, is used for Buzzard, and *vråk* is the modern Swedish term for Buzzard (Per Terje Smiseth, pers. comm., Cramp *et al.*, 1977–94). Though use of an Old Norse (ON) name on the Isle of Wight (where one Wroxhall, *Warochesselle* in 1086, is located) seems less likely than in northern England, some ancient common root seems to be indicated. Although Kitson (1998) argues for it meaning Marsh Harrier, because *halh* often implies a nook between two rivers or a projection of slightly higher land into a marsh, the places in question do not really seem to fit that notion; that on the Isle of Wight, for example, is indeed between two streams, but is high on the Downs above Ventnor, an unlikely place for marshes or Marsh Harriers. Perhaps, as the name is missing from the glossaries, it represented a rarer, shyer species that was unfamiliar to most people. Honey Buzzard (Swedish *bivråk*) comes to mind, and the sites all seem to be southern, but that is no more than a guess on our part.

This *wrocc* is not the only old element that has not been satisfactorily identified. The element *pur* in Purley, Berkshire and Purleigh, Essex is believed to be a water-bird of some sort, but Bittern, Pelican, Tern, and Black-headed Gull have all been suggested. Kitson (1998) plumps for Dunlin as its original meaning, suggesting that purring might be onomatopoeic for a wintering flock on the marshes, or for the male's song. The Essex marshes might be a suitable location for a Dunlin meadow, but the Thames bank in Berkshire fits less well. The *skraki* in Scargill, Yorkshire, is another name not explained by Gelling & Cole (2000); the Norwegian *skrike*, Swedish *skrika* might suggest Jay as an identity (Per Terje Smiseth pers. comm., HBWP), but the interpretation by Mills (2003) of Merganser (cf. Swedish *skrake*) is a better match. Whether a narrow ravine on the River Warfe would be better habitat for Jays or Mergansers is a moot point.

The appearance of the OE *amer*, bunting, in some placenames is an interesting reminder of the root of Yellowhammer; Parsons *et al.* (1996) argue that many more *amer* names than currently recognized might be present. Until recently, names such as Amberley (Sussex and Gloucestershire), have been postulated to include an otherwise unknown person, *Ambre*, *Ambra*, or *Amber*, but they argue that inserting a *b* into *amer* was a likely development in the English language. The OE *stint* probably meant any small wader, but given the inland locations of Stinchcombe and Stinsford, it seems very likely that they were in fact referring to Common Sandpipers. The distinction between two names for woodpeckers, and two names for swan, is intriguing. It is considered likely that the difference between Green (OE *speot*) and Spotted (OE *fina*) Woodpeckers was being noticed. Was this also true for 'wild' swans and Mute Swans? Archbishop Aelfric's vocabulary (Wright, 1884) equates the OE *swan* with

Latin *olor*, implying Mute Swan, and the OE *ylfete* with Latin *cignus*. The *swan* places listed in Table 6.1 are in southern England, more likely (at least nowadays) Mute Swan habitat, while the places derived from *elfete* do include the more northerly places where Whooper Swans might winter, and Kitson (1997) makes the same deduction.

The small body of placenames founded on hen, cock (*hana*), and goose (*gos*), and a couple of peacocks, in Table 6.1 suggests either that domestic birds played a small part in Anglo-Saxon economy, or that they were not particularly notable elements of contemporary farming – there seem to be far more names founded on domestic mammals. However, a much fuller list compiled by Iain Pickles (2002) contains 1,588 placenames that refer to birds in general, or domestic birds in particular. These include 576 that refer to goose and 685 that imply domestic fowl (417 to cock, 239 to hen and 29 more generally to poultry), but only 179 that refer to ducks. Only 24 of these are major placenames (hundreds, parishes, towns), 390 are minor places (farms, landscape features) but as many as 1,174 are field names. This suggests an association with the local farming economy, matching the expectation that domestic birds might be less noticeable, but the sheer number of goose names is an intriguing hint that they might have been at least as important as chickens in the contemporary economy. Does the archaeological record do anything to support this?

Archaeological Saxon birds

Among the earliest places to be settled by the Anglo-Saxons were small East Anglian settlements like West Stow and Mucking. Dating back to the fifth century, they suggest that small farms, rather than towns, were their earliest settlements. At West Stow, Crabtree (1985, 1989a) documented 431 fragments of Domestic Fowl and Domestic Goose bones, a very small fraction (0.95%) of the number of domestic mammal bones (15,988). However, they were numerous at all levels, and in the earliest, dating perhaps back to the fifth century, goose bones were almost as abundant as fowl. Fowl bones were very variable in size, ranging from those of a modern bantam and game fowl up to a few bones comparable in length, though not in thickness, to large modern breeds. Clearly selective breeding was undertaken, and some cocks were capons, surgically sterilized to gain greater size. The geese were mostly of the size of Domestic or Greylag Geese, though a few bones of smaller wild geese, probably White-fronted, were also present. At least a couple of sterna had deeper keels than wild geese, suggesting breeding for meat, but their size overlap with both Greylag and Domestic Goose is complete. While there is not enough information to claim that all are in fact domestic, the numbers involved strongly suggest that. There are a few bones of various wild species, including Swan (species?), Teal and another wild duck (perhaps Wigeon, on size), Crane, Moorhen, Heron, Grey Plover, Lapwing, Woodcock, Snipe, Herring/Lesser Black-backed Gull, Common Gull, thrushes, including Song Thrush, and Starling, most of which could have been eaten. Most are represented by only one or two bones, but there are 30 of Crane, some showing butchery marks. There is also a Buzzard, probably a scavenger, and Goshawk, which might have been used to catch some of the wild birds.

The mid-Saxon layers at North Elmham, Norfolk, also have a bird fauna dominated by Domestic Fowl and Domestic Goose. Bramwell (1980a) identified remains of 37 and 18 individuals respectively, along with five each of wild Mallard and Domesticated Duck, two

Cranes, and single individuals of White-fronted, Pink-footed, Barnacle and Brent Goose, Shelduck, Teal, Wigeon, Red Kite, Sparrowhawk, Buzzard, Golden Plover, Curlew, and Wood Pigeon. He remarks on the contrast with Roman sites, where goose remains are sparse, and argues that the Saxons were probably responsible for increasing the level of goose farming in England. Ducks by contrast are never as numerous as geese in Saxon sites (Albarella, 2005). By Mid-Saxon times, towns had redeveloped, and as at North Elmham, Domestic Fowl increasingly outnumbered Domestic Goose. At Thetford, 86 Fowl and 14 Goose accompanied three Mallard, three Wigeon, two Peacocks, a Greylag Goose, Shoveller, Pochard, Crane, Oystercatcher, and Raven (Jones, 1984). Possibly different breeds of Fowl had been developed by this time, because two size classes can be detected at Flaxengate (O'Connor, 1982), though this might reflect caponization, which produces longer legs. The largest bone collection from an urban Saxon site comes from Hamwic, better known now as Southampton (Bourdillon & Coy, 1980). Some 46,904 bones were divided between 45,704 mammal bones and 1,200 bird bones. An alternative division was into 46,823 domestic and only 81 from wild birds and mammals. The birds were resolved into an estimated 101 individuals, including 63 Domestic Fowl and 16 Domestic Geese. Mallard (three, wild or domestic?), Teal, ?Wigeon, Woodcock, two Starlings, ?Redwing, and Song Thrush, likely food species, along with a Great Black-backed and two Herring (or Lesser Black-backed) Gulls, Carrion/Hooded Crow, Jackdaw and Buzzard, likely scavengers or predators of the domestic birds, contributed the rest of the bird fauna, along with one more surprising species, a Great Northern Diver represented by an unmistakable humerus. Even more surprising than its presence were the cut-marks, implying that it had been butchered, though whether eaten or for its plumage can only be conjectured. As Coy (1997) remarks, discussing Saxon bird faunas in general, most meat was provided by domestic mammals, but wild birds as well as domestic ones added variety to the diet. Different Fowl/Goose ratios, ranging from 1.6:1 (at West Stow) to an extreme 19.8:1 (in the western suburbs of Winchester) may reflect differences between rural and urban sites, between earlier and later Saxon sites, or between sites where bones were hand-picked by the archaeologists and those where sieving was routinely undertaken; three different Saxon sites in Southampton cluster at 2.7, 3.3, and 3.6:1, a greater degree of consistency. Perhaps the most important conclusion is that the fauna looks not unlike what a modern bird-watcher to the Southampton area might expect to see; even Great Northern Divers are regularly recorded, in small numbers, off the modern Hampshire coast in winter.

Portchester, only 23 km to the south-east, was a 'high status' site, a Saxon hall or palace built on the site of the former Roman castle. An early–mid Saxon layer contains few birds, mostly Domestic Fowl and probable Domestic Geese – the size of Greylag or Bean, but rather variable, suggesting selective breeding (Eastham, 1976). Some nine to 11 Fowl and eight to nine Geese are implied by the bone collections, along with single individuals of Bittern, Mallard, Curlew, Partridge, and Wood Pigeon. The Late Saxon layers provide a much richer fauna. Domestic Fowl, about 84–87 individuals, and Geese, 39 individuals, are again dominant, but a wider range of wild species is present. There are some waterbirds, including Great Northern Diver again, with Shelduck, Mallard, Teal, Wigeon, and Pintail, but mostly only one or two of each. More striking is the variety and numbers of waders, including Golden Plover, Dunlin, Redshank, (Black-tailed?) Godwit, Curlew, Whimbrel and Woodcock. Indications of at least 13 Curlew and two Whimbrel suggest that these were taken in autumn, when the Whimbrel would have passed through on migration, and perhaps on into winter, when the Woodcock would arrive; none of these waders is likely to have bred locally, except the Redshank and possibly the Godwit. Other food species included Wood Pigeon and Rock/Domestic Dove, while the remains of six Common Terns are harder to explain; terns are not often eaten. A few scavengers are suggested by a Red Kite, two Herring Gulls, two Carrion Crows, and a Jackdaw. The abundance of Domestic Geese confirms the Saxon penchant for this species, while the preference here for waders seems to be a mark of the high status of this site (Eastham, 1976).

At Flixborough, Dobney *et al.* (1994) estimated some 5,000 goose bones, 4,700 fowl, and 350 crane, implying even greater emphasis on Domestic Geese at that eighth to ninth century site. A fuller analysis of this fauna has now been published (Dobson *et al.*, 2007). They confirm some 5,700 Domestic Fowl bones, 3,698 Domestic Goose, and 846 probable Barnacle Goose bones, as well as 228 Crane bones. The rich avifauna includes other wetland species such as Marsh Harrier, Grey Heron, ?Lapwing, Curlew, Brent and Pink-footed Goose, Mallard and Teal. ?Woodcock, Black Grouse, pigeon, and wader spp also contributed to the diet, while other predators included ?Red Kite, ?Common Buzzard, and Tawny and Barn Owl. The geese are particularly interesting in being the first case where archaeological birds have been identified by analysis of their DNA: this has confirmed the identity of Barnacle and Pink-footed Geese, as well as the presence of three different genotypes of Domestic Geese; interestingly, wild Greylags do not seem to have been present. (The Brent Goose was identified more simply, from its small size and morphology.) Domestic Geese and Fowl were a major item of diet at this site, numerically outnumbering even cattle and pigs in some layers (even if not so important in supplying the weight of food).

Among the rarer species in the Anglo-Saxon archaeological record, Quail at sixth to seventh century Viroconium (Hammon, 2005) and Pheasant at Flaxengate (O'Connor, 1982) are worth noting. Corvid bones in general are rarer than in the Roman period, perhaps because the Raven lost its symbolic importance. However, a Raven bone from late Saxon Chalk Lane has cut marks (on the humerus, suggesting feather removal) (Coy, 1981b). Butchery marks are rarely reported for Saxon sites, but are also present on a Grey Partridge bone from Viroconium (Hammon, 2005). Being located on the wing bones, they may, as for the Raven, reflect feather removal, but consumption of this species is very likely, even in the absence of direct evidence.

Rather different contemporary faunas have been recorded from more distant parts of the British Isles, well beyond the Saxon kingdoms. In Ireland, Stelfox (1938) reported a remarkably rich avifauna from the crannog of Lagore, in County Meath. This is also believed to have been a royal site, occupied from the eighth to tenth centuries. Over 1,000 bird bones were available to Stelfox, along with an extensive reference collection to assist in their identification in the National Collection, Dublin. Surprisingly, Domestic Fowl, with 176 bones, was not quite the largest contributor. Goose bones were more plentiful, but most belonged to White-fronted (124) and Barnacle Geese (202). A few seemed referable to Brent (seven bones), possibly Bean (three bones) and some 56 to Greylag Goose. Stelfox found no evidence for Domestic Goose, presumed they were probably absent this early in Ireland, and considered the largest goose bones to be wild Greylag. Among the rich diversity of wetland birds were Red- (or Black-?) throated Diver, Great Crested Grebe, Cormorant, Crane, Heron, Coot, Moorhen, Corncrake, Bewick's and Whooper Swan, Mallard, Pintail, Garganey, Teal, Wigeon, Scaup, Tufted Duck, Goldeneye and Red-breasted Merganser, most of which were probably eaten. Probable scavengers include White-tailed Eagle, Buzzard, Hooded/Carrion Crow, Raven and perhaps a large gull, probably Herring Gull. Other species present in the area included Barn Owl, Rook, and Chough. The absence of waders is surprising, and it is not clear whether they were absent, too small to be noticed, or too difficult to identify to species with the available reference collection; one would expect plovers, Woodcock and Curlew to figure in this list. Lack of small passerines surely reflects lack of sieving. Even so, the fauna is a rich one, and draws attention to the loss of Crane, White-tailed Eagle and (from most of Ireland) Corncrake and Buzzard since that time. A contemporary site at Raystown, County Meath, not yet published, has Corncrake more numerous than the domestic poultry, and ducks more common than geese or chickens. Other species reported there include Quail, Crane, Goshawk, Woodcock, Snipe, Nightjar, and Raven (Murray & Hamilton-Dyer, 2007). In Dublin, at Woods Quay, a slightly later (ten to eleventh century) Irish site, the avifauna has not been properly documented, but D'Arcy (1999) reports that 58 Domestic Fowl and 16 individual geese (wild and domestic) are represented, along with 21 Ravens and 10 Buzzards. Predators are well represented, by seven White-tailed Eagles, four Kites, two each of Peregrines, Ospreys, Hen Harriers and Marsh Harriers, and a Sparrowhawk. Other species represented by single individuals include Red-throated Diver, Crane, Gannet, Shag, Cormorant, Curlew, Bar-tailed Godwit, Guillemot, Great Black-backed Gull, Kittiwake, Crow, Rook, and Jackdaw, along with two Swans (species?), three Ducks (species?), and a number of unidentified wader and other bones.

A different fauna comes yet again from coastal sites. The monastery at Illaunloughan, County Kerry (seven to ninth centuries) has Domestic Fowl and Goose (species?) as well as Cormorant, Shag, Gannet, Godwit, Snipe, Kittiwake, Guillemot, Puffin, Wood Pigeon, Crow, but 70% of the bird bones are Manx Shearwater (Murray *et al.*, 2004). In the north, Scottish sites similarly yield bird faunas dominated by marine species. At Brough of Birsay, Orkney, a Viking-age site, Allison (1989, also Allison & Rackham, 1996) points out that 60% of the bird remains were of seabirds, and that Domestic Fowl were scarce (only four bones, in 305); most of the goose were probably Domestic Goose, but could have been or included wild Greylag. Manx Shearwater, Shag, Cormorant, Gannet, Little and Great Auk, Razorbill, Black Guillemot, Guillemot, Razorbill, and Puffin were the main food species, along with ?Crane, Oystercatcher, Curlew, Great and Lesser Black-backed Gull, and Starling. Bramwell (1977b) made very similar observations on the Norse (and Pictish) levels of Buckquoy: Domestic Fowl and Goose were scarce, and seabirds provided much of the meat to this community. As at Brough of Birsay, all the auks were taken, as were Fulmar, Manx Shearwater, Gannet, Cormorant, and Shag, but the species list was much longer. Great Northern Diver, Eider, and Common Scoter were among the other marine species, as well as Herring/Lesser Black-backed, Great Black-backed, and Black-headed Gull. Both Mute and Whooper Swan, along with Shelduck, Mallard, Teal, Wigeon, Shoveler, and Goldeneye contributed to the waterfowl, with Crane, Water Rail, and Corncrake also coming from freshwater habitats. Most surprising was the occurrence of both Red and Black Grouse; while the former still breed in Orkney, the latter is not known to have done so in recent times. Bramwell lists a number of waders – Oystercatcher, Golden Plover, Dunlin, Knot, Greenshank, Curlew, Whimbrel, Jack Snipe, and a phalarope, perhaps Grey – along with a very few terrestrial birds – Rock Dove, Ring Ouzel/Blackbird, Song Thrush/Redwing, Crow/Rook, and Raven. Merlins and Kestrels competed with Humans for the smaller prey. At Skaill, Deerness, Allison (1997b) found most of the auk species in the Viking levels, including Little and Great Auks (though not Black Guillemot or Puffin), along with other seabirds such as Manx Shearwater (but not Fulmar), Shag, Cormorant, and Gannet. Large gulls included Glaucous/Great Black-backed and Herring/Lesser Black-backed, and both Great Northern and Red-throated Divers were reported. Common Buzzard, White-tailed Eagle, Raven, and Short-eared Owl were among the predators and scavengers recorded. Modest numbers of Red Grouse, Domestic Fowl, and Domestic Goose bones were found, but seabirds were more numerous.

If the Fulmar seems to be much more numerous in these coastal faunas than modern knowledge would lead us to expect (cf. Chapter 4), even more surprising is the presence of a small gadfly petrel *Pterodroma* (Serjeantson, 2005). Its remains have been discovered at three widely scattered sites, The Udal on North Uist in the Outer Hebrides, Kilellan Farm, Islay in the Inner Hebrides, and Bretaness on Rousay, Orkney. All are thought to date from the first millenium AD, in what would elsewhere be termed Early Christian (in Ireland) or Anglo-Saxon (in England). Among the 11 bones represented are parts of both upper and lower bill, two coracoids, a pair of tibiotarsi, an ulna, radius, and two broken femora; collectively, at least six birds are represented by these bones. Comparing them carefully with those of all the North Atlantic petrels, Dale Serjeantson suggests that they are those of the Madeira Petrel *Pt. feae*, which breeds only on Bugio, Madeira, and in the Cape Verde islands; the bones of *Pt. madeira*, which breeds in small numbers at high altitude on Madeira itself is a smaller bird, while the Cahow *Pt. cahow* of Bermuda is much the same size, but has a shorter coracoid. Of course all these are rare species, and only a limited range of comparative specimens is available. Given the sad record of these species throughout the world, it is also possible that it is an entirely extinct species, but the most plausible identification is *Pt. feae*. Apparently the same species has also been reported from two archaeological sites in Sweden (Ericson & Tyrberg, 2004). Intriguingly, a single bone, a furcula, of another *Pterodroma* species, smaller even than *Pt. madeira*, was also found at The Udal. No identification has yet been suggested for this, and more material is needed.

Even further north, Platt (1956) produced a succinct account for Jarlshof on Shetland, lacking details of numbers, but describing a ninth century Viking-age fauna like those from Orkney. Among the more numerous birds were Great Black-backed and Herring Gull, Shag, Gannet, Cormorant, and Eider. Less numerous were Guillemot, Hooded Crow, Heron, Shelduck and Black-headed Gull, while Red-throated Diver, Shoveler, Velvet Scoter, Whooper Swan, Bittern, Oystercatcher, Kittiwake, Curlew, Leach's Petrel, Black Grouse, Peregrine, Magpie, and Raven were also present. Presumed Domestic Ducks, Geese, and Fowl were evidently present, but not numerous. A more recent excavation on Shetland, at the Broch of Scalloway, produced a similar, though smaller, fauna of late Iron Age and Viking dates (AD 500–1000): Gannet and Puffin were most numerous, along with Domestic Fowl. Red-throated Diver, Grey Heron, Cormorant, Shag, Mute Swan, Greylag Goose, Teal, Mallard, Curlew, Snipe, Bar-tailed Godwit, Guillemot, Kittiwake, Herring Gull, Raven, and Hooded Crow were also recorded (O'Sullivan, 1998).

Norman birds – castles, feasts, and falconry

The conquest of England by William of Normandy resulted in a series of linked changes. The towns, boroughs, and counties of the English kingdom were retained and exploited, but a new language was imported, giving some changes of names – ernes became eagles under the influence of French *aigle*, and the Anglo-Saxon *hragra* was replaced by Heron (cf. French *héron*). Castles, the epitome of conquest and the ruling hierarchy, were also sites that have endured, to allow rich archaeological investigations. They provided sites for, and provide evidence of, banquets in which wild birds figured not only as items of diet but as status indicators. Falconry, one means by which the wild birds were obtained, joined hunting, forests and parks with the notion that the right to hunt was the prerogative of the King and his nobles, and another symbol of rule (Rackham, 1986, Yalden, 1999). A written record of the state, of the organization of hunts and feasts, and thus indirectly of the wild birds and mammals that were hunted, also begins with Norman conquest, and with that remarkable record, the Domesday Book of 1086.

Domesday Book was essentially a tax return, an attempt by the Conqueror to ascertain what he and his barons owned, and what he could expect to get in taxation. As such, it is mostly a list of manors and ownership, with a concentration on the agricultural wealth – areas of ploughland, numbers of ploughs, and villagers to operate them. However, woodland was an important commodity, providing firewood, timber for construction and pannage for pigs, and was also well documented. Rackham (1986) estimates that England was only about 15% wooded and moreover points out that the woodland was irregularly distributed – as now, the Weald was well wooded, but large areas of the Midlands and Fenland lacked any woods. For present purposes, it is the woodland entries for some of the counties, mostly in the western Midlands, that are of interest – they contain the first serious written ornithological records for England. Hawk's nests, usually in the phrase *airae accipitru*, hawk's eyries, immediately follow the entries for woodland in some entries for Buckinghamshire (one), Cheshire (24), Gloucestershire (two), Herefordshire (one), south Lancashire (but not enumerated), Shropshire (three), Surrey (one), Worcestershire (two), and also north Wales (four). For Limpsfield, Surrey, the phrase *nidi accipitris*, hawk's nest, is used. The use of *accipitris* implies hawks, rather than falcons, as does the association with woodland. (Surprisingly, no *airae falcones* are listed in Domesday, though Yapp (1982a) points out that the early Medieval accounts differentiate hawks from falcons.) Moreover, these were surely Goshawks, not Sparrowhawks, for they were too sparsely distributed to have been the much more numerous Sparrowhawk. Only Cheshire seems to have had a fairly complete listing, and the eyries were confined to the better wooded centre and east of the county. The 24 pairs would have had about 50 km² each, about the size of modern Goshawk territories, whereas the same area of woodland should have supported some 1,300 pairs of Sparrowhawks, too numerous to have been noted. Their value, too, indicates that these Domesday *accipitres* were Goshawks – £10, at a time when the whole of Macclesfield was only worth £1 (Yalden, 1987). Just to clinch the matter, one Cheshire manor, Hampton, paid a rent of 2 shillings and one Sparrowhawk – *spreuariu*. This was clearly different from, and less valuable than, the *accipitres* whose eyries were recorded elsewhere in Cheshire.

Falconry in archaeology

The Romans apparently did not indulge in falconry, and Toynbee (1973) does not give any indication that they kept, or illustrated, tamed hawks. They might have used them, like owls, as decoys, as one hunting scene from a fourth century mosaic in Sicily shows a man with a falcon (Wilson, 1983). By Late Saxon times, falconry was certainly practised in England, and in Archbishop Aelfric's eleventh century Colloquies (Garmonsway, 1947; Swanton, 1975), the fowler, asked how he caught birds, replied that he used nets, snares, lime, whistling, hawks (*hafoce* in the OE text, *accipitre* in Latin), and traps. Asked how he fed the hawks, he replied that in winter they fed themselves, as well as himself, but in spring he let them go, and in autumn he took young birds and tamed them. However, it was the Norman and later kings who indulged in falconry to the full. Yapp (1982a) discusses the disinformation perpetrated by the *Boke of St. Albans*, in which the appropriate raptors for people of different ranks are supposedly allocated (from 'an eagle for an emperor' to 'a muskett – male sparrowhawk – for a holywater clerk'), pointing out both the internal inconsistencies and the ornithological nonsense implied by the list (Table 6.2). For instance, a yeoman would hardly have afforded a Goshawk, though the Emperor Frederick II certainly used them. While vultures and melawnes (?kites), were assigned, with eagles, to an emperor, they have never been useful birds for hawking. Abbots and bishops are missing, Peregrine Falcons appear at least three times, and the notion that fifteenth century hawkers could distinguish Lanners, Sakers, and Peregrines, is to project back nineteenth century knowledge well beyond any reality; indeed, from the evidence of Frederick II's account of falconry, sacer and lanner were then simply other names for Peregrines (Yapp, 1983). Turning to the more substantial evidence, of birds in illustrated manuscripts, which often depict hawking scenes, Yapp (1981b) points out that even distinguishing between hawks and falcons is difficult in poor drawings. Of the better illustrations, only one seems to be a Peregrine, while five good Goshawks are shown. Yapp also points out the level of confusion in the use of names in Medieval glossaries – while hawks (short-winged) and falcons (long-winged) are usually distinguished (as *accipiter=hafoc* and *herodius* or *falco=wealhafoc*, Welsh hawk), species are not. The Latin *peregrinus* is used for Hobby as well as for Peregrine (= *Faucon Pelryn*), and even Goshawk and Sparrowhawk are sometimes confused. As he remarks, Peregrines

Table 6.2 Supposed allocation of raptors for hawking, according to the "Boke of St Albans", reputedly dating from 1486, but probably a much later fabrication (after Yapp 1982a, Table 1). Attributions in brackets appear in different versions of this list.

Emperor	eagle, bawtere (vulture), melawne (kite?)
King	gyrfalcon and its tiercel (i.e.male)
Prince	falcon gentle and its tiercel
Duke	falcon of the rock
Earl	falcon peregrine
Baron (Lord)	bastard
Knight	sacer, sacret (i.e. male)
Squire	lanner, lanret
Lady	merlin
Young man (young squire, squire of the first head, infant)	hobby
Yeoman (gentleman, poor gentleman, poor man)	goshawk
Poor man (gentleman, yeoman)	tiercel (of goshawk, implied)
Priest	sparrowhawk
Holywater clerk	muskett (i.e.male sparrowhawk)
(Knave)	kestrel

are rare in southern England, and, in a more wooded countryside, Goshawks would have been more available. The attention given to them in Domesday clearly points to their importance. Both Frederick II and more recent accounts of hawking stress the value of Peregrines and Goshawks as the prime hunters. How does the archaeological record illuminate this interpretation?

The archaeological records of major birds of prey are summarized in Table 6.3. If falconry was, as surmised above, a more frequent sport in Medieval times, Goshawk remains should be more numerous than those of Peregrines, and both should show a noticeable increase from Norman times onwards. It is evident that Peregrines have indeed been less numerous than Goshawks throughout. It is also evident that in Iron Age and Roman times, when towns first provided scavenging habitats, scavengers were the common ecological group, and 'others' were as numerous as the falconry species, but this is no longer true from Anglo-Saxon times onwards. The sharp increase in numbers of Sparrowhawks in Medieval times is particularly notable, but the increase in numbers of Goshawks and Peregrines is also marked. The numbers of scavengers, particularly Red Kites and Common Buzzards, also seems high in Medieval times, and this might be a comment on the lack of hygiene. However, there is also documentary evidence that Gyr Falcons were sometimes flown against Red Kites, so their presence might alternatively be another indication of the importance of falconry (Dobney & Jaques, 2002). One technique was to release an Eagle Owl with the tail of a Fox attached to its

Table 6.3 Raptors in the archaeological record.
The number of sites recording each species is listed. Note that the raw numbers are biassed because some periods have produced far more records than others. To consider the effects of falconry on this accumulation of records, the probable falconers' birds (Goshawk, Sparrowhawk, Peregrine, Merlin, Hobby) are contrasted with the likely scavengers (Red Kite, Common Buzzard, White-tailed and Golden Eagle) and the others (Osprey, Rough-legged Buzzard, harriers, Kestrel) in the summary rows at the bottom of the table.

Species	Pleist	LatePleist	Mes	Neo	B.A.	Iron	Rom	A-S	Nor	Med	PostMed
Red Kite	1		1	1		4	14	11	2	28	9
Osprey		2				1	1	2			1
Com. Buzzard	2		6	5	2	12	19	19	5	33	7
R-l. Buzzard	1	1		1		1					
Goshawk	1	3	3	3		5	3	6	6	11	2
Sparrowhawk			4				3	9	3	17	7
W-tailed Eagle	3	5	5	7	3	7	18	6		4	
Hen Harrier						1	1	4			1
Mont's Harrier						1					
Marsh Harrier			1			3		6	1	3	1
Golden Eagle	1	4		1		2	3	1		3	
Peregrine	2	1	2			4	1	5	2	6	2
Kestrel	5	12	1	3	2	3	6	2		5	4
Hobby	1	1								1	
Merlin		4	1			2		1		2	
"Scavenger" Total	7	9	12	14	5	25	54	37	7	68	16
"Falconry" Total	4	9	10	3		11	7	21	11	37	11
"Others" Total	6	15	2	4	2	10	8	14	1	8	7

leg in view of the Kite, which would swoop down to ambush the Eagle Owl and deprive it of its 'prey', at which point the Falcon would be released to attack the Kite (Salvin & Brodrick, 1855). Eagle Owls are not represented in Medieval British archaeology (a strong argument that they had become extinct well before then) but Tawny Owls are surprisingly frequent. Perhaps they were the local substitute, though they were themselves surely also used to decoy smaller birds nearer to nets or bird-lime. It is also possible that Red Kites and Buzzards were themselves used in falconry; Kites to prevent gamebirds or waterfowl from taking flight prematurely, Buzzards as a robust species that could be used by novices to falconry (Dobney & Jacques, 2002). Gyr Falcons are very rarely seen in Britain, though they do sometimes occur as winter visitors. Formerly they were obtained for falconry from Norway, Iceland, or Greenland. There are only two possible archaeological records from Britain; one is certainly not a falconer's bird – an uncertain Peregrine/Gyr Falcon from a Late Glacial site, Potter's Cave, Pembrokeshire (David, 1991), but the other, from Winchester, certainly a high-status site, probably was a falconer's bird (Serjeantson, 2006). There is a well-recorded letter from King Ethelbert of Kent (AD 748–755), asking St Boniface in Germany to procure two hawks for crane-hawking, and it is presumed that he was seeking Gyr Falcons, though Peregrines were also flown at Cranes (Salvin & Brodrick, 1855; Dobney & Jacques, 2002). The Bayeux tapestry shows King Harold out hawking, with what appears to be a Goshawk. Evidently hawking started in Britain in Saxon times, and this is confirmed by the increased number of falconers' species from somewhat before Norman times (Table 6.3).

The fact that raptors are found in archaeological sites does not of itself indicate that they were used for falconry, though it is consonant with that thesis. Several other facts point the same way. Most often, the sites with raptors are high status sites, including castles and abbeys – for example, Peregrines are recorded at Loughor, Castle Rising, and Baynard's Castles, as well as Faccombe Netherton, Ilchester, Beverley, and King's Lynn, while sites yielding Goshawks included Scarborough, Stafford, Hen Domen, Portchester, and Castle Rising Castles, as well as Battle Abbey, Ilchester, Faccombe Netherton, King's Lynn, Norwich, and York. Most emphatically, Goshawk bones from Hen Domen carried a faint green stain, thought by Browne (2000) to indicate a falconer's ring, while Cherryson (2002) reported hawk rings found (but without bones) at two other sites, Heddingham Castle and Biggleswade. At Faccombe Netherton, almost complete skeletons of a Goshawk, Sparrowhawk, and Peregrine were found in one pit, sure evidence of their use in falconry (Sadler, 1990). Often the sort of 'high status' prey, notably Crane, Grey Heron, and Bittern, which are known to have been hunted with hawks or falcons, are also found at these sites. At Baynard's Castle, London, all six of the species that Sykes (2004) investigates as possible indicators of high status are present, i.e. Grey Heron, Bittern, Common Crane, Mute Swan, Grey Partridge, and Woodcock, along with Peregrine (Bramwell, 1975a). Of these, Woodcock, being nocturnal, were rarely hunted by hawks, but were usually taken by specialized netting techniques during their crepuscular display (roding) flights. Swans, Mute or Whooper, are too large even for a Goshawk to tackle, and would have been taken with arrows, or (in the case of Mute Swans) harvested during their moult in late summer, when they become flightless. Hawking was a usual way of taking the other four. As Serjeantson (2006) documents, wild birds (other, that is, than geese and chickens) that were probably taken by hawking, such as thrushes, larks, and waders, become much more numerous in archaeological sites of post-Norman age.

Table 6.4 Numbers of Anglo-Saxon to Medieval sites from which Peregrine, Goshawk and their "high status" prey are recorded, compared with the total number of sites of these ages. The figures for the total number of sites is approximate, because several sites span a range of ages.

Site	Anglo-Saxon	Norman	Medieval	Total
Peregrine	5	2	6	13
Goshawk	6	3	14	23
Common Crane	19	5	34	58
Grey Heron	14	3	22	39
Bittern	4	–	5	9
Black Grouse	7	–	11	18
Total Sites	46	9	144	199

Of the 13 sites with Peregrines that date to Anglo-Saxon, Norman, or Medieval date (Table 6.4), seven also contain bones of Crane and six also have Heron bones. As six of them also contain Woodcock remains, the coincidences might not seem to strengthen the evidence for falconry, though they do confirm the high status of these sites. However, in both cases, this is a significant set of coincidences (Peregrines and Cranes coincide more often than expected by chance, with $\chi^2 = 4.1$, $P = 0.04$; similarly, so do Peregrines and Herons, $\chi^2 = 6.1$, $P = 0.01$). It is interesting to note that nearly all these Peregrine sites are in southern and eastern England, well outside their breeding range, another strong indication that they had been caught for falconry; only Jarlshof, Shetland, and Lougher Castle at the base of the Gower Peninsula might be considered to lie within their natural breeding range.

Of the 23 sites with Goshawk, nine also contained Crane and six Heron, but in these cases the coincidence is no more than expected by chance. Perhaps the Goshawks were not used to catch these prey (they might have been more usefully flown at grouse, hares, or other game), and as Herons can even out-fly Peregrines, the slower Goshawk would have found them difficult. Goshawks, by contrast with Peregrines, could probably have been obtained in any relatively well-wooded part of the country, and it is harder to make any geographical point with them. Their sites are well spread across southern Britain.

Cranes, Ernes, Brewes, and other Mediaeval birds

Mention of Cranes as targets for falconry prompts consideration of their status and distribution in Mediaeval Britain in their own right and, with them, other species, now lost. What does the Mediaeval record tell us about the status of some of these charismatic species? Were they widespread, or were they already showing evidence of decline?

Cranes remained apparently numerous and widespread through Anglo-Saxon to Mediaeval times, according to their archaeological record. Not only are there some 59 records, to add to the 60 earlier and eight later records (plus another seven of uncertain date, Table 6.5), but they remained widespread (Figure 6.4). Their absence from Pleistocene and Late Glacial sites is interesting. It presumably reflects their preference for open wet marshy (not frozen) areas, and the fact that their large size inhibited cave-dwelling raptors from carrying their remains into such sites. Humans of all periods from Mesolithic to Modern

Table 6.5 Archaeological records of Crane *Grus grus* from the British Isles. Records marked + have been assigned to *Grus primigenia*, but that is discounted as a distinct species (see p. 37). Records marked * assigned only to *Grus* sp., but can hardly have been any other species.

Site	Grid Ref	Date	Citation
+Ilford	TQ 45 85	Ipswichian	Harrison & Cowles (1977)
Hackney Marshes	TQ 36 86	Pleistocene/Holocene	Harrison (1980a)
Thatcham	SU 50 66	Mesolithic	King (1962)
Star Carr	TA 02 81	Mesolithic	Fraser & King (1954); Harrison (1987a)
Formby Point	SD 26 96	Neolithic	Roberts *et al.* (1996)
Mount Pleasant, Dorset	SY 71 89	Neolithic	Harcourt (1971b)
Lough Gur, Co Limerick	R 64 41	Neo/Bronze	D'Arcy (1999)
Shap – Hardendale Quarry	NY 58 14	Beaker	Allison (1988b)
Barton Mere	TL 91 66	Bronze Age	Fisher (1966)
Burwell Fen	TL 59 67	Bronze Age	Northcote (1980)
Ballinderry crannog, Co Westmeath	N 22 39	Bronze Age	Stelfox (1942)
Ballycotton, Co Cork	W 98 64	Bronze Age	Harkness (1871); Newton (1923)
Norwich	TG 23 08	Bronze Age	Bell (1922)
Islay – Ardnave	NR 28 74	Bronze Age	Harman (1983)
Caldicot	ST 48 88	Bronze Age	McCormick *et al.* (1997)
*Brean Down	ST 29 58	Bronze Age	Levitan (1990)
West Harling-Micklemoor Hill	TL 87 95	Early Iron Age	Clarke & Fell (1953)
Dun an Fheurain, Gallanach	NM 82 26	Iron Age	Ritchie (1974)
+Meare Lake Village	ST 44 42	Iron Age	Gray (1966); Harrison (1987b)
Meare East	ST 45 41	Iron Age	Levine (1986)
Howe, Orkney	HY 27 10	Iron Age	Bramwell (1994)
+Glastonbury Lake Village	ST 49 38	Iron Age	Andrews (1917); Harrison (1980a, 1987b)
+Longthorpe	TL 15 97	Iron Age	King (1987)
Cat's Water, Fengate, Peterborough	TL 20 98	Iron Age	Biddick (1984)
Gussage All Saints	SU 00 10	Iron Age	Harcourt (1979a)
Blunsdon St Andrews	SU 15 58	Iron Age	Coy (1982)
Ower – Cleavel Point	SZ 00 86	Iron Age	Coy (1981a)
Dragonby	SE 90 12	Iron Age	Harman (1996a)
Haddenham	TL 46 75	Iron Age	Evans & Serjeantson (1988)
*West Stow	TL 81 70	Iron Age	Crabtree (1989b)
Woodbury, Devon	SX 84 51	Iron Age	Harrison (1980a, 1987b)
Wakerley	SP 95 99	Iron Age	Jones (1978)
North Uist – Bac Mhic Connain	NF 80 70	Iron Age	Hallen (1994)
*Burgh	TM 22 52	Iron Age	Jones *et al.* (1988)
Harston Mill	TL418507	Iron Age	R. Jones, pers. comm.
*York – 9 Blake Street	SE 60 52	Early Roman	O'Connor (1987b)
*Worcester – Sidbury	SO 68 85	Early Roman	Scott (1992b)
*Carlisle – The Lanes	NY 39 56	Early Roman	Connell & Davis unpub.
Gorhambury	TL 11 07	Roman	Parker (1988); Locker (1990)
Colchester	TL 99 25	Roman	Luff (1982, 1993)
Claydon Pike	SU 19 99	Roman	Locker unpub, Parker 1988
Exeter	SX 9192	Roman	Bell (1915), Maltby (1979), Bidwell, (1980)

Site	Grid Ref	Date	Citation
Lincoln	SK 97 91	Roman	Cowles (1973), Dobney *et al.* (1996)
Silbury Hill	SU 10 68	Roman	Gardner (1997)
Camulodunum	TL 98 25	Roman	Luff (1982, 1985)
Caerleon	ST 33 90	Roman	Hamilton-Dyer (1993)
Carlisle, Annetwell Street	NY 39 56	Roman, phase 3	Allison (1991)
London – St Mildred's	TQ 32 80	Roman	Bramwell (1975f); Parker (1988)
Carlisle, Annetwell Street	NY 39 56	Roman, phase 5	Allison (1991)
Newstead	NT 57 34	Roman	Ewart (1911); Parker (1988)
Papcastle	NY 10 31	Roman	Mainland & Stallibrass (1990)
Housesteads	NY 78 69	Roman	Gidney (1996)
York – Blake Street	SE 60 52	Roman	Allison (1986); Parker (1988)
York – colonia	SE 60 52	Roman	O'Connor unpub.; Parker (1988)
York – Minster	SE 60 52	Roman	Allison (1986); Parker (1988)
Shiptonthorpe	SE 81 38	Roman	Mainland & Stallibrass (1990)
Barnsley Park, Gloucs.	SP 08 06	Roman	Bramwell (1985)
Corbridge	NY 98 64	Roman	Bell (1922); Parker (1988)
Dorchester	SY 68 90	Roman	Maltby (1993)
Wookey Hole, Somerset	ST 53 47	Roman	Balch & Troup (1910)
*Plants Farm, Maxey	TF 11 08	Roman	Harman (1993a)
Caister-on-Sea	TG 51 12	Roman	Harman (1993b)
Dragonby	SE 90 12	Roman	Harman (1996a)
Flint – Pentre Farm	SJ 25 72	Roman	King & Westley (1989)
Silchester	SU 64 62	Roman	Newton (1906b), Maltby (1984); Parker (1988)
Wroxeter	SJ 56 08	Roman	Meddens (1987); Parker (1988)
London Wall	TQ 29 79	Roman	Harrison (1980a); Parker (1988)
Chester	SJ 40 66	Roman	Fisher unpub.; Parker (1988)
Ballinderry crannog, Co Westmeath	N 22 39	Early Christian	Stelfox (1942)
St Alban's Abbey	TL 14 07	Early/Mid Saxon	Crabtree (1983) (Unpublished)
St Alban's Abbey	TL 14 07	Mid Saxon	Crabtree (1983) (Unpublished)
Ipswich – St Peter's Street	TM 16 44	Saxon	Crabtree (1994)
Walton, Aylesbury	SP 82 13	Saxon	Bramwell (1976b)
Ipswich – St Nicholas Street	TM 16 44	Saxon	Crabtree (1994)
Ipswich – Buttermarket/St Stephens Lane	TM 16 44	Saxon	Crabtree (1994)
+London – Barking Abbey	TQ 29 79	Saxon	West (1994)
London – Westminster Abbey	TQ 30 79	Saxon	West (1994)
*West Stow	TL 81 70	Anglo-Saxon	Crabtree (1985, 1989a)
Ipswich	TM 16 44	Anglo-Saxon	Jones & Serjeantson (1983)
Flixborough	SE 87 15	Anglo-Saxon	Dobney *et al* (1994)
North Elmham Park	TF 98 20	Anglo-Saxon	Bramwell (1980a)
London – Shorts Gardens	TQ 30 81	Anglo-Saxon	J Stewart pers comm (quoted in Boisseau & Yalden 1999)
Thetford	TL 87 83	Anglo-Saxon	Jones (1984, 1993)
Raystown, Co Meath	O 04 51	6–7th C	Murray & Hamilton-Dyer (2007)
*Flixborough	SE 87 15	7th C	Dobney *et al.* (2007)
*Flixborough	SE 87 15	7–8th C	Dobney *et al.* (2007)
*Flixborough	SE 87 15	8–9th C	Dobney *et al.* (2007)

Table 6.5 (*Continued*)

Site	Grid Ref	Date	Citation
*Flixborough	SE 87 15	9th C	Dobney *et al.* (2007)
*Flixborough	SE 87 15	10th C	Dobney *et al.* (2007)
Lagore	N 98 52	Late Christian	Stelfox (1938), Henken (1950)
Buckquoy	HY 36 27	Norse	Bramwell (1977b)
Brough of Birsay	HY 23 28	Viking	Allison (1989)
York – Coppergate	SE 60 52	Anglo-Scand	O'Connor (1989)
Castle Rising Castle	TF 66 24	Saxo-Norman	Jones *et al.* (1997)
Maynooth Castle	N 934375	pre-Anglo-Norman	Hamilton-Dyer pers. comm.
Trim Castle	N 79 56	Anglo-Norman	Hamilton-Dyer pers. comm.
Lindisfarne – Holy Island	NU 13 41	AD 850–1100	Allison *et al.* (1985)
York – Minster – Contubernia	SE 60 52	9th–11th C	Rackham (1995)
Dublin – Fishamble Street	O 15 35	10th–11th C	T O'Sullivan, in D'Arcy (1999)
Dublin Castle	O 15 35	10th C	McCarthy (1995)
Oxford – St Ebbes	SP 51 06	11th–12th C	Wilson *et al.* (1989)
York – Parliament St	SE 60 52	11th–13th C	Carrott *et al.* (1995)
Beverley – Lurk Lane	TA 04 40	11th–13th C	Scott (1991)
Stafford Castle	SJ 92 23	11th C.	Sadler (2007)
Dublin – Woods Quay	O 15 35	10th–11th C	D'Arcy (1999)
Stafford Castle	SJ 92 23	12th C.	Sadler (2007)
Scarborough Castle, Kitchen	TA 05 89	12th–13th C	Weinstock (2002b)
York – Tanner Row	SE 60 52	12th–13th C	O'Connor (1988); O'Connor & Bond (1999)
Scarborough Castle	TA 05 89	13th C.	Weinstock (2002b)
Dublin Castle	O 15 35	13th C	McCarthy (1995)
Dragon Hall, Norwich	TM 23 08	13th–14th C	Murray & Albarella (2005)
Kings Lynn	TF 61 20	13th–14th C	Bramwell (1977a)
Dublin – Cornmarket	O 15 34	13–15th Cent	Hamilton-Dyer pers. comm.
Battle Abbey	TQ 74 15	Mediaeval	Hare (1985)
Chester – Dominican Friary	SJ 40 66	Mediaeval	Morris (1990)
Launceston Castle	SX 33 84	Mediaeval	Albarella & Davies (1996)
Newcastle – Quayside	NZ 25 64	Mediaeval	Allison (1987, 1988)
Beverley – Eastbrake	TA 03 39	Mediaeval	Scott (1984, 1992a)
Dublin – Back Lane	O 15 34	Mediaeval	Hamilton-Dyer pers. comm.
Clonmacnoise	N 01 30	Mediaeval	Hamilton-Dyer pers. comm.
Carlisle – Southern Lanes	NY 39 55	Mediaeval	Allison (2000)
York – Walmgate	SE 60 52	Mediaeval	O'Connor (1984b)
Walton Abbey	SE 46 48	Mediaeval	Newton (1923)
Lincoln – Flaxengate	SK 97 71	Mediaeval	O'Connor (1982)
Northampton – St Peters Street	SP 75 61	Mediaeval	Bramwell (1979e)
Walton, Aylesbury	SP 82 13	Mediaeval	Bramwell (1976b)
Loughor Castle, W Glamorgan.	SS 57 98	Mediaeval	Brothwell (1993)
Galway	M 29 24	Mediaeval	Hamilton-Dyer pers. comm.
York – General Accident Site	SE 60 52	Mediaeval	O'Connor (1985)
*Carlisle – Fisher Street	NY 39 56	Mediaeval	Rackham (1980)
*Leicester – Austin Friars	SK 58 06	Mediaeval	Thawley (1981)
Southampton, Cuckoo Lane	SU 42 13	14th C	Bramwell (1975c)
Southampton, Westgate	SU 42 13	14th–15th C	Coy (1980b)
Winchester	SU 48 29	14th–15th C	Coy (1984b)

Site	Grid Ref	Date	Citation
*Newcastle – Queen Street	NZ 25 63	14th-16th C	Rackham (1988)
Okehampton Castle	SX 58 95	Late Mediaeval	Maltby (1982)
London – Baynard's Castle	TQ 32 80	1500	Bramwell (1975a)
London – Baynard's Castle	TQ 32 80	1520	Bramwell (1975a)
Roscrea Castle, Co Tipperary	S 13 89	17th C	McCarthy (1995)
Castle Rising Castle	TF 66 24	Post-Med	Jones et al. (1997)
Kings Langley	TL 06 02	Post-Med	Locker (1977)
Carrickfergus	J 4187	Post-Med	Hamilton-Dyer pers. comm.
Peel – Isle of Man	SC 24 84	Post-Med	Fisher (2002)
Norton Priory	SJ 55 85	Post-Med	Greene (1989)
*Durham Cathedral	NY 27 42	Post-med	Gidney (1995a)
*Womersley – Wood Hall	SE 53 19	Post-med	Mulville (1995)
Galway	M 29 24	Post-Med	Hamilton-Dyer pers. comm.
*Hull – Magistrates Court	TA 10 28	Late Post-med	Carrott et al. (1995)
York – General Accident Site	SE 60 52	Modern	O'Connor (1985)
London – Cannon Street	TQ 32 80	Unknown	J Stewart pers comm (quoted in Boisseau & Yalden 1999)
London – Borough High Street	TQ 32 79	Unknown	J Stewart pers comm (quoted in Boisseau & Yalden 1999)
London – Rangoon Street	TQ 33 80	Unknown	J Stewart pers comm (quoted in Boisseau & Yalden 1999)
Catacomb Cave – Co Clare	R 33 73	Unknown	Newton (1906a)
Cambridge Fens	TL 4 6		Harrison (1980a)
London – East Cheap	TQ 33 80	Unknown	J Stewart pers. comm. in Boisseau & Yalden (1999)

times evidently hunted them, and their remains are among the most frequently recorded of all species of wild birds in later sites. They are relatively numerous at high status sites in the later Mediaeval period, and rare or absent then in villages (Albarella & Thomas, 2002; Sykes, 2004), but this difference is not evident in Anglo-Saxon to earlier Mediaeval times. When they were still common, it seems as though everyone ate them. In many cases, the bones themselves carry cut marks, which tallies with the frequent references to them in reports of meals, cooking, and game ordered for high class feasts. Sykes (2007) illustrates the distal end of a tarsometatarsus from Lincoln that had evidently been skinned very carefully; as she remarks, it would have been easier to cut the leg at the ankle, below which there is no meat anyway, and this indicates careful preparation of a high class dish. Henry III's Christmas dinner in 1251 apparently included 115 Cranes, along with 2,100 Partridge, 290 Pheasants, 395 swans, 7,000 hens, and 120 Peafowl (Rackham, 1986). Cranes were also kept in captivity for, as a grain-eating species, they were reasonably easy to feed, indeed to force-feed, like geese. They may have been held for falconry, as well as for eating (Yapp, 1982a). By the end of the Medieval period, they were evidently becoming scarce (Sykes, 2004), as indicated by a drop in representation among archaeological faunas. Henry VIII protected their eggs with a fine of 20d per egg in 1534, the highest fine imposed for any infringement. Perhaps the last indication of them breeding in England is the reference given by Gurney (1921) to a young one being obtained in Norfolk in June 1542, and the less precise report

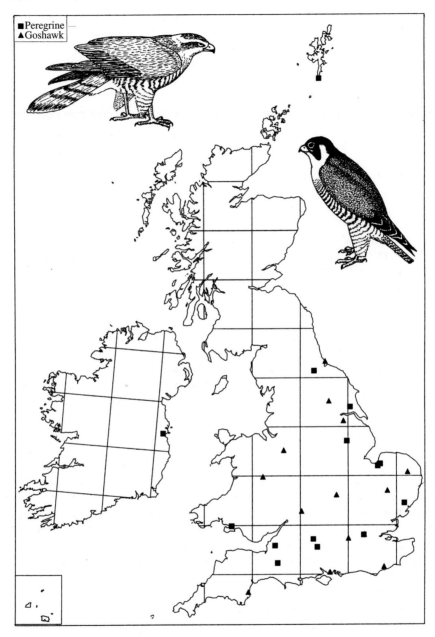

Fig 6.4 Map of Peregrine and Goshawk sites from Medieval Britain (cf. Tables 6.3, 6.4).

that William Turner, writing in 1544, had seen their young himself in England also during the sixteenth century (Boisseau & Yalden, 1999). The great third draining of the fens during the seventeenth century is likely to have resulted in the final loss of their last breeding habitat. Cranes still appeared in menus later than this, but they would have been wintering birds, immigrants from Scandinavia, as indeed is implied by their appearance at Christmas banquets. The latest archaeological records are Post-Mediaeval, probably indicating human consumption of these wintering birds. The menus that list Cranes, Herons, and Bitterns also often include Brewes. The identity of these has been uncertain, but Bourne (2003) has argued convincingly that the reference is to Night Herons. Egrets, presumably Little Egrets, are also specified in these menus, making it surprising (see also Chapter 4) that there is no Mediaeval archaeological record, and while there is evidence that both were imported, they must surely, given the numbers involved, also have bred in Britain. The tarsus lengths of Little Egret and Bittern overlap completely, and it seems possible that, with reference skeletons also scarce, Egrets have been overlooked/misidentified by archaeologists.

By contrast with the Crane, the White Stork, a similar bird ecologically but with a much more southern European distribution, is recorded from only 11 sites. In further contrast with Crane, there are two Pleistocene cave records, from Pinhole Cave and Robin Hood's Cave at Creswell Crags (Jenkinson & Bramwell, 1984; Jenkinson, 1984). The best known record is very doubtful; the Mesolithic record from Star Carr (Fraser & King, 1954) was thought by Harrison (1987a) not to be White Stork at all. Two Bronze Age (Jarslhof: Platt, 1933a, 1956; Nornour: Turk, 1971, 1978) and two Iron Age records (Dragonby: Harman, 1996a; Harston Mill, R. Jones pers. comm.) are followed by single Roman (Silchester: Newton, 1908; Maltby, 1984) and Saxon (Westminster Abbey: West, 1991) records, and then by a single Medieval record, from St Ebbes in Oxford (Wilson *et al.*, 1989). This is strange, as the species is recorded as having bred in Edinburgh in the fifteenth century. More records, especially from warmer periods and further south in England, would have been predicted. Presumably its rarity explains why it was not hawked, and is not discussed in regard to banquets either. It may be relevant to note that there is no Swedish archaeological record of White Storks, either (though there are a couple of records of Black Stork): it is presumed to have always been a more southerly species (Ericson & Tyrberg, 2004).

Another species which was eaten, and is even more famous as a former breeder, is the Great Bustard. This is well known as a former inhabitant of open downland and the Breckland in Wiltshire, Suffolk, Norfolk, and elsewhere. Like the White Stork, the interesting thing about this species is its archaeological rarity. Beyond the Late Glacial records (see Chapter 3, pp. 50, 58), it is recorded only from Roman Fishbourne (Eastham, 1971)) and Medieval Baynard's Castle, London (Bramwell, 1975a). Given the large numbers reputedly eaten in Medieval banquets – apparently Henry VIII included 4 dozen in his menu for October 1539 – it ought to have been found more often. It is possible that it was imported for such meals, though Yapp (1982a) does not mention it among birds that might have been kept captive in Medieval Europe. It cannot have been present in Britain during wooded Mesolithic times, but must have depended on the creation of open farmland by humans. It seems likely that it was never very common, perhaps difficult to catch. Perhaps the farmed landscape of Roman times was sufficiently open for it, if the Fishbourne specimen is correctly identified (rumour suggests it is actually a Crane bone), or that was imported, and it only established itself in England during the period of open field systems in Medieval times. It may be relevant that there is no native Old English (Anglo-Saxon) name for it. Various recent accounts, in particular those considering the recent or current reintroduction programmes, imply that it was once common (e.g. Waters & Waters, 2005), but the archaeological record suggests otherwise. Interestingly, there is no Swedish archaeological record, and Ericson &

Tyrberg (2004) suppose that it colonized Sweden only for a limited period from about 1780 to 1860. In the detailed accounts of its demise in England during the nineteenth century, overhunting is certainly documented, but enclosure of the open fields, and the planting of hedges and estate woodlands, produced a landscape which was no longer really suitable for it (cf. Chapter 7).

Another interesting archaeological record that ends in Mediaeval times, so far as England is concerned, is that for the Capercaillie. It has the additional interest of documenting a well established, largely contemporary, population, in Ireland (Table 6.6). The history of Capercaillie in Scotland in historical times is well known. It was assumed that it survived there because of the extent of Scot's Pine, its main food and the main component of the Caledonian Forest. Further, its extinction in the late eighteenth century parallels the deforestation at that time, while its reintroduction in 1835 followed an era of much replanting of conifers. Given the scarcity of woodland in Ireland, and the nature of deciduous woodlands that predominated in England, it had always seemed unlikely that Capercaillie could have survived into historical times in either. Thus the strong archaeological record is

Table 6.6 Archaeological records of Capercaillie in the British Isles.

Site	Grid Ref	Date	Citation
Swanscombe, Kent	TQ 60 74	Wolstonian	Harrison (1979), 1985
Kirkdale Cave, Yorkshire	SE 67 85	Pleistocene	Salcho Jones quoted by Bramwell (1971a)
Ravencliffe Cave, Derbyshire	SK 17 73	Pleist-Post Glacial	Harrison (1980a)
Kent's Cavern	SX 93 64	Late-Post Glacial	Bell (1915, 1922); Bramwell (1960b, 1971a)
Teesdale Fissure	NY 86 31	Flandrian	Simms (1974); Bramwell (1971a)
Mother Grundy's Parlour, Creswell	SK 53 74	10–7,000 BP	Bramwell (1971a)
Wetton Mill Rock Shelter	SK 09 56	Mesolithic	Bramwell (1976a)
Dowel Cave, Derbyshire	SK 07 67	Mesolithic	Bramwell (1971a)
Mount Sandel, nr Coleraine	C 86 32	Mesolithic	Van Wijngaarden-Bakker (1985)
Fox Hole Cave, Derbyshire	SK 10 66	Neolithic	Bramwell (1971a, 1978c)
Embo, Sutherland	NH 82 92	Neolithic	Henshall & Ritchie (1995)
Heathery Burn Cave, Devon	NY 99 39	Bronze Age?	Harrison (1980a)
Wookey Hole, Somerset	ST 53 47	Roman	Balch & Troup (1910), Bramwell (1971a)
Durham – Saddler Street	NZ 27 42	Anglo-Saxon	Rackham (1979a)
York – General Accident Site	SE 60 52	Anglo-Scand	O'Connor (1985b)
Dublin – Fishamble Street	O 15 35	10th–11th C	T O'Sullivan in D'Arcy (1999)
Dublin Castle	O 15 35	10th C	McCarthy (1995)
Leicester – St Peters	SK 58 04	12th C.	Gidney (1991a, 1993)
Wexford	T 05 23	12th C	D'Arcy (1999)
Waterford	S 60 12	12th C	D'Arcy (1999)
York – Tanner Row	SE 60 52	12th–13th C	O'Connor & Bond (1999)
Dublin Castle	O 15 35	13th C	McCarthy (1995)
Waterford	S 60 12	13th C	D'Arcy (1999)
Trim Castle, Co. Meath	N 79 56	Anglo-Norman	Hamilton-Dyer pers. comm.
Galway	M 29 24	Post-Med	Hamilton-Dyer pers. comm.
Carrickfergus, Co. Antrim	J 4187	Post-Med	Hamilton-Dyer pers. comm.
York – Aldwark	SE 60 52	Post-Med	O'Connor (1984a)

as surprising as the late date of the last records. The pollen record does suggest a fringe of pine around the upper slopes of the Pennines and the North York Moors in the Mesolithic period, as well as more extensively in Ireland, and it is not clear how late in time this lasted. It is notable that the later records of Capercaillie in England are mostly in the north, while in Ireland they are more widespread. The continuous record from Anglo-Saxon through Mediaeval times, and the Post-Mediaeval records from York, Galway, and Carrickfergus, support the notion that the species survived more widely than just in Scotland, well into the historical period. Giraldus Cambrensis has been ridiculed for suggesting in the fourteenth century that there were *pavones sylvestres*, wood peacocks, in Ireland, but this would not be a poor description of a Capercaillie for someone unfamiliar with the bird. Deane (1979) argued strongly that the Capercaillie was never an Irish bird, particularly because there was no record of an Irish Gaelic name for it. Hall (1982) countered that it was known as cock of the woods, *coileach feadha*. (The Scots Gaelic *capall coille*, horse of the woods, is the probable source of the modern name, horse here being, as in horse-fly and horse-mushroom, an expression of size.) The discovery of archaeological remains, first at Mount Sandel in 1982 and more recently in Dublin and elsewhere, resolves the question unequivocally. Moreover, the later records validate the late literary and culinary records which Hall (1982) assembled, suggesting a survival in the west of Ireland as late as the early eighteenth century; Dr Charles Smith (1749) remarks in describing County Cork that 'the bird is not found in England and now rarely in Ireland since our woods have been destroyed' (as quoted by D'Arcy, 1999). It appears that memory of the bird itself in England did not quite linger long enough for the bird to be listed by the early ornithologists as an English species; Willughby (1676) confirmed its absence from England, while commenting on the delicate taste and rarity of the bird (D'Arcy, 1999).

One other species with a strong archaeological record is worth discussing at this point; it was argued above that the White-tailed Eagle was well-enough known to the Anglo-Saxon settlers to have given its name to a number of places in England. With 58 records, as against only 15 of Golden Eagle (see Table 6.3), it was always much more widespread, and probably more common (or at least more available to Humans) (Yalden, 2007). The archaeological record confirms that this was true as late as Roman times, but it was becoming scarce by Anglo-Saxon times, and apparently restricted to a more northern and western range by Mediaeval times (Table 6.7). The few Mediaeval records, from Waterford and Nantwich outside its modern range, and Iona, well within its historical range, suggest that the last ought to have been somewhere in the Scottish islands. It is interesting therefore to find the latest record appears to be from Cumbria, as the latest known nesting in England is believed to have been at Haweswater in the Lake District in 1787 (Love, 1983), though other eighteenth century nests remembered on the Isle of Wight, near Plymouth, on Lundy and on the Isle of Man bear testimony to its once wide range. Despite much persecution, in 1894 the White-tailed Eagle was described by Ussher as 'still one or two pairs in Mayo and Kerry', but the last nested in 1898. In Scotland, too, persecuted to extinction, the last nested on Skye in 1916 (Love, 1983).

In general, the diversity of wild birds seems to increase from Anglo-Saxon through Norman to later medieval times (Albarella & Thomas, 2002). Herons, Woodcock, and Grey Partridge as well as swans register marked increases in the later Middle Ages (Sykes, 2004). This must be partly the consequence of the fact that swans were increasingly kept as tamed

Table 6.7 Archaeological records of White-tailed Eagle in the British Isles.

Site	Grid Ref	Date	Citation
Walton, nr Clevedon	ST 42 74	Ipswichian?	Reynolds 1907, Palmer & Hinton (1928)
Tornewton Cave	SX 81 67	Wolstonian	Harrison (1980a, b, 1987b)
Soldier's Hole, Cheddar	ST 46 54	Middle/early Late Devensian	Harrison (1988)
Cat Hole, Gower	SS 53 90	Devensian	Harrison (1980a)
London Basin	TQ 2 7	Upper Devensian	Harrison (1985)
Walton Cave, Somerset	ST 41 72	Late Glacial	Reynolds (1907)
Walthamstow, Essex	TQ 37 88	Late Glacial	Harrison & Walker (1977), Bell (1922)
Soldier's Hole, Cheddar	ST 46 54	Late Glacial	Harrison (1988)
Rousay, Orkney	HY 40 30	Post-Glacial	Bramwell (1960a)
Church Hole Cave	SK 53 74	Flandrian	Jenkinson (1984)
Hornsea	TA 21 47	Holocene	Bell (1922)
Skipsea	TA 16 55	Mesolithic	Sheppard (1922)
Port Eynon Cave, Gower	SS 47 85	9,000–6,000 BP	Harrison (1987b)
Carding Mill Bay	NM 84 29	5000 BP	Hamilton-Dyer & McCormick (1993)
Lough Gur, Co Limerick	R 64 41	Neolithic	D'Arcy (1999)
Links of Noltland, Orkney	HY 42 49	Neolithic	Armour-Chelu (1988)
Rousay – Knowe of Ramsay	HY 40 28	Neolithic	Davidson & Henshall (1989)
Isbister	HY 40 18	Neolithic	Bramwell (1983a)
Westray – Point of Cott	HY 46 47	Neolithic	Harman (1997)
Dublin – Dalkey Island	O 27 26	Neolithic	Hatting (1968)
Burwell Fen	TL 59 67	Bronze Age	Northcote (1980)
Potterne	ST 99 59	Bronze Age	Locker (2000)
Coneybury Henge, nr Stonehenge	SU 13 41	Bronze Age	Maltby (1990)
Dragonby	SE 90 12	Iron Age	Harman (1996a)
Meare Lake Village	ST 44 42	Iron Age	Gray (1966)
Glastonbury Lake Village	ST 49 38	Iron Age	Andrews (1917); Harrison (1980a, 1987b)
Deerness – Skail	HY 58 06	Iron Age	Allison (1997b)
Howe, Orkney	HY 27 10	Iron Age	Bramwell (1994)
Cats Water, Fengate, Peterborough	TL 20 98	Iron Age	Biddick (1984)
Carlisle – The Lanes	NY 39 56	Early Roman	Connell & Davis unpublished
Leicester – High Street	SK 58 04	Roman	Baxter (1993); Mulkeen & O'Connor, (1997)
Stanwick – Redlands Farm	SP 96 70	Roman	Davis (1997)
Uley Shrines, Gloucs.	ST 78 99	Roman	Cowles (1993)
Ower	SZ 00 85	Roman	Coy (1987b)
Stonea, Cambridgeshire	TL 44 93	Roman	Stallibrass (1996)
London – Billingsgate Buildings	TQ 32 80	Roman	Cowles (1980a); Parker (1988)
London – Southwark	TQ 31 79	Roman	Cowles (1980b); Parker (1988)
Caerleon	ST 33 90	Roman	Hamilton-Dyer (1993b)
Camulodunum	TL 98 25	Roman	Luff (1982, 1985); Parker (1988)
Long Bennington	SK 82 47	Roman	Harman (1994)
Dragonby	SE 90 12	Roman	Harman (1996a)
Tolpuddle Ball	SY 81 94	Roman	Hamilton-Dyer (1999)

Site	Grid Ref	Date	Citation
Longthorpe	TL 15 97	Roman	King (1987)
Binchester	NZ 21 31	Roman	Mulkeen & O'Connor (1997)
Segontium	SH 48 64	Roman	O'Connor (1993); Mulkeen & O'Connor (1997)
Dunstable	TL 01 21	Roman	Jones & Horne (1981); Parker (1988)
Droitwich	SO 89 63	Roman	Cowles (1980); Parker (1988)
Scole-Dickleburgh	TM 16 80	Late Roman	Baker (1998)
York – Minster – SE	SE 60 52	5th–8th C	Rackham (1995)
Lagore	N 98 52	Late Christian	Stelfox (1938), Hencken (1950)
York – Coppergate	SE 60 52	Anglo-Scand	O'Connor (1989)
Deerness – Skail	HY 58 06	Viking	Allison (1997b)
York – Minster – Contubernia	SE 60 52	9th–11th C	Rackham (1995)
Dublin – Woods Quay	O 15 35	10th–11th C	D'Arcy (1999)
Nantwich	SJ 65 52	Medieval	Fisher (1986)
Waterford	S 60 12	Medieval	D'Arcy (1999)
Iona – Abbey	NM 28 24	Medieval	Coy & Hamilton-Dyer (1993)
Brougham Castle, Cumbria	NY 53 28	14th–16th C	Gidney (1992c)

birds in parks (MacGregor, 1996); parks would have provided a suitable habitat too for partridges, and estate woodlands might have suited the operation of nets for Woodcock. This increase in wild (or, in some cases, tamed) birds is unlikely to reflect their genuine increase in the countryside, but reflects their enhanced importance as symbols of social status.

Birds in early literature and art

A third line of evidence concerning bird life in the earlier part of the Middle Ages comes from the work of early monks, illustrating their bibles, psalters, and hymn books, and monarchs describing the art of hunting. Roughly contemporary, the bestiaries, sometimes based on fact and sometimes on legend, also give an interesting insight into what might, generously, be construed as early ornithology.

Illustrated bibles and similar religious works vary in quality, in ways that seem to have little to do with artistic ability, but may have more to do with style. In most, eagles, symbolic of St John, and doves, symbolic of the Holy Ghost, are frequent. More rarely, other birds are depicted. Fortunately, Yapp (1982a,b, 1983, 1987) investigated these sources in some depth, and his account of birds in illustrated manuscripts (Yapp, 1981b) deserves particular attention. Illustrations are not always readily recognizable, and rarely have the details that we would expect in a modern field guide. Nor should they be expected to be so accurate: in most cases, they are intended as incidental decoration, and are rarely depicted as subjects in themselves. Even so, some species are evident, either because of the accompanying text, or because they are so distinctive. Cranes, as already remarked, are a familiar subject. The 'bustle' of secondary feathers is a sure indication, and most show also the red crown characteristic of adults. The story of Tobit (in the biblical Apocrypha) being blinded by bird dung is illustrated by a swallow flying up to a characteristic nest, though the nest sometimes appears to be that of a Red-rumped Swallow or of a House Martin, while the bird seems usually to be a (Barn) Swallow (Yapp, 1981b). Some biblical passages lent themselves to bird illustrations, notably the Creation, Adam naming the birds, and the passage in Revelations describing the Apocalypse, in which the birds are called by an angel and then invited to feast on the corpses of the damned. Psalters and hymn books provided an annual litany, and certain months were typically illustrated by reference to particular seasonal pursuits. Hawking often illustrated May.

Yapp (1982b) drew particular attention to the variety of birds illustrated in the Sherborne Missal, an illustrated prayer book containing the text for mass on every day of the year. Written by a scribe called John Whas, of whom we know nothing more, and illustrated by a Dominican friar John Siferwas, it was created for Robert Brunyng, Abbot of the Benedictine Abbey of Sherborne in Dorset from 1385 to 1415. Among the many decorations are a remarkable series of 48 birds, illustrated in colour, 41 of which also have Middle English names appended. While Yapp (1981b, 1982b) could only illustrate a few, or only in monochrome, Backhouse (2001) has provided a full complement. Some (Crane, Peahen, Peacock, Pheasant, Robin, Goldfinch) feature regularly in other illustrated manuscripts, but others are unique to this one – a juvenile Gannet (*ganett*), Shelduck (*bergandir*), Common Gull (*mew*) and Moorhen (*more hen*), Long-tailed Tit (*tayl mose*), a Cormorant (*cormerant*), and Barnacle Goose (*bornet*). Others are rarely illustrated: Fieldfare (*vuelduare*), Great Grey Shrike (*waryghanger*), Green Woodpecker (*wodewale*), Kingfisher (*kyngefystere*), for examples. Lark, Wagtail, Coal Tit, female and male House Sparrow and Chaffinch, Bullfinch, Blackbird, Quail, Woodcock, and Snipe are also illustrated. Some errors are evident – the *grene fynch* is clearly a Goldfinch, and the *linet*, with a white rump, might have been drawn from a Brambling (despite its yellow bill, it is not streaky enough to have been drawn from a Twite). Some puzzles remain. One bird, labelled *morcoc*, is a short-tailed but long-billed bird with a striking zig-zag pattern on its flanks. It is surely intended to be a Water Rail, and its name, paired with *more hen*, recalls a time when Wrens were thought to be the female partners of Robins, and Great and Blue Tits were similarly thought to be mates (*mose cok* and *mose hen* in the Sherborne Missal). More problematical are the *vinene coc* and *fyne hen*. These are both black and white birds, with some red on the head, and the *vinene coc* (though not the *fyne hen*) shows the zygodactylous feet (two toes forward, two back) of a woodpecker, recalling (see Table 6.1) the OE name *fina* for a spotted woodpecker. Were Great and Lesser Spotted Woodpeckers also thought to be male and female of the same species? Yapp (1982b) suggests that Siferwas had a northern assistant, the inclusion of, for example, Barnacle Goose, juvenile Gannet, Fieldfare, and Great Grey Shrike recalling his northern roots.

The Mediaeval bestiaries were accounts, in Latin, of the 'known' mammals and birds, some real and others imaginary. Thus unicorns and phoenixes were given similar treatment to wolves and eagles. Their origin was a Greek text, the *Physiologus*, dating back to fourth century Alexandria, but new knowledge was added. Unfortunately, old knowledge was also sometimes corrupted, as the meaning of original Greek names was lost. Thus *fulica*, which to us translates as Coot, was equated in an early Latin version of the *Physiologus* with *erodios*. However, while the Latinized *herodius* later became the Mediaeval name for a falcon, or more specifically for the Gyr Falcon, *fulica* was a bird that nested in the middle of a lake or sea, and only later became the Coot. The only waterbird consistently listed is *pelicanus*, which probably was the familiar pelican to the original author of the *Physiologus*. By Mediaeval times in northern Europe, it had become a mythical bird that killed its own young and then resuscitated them by pecking at its own breast; appropriately, the illustrations sometimes show it with a hooked bill, doing just this. The Latin *coturnix* is, as we now regard it, the quail, and evidently an onomatopoeic name, but by the Middle Ages it had somehow become the Curlew, and was sometimes quite well illustrated as such, with a decurved bill and long legs. Among other bestiary birds, of which about 65 in all are listed, described and illustrated, *grus*, *ciconia*, *ardea*, *milvus*, *accipiter*, *corvus*, *pica*, *graculus*, *passer*, *alauda*, *hirundo*, and *merula* have familiar modern meanings (Crane, White Stork, Heron, Kite, Hawk, Crow/Raven, Magpie, Jay, Sparrow, Lark, Swallow, and Blackbird); the domestic/semidomestic *gallus*, *pavo*, *fasianus*, *anser*, *anas*, *olor*, and *psittacus* were also of course well known. Perhaps most informative are the comments by Yapp (1987) on the Greek *epops*, in the *Physiologus*, later Latinized as *epopus* or *upupa*, all onomatopoeic for the Hoopoe. This was a familiar bird in southern Europe, and well illustrated in French and Italian bestiaries, with the correct pink, black and white colouring, and the characteristic crest. English bestiaries, however, have poor drawings of indeterminate birds: evidently it was not a familiar species, despite the fact that it was more likely to have occurred here in slightly warmer times. English bestiaries are similarly uncertain about *vultur*, which must also have been unfamiliar, and very confused about owls. The original *Physiologus* had only one owl, *nocticorax* equated in later bestiaries with *noctua*. However the latest bestiaries attempt to distinguish four species, *noctua* (=*nycticorax*), *strix*, *bubo*, and *ulula*. The original *noctua* was the Little Owl, not familiar to English Mediaeval writers or illustrators, and the Eagle Owl *bubo* was also unknown to them. Sometimes *noctua* is shown with small ears, and *bubo* without them. Both *strix* and *bubo* are said to refer to their calls, which might suggest an attempt to distinguish screech owls *strix* (Barn Owls?) from owls that hoot *bubo* (Tawny Owls?), but the illustrations do not help (Yapp, 1987). The small birds too are rather confusing. The Nightingale *lucinia* is generally described, and sometimes equated with *acredula*. *Ficedula* (literally 'fig-eater') may have been used to refer to one or more of the *Sylvia* warblers, but the illustrations are unhelpful. One bestiary appears to illustrate a Grasshopper Warbler under the name *ciconia* following the text which says, presumably referring to the bill rattling of Storks, that they are so called from the crackling noise that they make; this would, as Yapp remarks, equally well describe the song of this warbler.

Conclusions

The archaeological record of British birds provides a direct, but very patchy and selective, record: biassed to larger species, edible species or other human interests. There is a wealth of wild birds found in Mediaeval sites, especially those of high status. Among the more charismatic species, Crane, White-tailed Eagle, and Capercaillie are widespread. The bestiaries, like placenames, represent a hesitant first step in recording the bird life of the British Isles. By Mediaeval times, however, a proper written record of British birds had begun, and this direct recording increasingly replaces the hesitant proxy records. It is necessary to step back in time a little to pick up this record.

7
From Elizabeth to Victoria

Assembling a list of British birds

One of the many interesting features in *The Shell Bird Book* (Fisher, 1966) was the account of the assembling of the British bird list. Fisher suggested that it might be started with the Anglo-Saxon poem, *The Seafarer*,

Þaer ic ne gehyrde	butan hlimman sae,
iscaldne waeg	hwilum yfelte song.
Dyde ic me to gomene	ganetes hleoþor
& huilpen sweg	fore hleahtor wera
maew singende	fore medodrince
stormas þaer stanclifu beotan	þaer him stearn oncaeð
isigfeþera;	ful oft þaet earn bigael
urigfeþra...	

which he translated as:

There heard I naught	but seething sea,
Ice-cold wave,	awhile a song of swan,
There came to charm me	gannets' pother,
And whimbrels' trill	for the laughter of men,
Kittiwake singing	instead of mead,
Storms there the stacks thrashed,	there answered them the tern,
With icey feathers	full oft the erne wailed round,
Spray-feathered...	

suggesting that this described a scene near Bass Rock (a known early gannetry, whence *Morus bassanus*) sometime before AD 685, in spring as the Common Terns and Whimbrels returned north to breed, as the Whooper Swans departed for the north, and as the others settled on their nesting cliffs. He regarded this as a first indication of a list of British birds, even if the specific identity of the gulls (*maew*) and terns (*stearn*), let alone *huilpe* (Whimbrel, Curlew, or other wader?) was uncertain. Whimbrels trill better than Curlews, and *medodrince*, he suggests, is an attempt, like Kittiwake, at onomatopoeia. The lives of the early saints had already contributed a few other birds – the tame Robin restored to life by Kentigern (later St Mungo) in about AD 530 is perhaps the first recorded British bird (after Caesar's mentions of Domestic Fowl and Goose). The migrating Crane restored to health by St Columba on Iona in about AD 570, the White-tailed Eagle, Carrion Crow, and perhaps Eider reported by St Cuthbert on the Farne Isles about a century later (contemporary with The Seafarer?), the Woodpigeon, Nightingale, Swallow, and Chaffinch alluded to by St Adhelm's riddles, and the Cuckoo, Swallow, and Raven noted by St Guthlac, in his fenland retreat about AD 700, combine to produce a list of about 16 species recorded by that date. In the eighth and ninth centuries, the Anglo-Saxon glossaries, giving Latin equivalents to Anglo-Saxon bird names, give about 70 species, though this total includes domestic birds and such exotics as *pawe* (Peacock), *earngeap* (Vulture), and *geolna* (Ibis). Among the more interesting additions are *hegesugge* (Hedge Sparrow), *swertling*, probably Blackcap (as Kitson (1997) remarks, *swert* is the same as German, *swart*, black), *colmase* (Coal Tit), *snite*, *wudecocc*, and *hulfestre* (Plover). Probably the lists recognize about 57 British wild birds. Placename evidence (see Chapter 6) suggests a number more, so probably about 60 or 70 species were recognized in Saxon times. Giraldus Cambrensis referred to Capercaillie (*pavones sylvestres* – Wood Peacocks), Hobby and Merlin, while travelling in Ireland in 1183–86. He managed an interesting confusion of Dipper and Kingfisher under the name *martinetas*, which were described as short quail-like birds that plunged into water to catch fish, and were variously white with black backs or, elsewhere, had red breast, beak and legs but back and wings gleaming bright green (Yapp, 1981b).

Geoffrey Chaucer (approximately 1340–1400) mentioned some 43 wild British birds (Table 7.1), including a reference to a shrike (Great Grey rather than Red-backed?) as waryangel, adding four more species and bringing the list overall to 100. He was perhaps not as original or accurate an ornithologist as Fisher (1966) supposed, for Romaunt of the Rose is a translation of a French work, and includes some bird names (Chalaundre, Calandra Lark, for example) which he translated with perhaps no understanding – though he did spend some time fighting in France. He may have confused Popinjay (usually Parrot, but perhaps Green Woodpecker here) and Wodewal (later usually Green Woodpecker, but originally, as perhaps here, Golden Oriole, cf. Lockwood, 1993). Never-the-less, he certainly referred to a wide variety of birds, as Table 7.1 makes clear, and in a very imaginative way, using a mixture of real and supposed (as by the bestiaries) characteristics of the birds. Thus his Heron was the eels' foe, the Swallow the murderer of bees that make honey, his Nightingale called the green leaves of spring to unfold, his Robin was tame, as British Robins are still. On the other hand, his Stork abhorred adultery, the Swan was proud, the Owl portended ill, the Raven was wise, the Crow had a careful voice, the Kite was a coward. Chaucer's listing is about contemporary with the birds shown and named in the Sherborne Missal (Yapp, 1982b; Backhouse, 2001), discussed in Chapter 6, and some of the names he used recall those used there, notably waryangle for shrike.

By the end of Mediaeval times, about 1500, the list had reached 118 species, by Fisher's reckoning. Serious ornithological writing begins at about this time. In 1544, William Turner published *Avium Praecipuarum* in Cologne, and in Latin, regarded by Fisher as the first printed bird book. Later Dean of Wells, Turner had an erratic life as a clergyman, his spells in Cologne and elsewhere being the result of banishment and discrete self-exile as that ecclesiastically tortured century progressed. In *Avium Praecipuarum*, he attempted to identify the birds mentioned by Pliny and Aristotle (Gurney, 1921). He described Cormorants nesting in East Anglian heronries, mentioned white Herons nesting in England (presumed to be Little Egrets), knew that Hobbies were missing in winter, that both Peregrines and Gyr Falcons

Table 7.1 Wild birds mentioned by Chaucer. Spellings taken from Fisher (1977), with examples of locations in Chaucer's works mostly from Harrison (1956). These are line numbers in Co (Cook's Tale), Fkl (Franklin's Tale), FT (Friar's Tale), HF (House of Fame), Kn (Knight's Tale), LGW (Legend of Good Women), PF (Parliament of Fowles), Prol (General Prologue), RR (Romaunt of the Rose), Sq (Squire's Tale), Su (Summoner's Tale), TC (Troylus and Criseyde), Th (Tale of Sir Thopas), WB (Wife of Bath's Tale). He also mentions, of domestic birds, Hen, Cock, Drake, Goose and Peacock, as well as the generic Dove, Finch and Hawk.

Alp (Bullfinch)	RR 658
Bitore (Bittern)	WB 972
Bosarde (Buzzard)	RR 4033
Chalaundre (Calandra Lark)	RR 81, 663, 914
Chough, Cow (Jackdaw)	PF 345, WB
Cokkow (Cuckoo)	7 refs, inc. PF 358, 498, 505, etc
Colver, Wodedowe (Wood Pigeon)	LGW 2319, Th 770
Cormeraunt	PF 362
Crane	PF 344
Crowe	4 refs, inc. PF 345, 363
Eagle (and Tercel Eagle – male)	22 refs, inc. PF 330, 332, 373, 393 etc.
Faucon peregryn (Peregrine Falcon)	Sq 428
Feldefare	PF 364, RR 5510, TC iii: 861
Goldfynch	Co 4367
Goshauk	PF 335, Th 1928
Heroun, Heronsewe	Fkl 1197, PF 346, Sq 68
Heysoge (Hedge Sparrow)	PF 612
Jay	6 refs, inc. PF 356
Kyte	5 refs, inc. PF 349, Kn 1179
Lapwynge	PF 347
Larke, Laverokkes ((Sky)larks)	9 refs, inc. PF340, RR 662
Mavys, Mavise (Song Thrush)	RR 619, 665
Merlioun (Merlin)	PF 339, 611
Nyghtyngale	15 refs, inc. PF 351.
Partridge	HF iii: 302, Prol 359
Oule (?Barn, ?Tawny Owl)	8 refs, inc. PF 434
Quayle	Cl 1206, PF 339, RR 7259
Papejay, Popingeie (both Parrot and Green Woodpecker)	6 refs, inc. PF359, RR 81, 913, Th 767
Pye (Magpie)	10 refs, inc. PF 345
Raven	4 refs, inc. PF 363
Roddok (Robin)	PF 349
Roke (Rook)	HF iii:1516
Sparwe (Sparrow)	PF 351, Prol 626, Sum 1804
Sperhauk	6 refs, inc. PF338, 569
Swan	7 refs, inc PF 342
Stare (Starling)	PF 348
Stork	PF 361
Swalwe (Swallow)	3 refs, inc. PF 353
Terin (?Serin)	RR 665
Thrustel, Thrustelcok (Mistle Thrush) 1963	5 refs, inc. PF 364, RR 665, T 1959,
Turtil (Turtle Dove)	8 refs, inc PF 355, 510, 577, 583
Tydif (Great Tit)	LGW 154
Waryangle ((?Great Grey) Shrike)	FT 1408
Wodewales (?Golden Orioles)	RR 658

were difficult to obtain in England, and that Marsh Harriers took ducks and coots, but that Hen Harriers, living up to their name, were serious raiders of farmyard poultry. He seems to have confused Goshawks and Sparrowhawks, remarking that the latter took 'doves, pigeons, partridges and the larger sorts of birds', which sounds more like the former (Gurney, 1921). He reported that (Red) Kites in England were larger and more numerous than (Black) Kites in Germany, that they tended to be whiter, and that they were likely to snatch bread from children, fish from women, and drying linen from hedgerows. Evidently they were still very common in Britain, matching earlier comments from the Venetian ambassador Capello, in winter 1496–97, who remarked on the abundance of both Kites and Ravens in London. Fisher suggests that Turner recognized 130 species, of which some were domestic and others European; perhaps 105 wild British birds were indicated in his list, although that included two gulls and two grey geese whose current identity is uncertain. Among these, 15 species seem to have been recognized for the first time by him, including such more difficult identifications as Woodlark, Meadow Pipit, Whitethroat, and Brambling, as well as the two harriers.

Like Chaucer, William Shakespeare (1564–1616) made frequent use of birds' attributes in his plays and poetry, and seems to have acknowledged some 50 British wild species (Table 7.2). Again like Chaucer, his poetical writing alluded to foreign birds, and to bestiary birds as well as real ones – ostrich, parrot, and pelican, as well as griffin and phoenix (Harting, 1864; Acobas, 1993). His avian allusions are dominated by references to hawking, to food species and to familiar song-birds. Thus, there are some 108 mentions of raptors, including 35 unspecified hawks or falcons not listed in Table 7.2. As well as wild ducks and geese, Partridges, Quail, and Pheasant, there are 31 references to cocks, 38 Geese, a Guinea

Table 7.2 Birds in Shakespeare. A list of the British wild birds mentioned by Shakespeare (after Acobas 1993). The number of mentions of each species, and a descriptive example, are given. (AWW, *All's Well That Ends Well*; COR, *Coriolanus*; CYM, *Cymbeline*; ERR, *Comedy of Errors*; HAM, *Hamlet*; LLL, *Love's Labours Lost*; LR, *King Lear*; MAC, *Macbeth*; MM, *Measure for Measure*; MND, *A Midsummer Night's Dream*; ROM, *Romeo and Juliet*; SHR, *Taming of the Shrew*; SON, *Sonnets*; TIT, *Titus Andronicus*; TMP, *Tempest*; TN *Twelfth Night*; VEN, *Venus and Adonis*; WIV *Merry Wives of Windsor*; WT, *A Winter's Tale*; 1H4, *Henry IV, part 1*; 1H6, 2H6, 3H6, *Henry VI, Parts 1, 2 and 3*.)

Barnacle (Goose)	1	TMP	"and all be turned to barnacles or apes"
(Corn) Bunting	1	AWW	"I took this lark for a bunting"
Buzzard	4	SHR	"O slow-winged turtle, shall a buzzard take thee?"
Cormorant	4	LLL	"When, spite of cormorant devouring Time,"
Chough (Jackdaw?)	8	MAC	"By maggot-pies, and choughs, and rooks, brought forth"
Crow	36	2H6	"And made a prey for carrion kites and crows"
Cuckoo	22	MND	"The plainsong cuckoo grey"
Daw (Jackdaw)	8	1H6	"Good faith, I am no wiser than a daw"
Dive-dapper (Dabchick)	1	VEN	"Like a dive-dapper peering through a wave"
Turtle (Dove)	14	WIV	"We'll teach him to tell turtles from jays"
(Domestic) Dove	46	WT	" As soft as dove's down and as white as it"
Duck (Mallard?)	11	1H4	"Worse than a struck fowl or a hurt wild duck"
Eagle	40	CYM	"I chose an eagle, And did avoid a puttock"
Estridge (?Goshawk)	2	1H4	"The dove will peck the estridge; and I see still"
Eyas-musket (Sparrowhawk)	1	WIV	"How now, my eyas-musket, what news with you?"
Finch (Chaffinch?)	2	MND	"The finch, the sparrow, and the lark"

(Grey-lag?) Goose	6	MND	"As wild geese that the creeping fowler eye"
Halcyon (Kingfisher?)	2	LR	" Renege, affirm, and turn their halcyon beaks"
Handsaw (young Heron)	1	HAM	" I know a hawk from a handsaw"
Hedge-sparrow	3	LR	"The hedge-sparrow fed the cuckoo so long"
Jay	5	SHR	"What, is the jay more precious than the lark?"
(Red) Kite	18	2H6	"To guard the chicken from a hungry kite"
Puttock (Kite/Buzzard)	3	2H6	"Who finds the partridge in a puttock's nest"
Lapwing	4	ERR	"Far from her nest, the lapwing cries away"
Lark (Skylark)	29	SON	" Like to the lark at break of day arising"
Loon (Grebe?)	1	MAC	"The devil damn thee black, thow cream-faced loon"
Martlet (House Martin)	2	MV	" the martlet, builds in the weather on the outside wall"
Nightingale (Philomel)	30	ROM	"It was the nightingale and not the lark"
Night-Heron?	2	3H6	"The night-crow cried, aboding luckless time"
Osprey	2	COR	"As is the osprey to the fish, who takes it"
Ousel (Blackbird)	2	MND	"The ousel cock, so black of hue"
(Barn) Owl	{36	2H6	"The time when screech-owls cry, and ban-dogs howl"
(Tawny) Owl	{	LLL	"Then nightly sings the staring owl, To-who; Tu-whit, to who…"
Partridge	2	ADO	"and then there's a partridge wing saved"
Pheasant	2	WT	"None sir, I have no pheasant, cock or hen"
Pie (Magpie)	5	3H6	"And chattering pies in dismal discords sung"
Pigeon	15	AYL	"Which he will put on us, as pigeons feed their young"
Quail	2	TRO	"an honest fellow enough, and one that loves quails"
Raven	31	JN	"As doth a raven on a sick-fall'n beast"
Rook	8	MAC	"Makes wing to th'rooky wood"
Robin (Ruddock)	3	TGV	"to relish a love-song, like a robin-redbreast"
Snipe	1	OTH	"If I would time expend with such a snipe"
Sparrow	12	MM	"Sparrows must not build in his house-eaves"
Staniel (Kestrel)	1	TN	" And with what wing the staniel checks at it"
Starling	1	1H4	"Nay, I'll have a starling shall be taught to speak"
Swallow	7	WT	"That comes before the swallow dares.."
(?Mute)Swan	17	TIT	"Can never turn the swan's black legs to white"
Tercel (male Peregrine)	2	TRO	"the falcon as the tercel, for all the ducks i'th'river"
Throstle (Song Thrush)	3	MND	"The throstle with his note so true"
Wagtail	1	LR	"Spare my grey beard, you wagtail?"
Woodcock	10	HAM	"Ay, springes to catch woodcocks…."
Wren	10	TN	"Look where the youngest wren of nine comes."
Wryneck	1	MV	"And the vile squealing of the wry-necked fife"

Fowl and five Turkeys. Song-birds include 29 references to Larks and 30 to Nightingale or its synonym Philomel, though the familiar cage-birds, Goldfinch, Linnet and Canary, seem surprisingly absent. There is a similar mix of acute observation with use of mythological, fanciful, or bestiary information. His sparrow is partly biblical (the fall of a sparrow), partly popular – a lecherous species but a popular cage bird and pet. The species are not always obvious. Choughs and Daws are surely both Jackdaws, although the notion that Cornish Choughs were implicated in the scenes at Dover in *King Lear* has been widely promulgated. Was the night-crow or night-raven a Night Heron, which is nocturnal and has a raven-like croak, or a Bittern, as Acobas suggests? (Booming is more usually associated with Bitterns, in poetry as in marshland). The 'cream-faced loon' sounds more like a wintering grebe than a diver. His reference to the 'vile squealing of the wry-neck'd fife' in *Merchant of Venice* is one of the first allusions to the Wryneck as a British bird, one moreover that does squeal in spring. Among apparently foreign birds, the Scamels or Sea-mells mentioned in *The Tempest* ('...I'll get thee, Young scamels from the rock. Wilt thou go with me...') are among the most intriguing. The name scamel is applied in Norfolk to Bar-tailed Godwits, according to Greenoak (1979), while the alternative reading of sea-mell sounds like sea-mall, a dialect form of sea-mew or gull (Lockwood, 1993). However, Acobas (1993) produces a convincing argument that this alludes to a contemporary account of harvesting Bermuda Petrels *Pterodroma cahow*.

By 1600, towards the end of Shakespeare's life, the British list had reached 150 species (Fisher, 1966).

In the following century, John Ray (1627–1705) and his companion Francis Willughby (1635–72) undertook a series of field trips between 1658 and 1671, which were journeys of botanical and ornithological discovery, covering most counties of England, Wales, and southern Scotland. They also made an extensive trip through the Netherlands and Germany to Austria, Italy and France in 1663–64 (Gurney, 1921). The results of their travels were used to compile *The Ornithologia*, published in Latin in 1676 and in English 2 years later. Started by Willughby, and completed by Ray after his friend's early death at the age of 37, this was a serious synthesis of what was known about British birds, backed by good field work, collected specimens and field notes, which added 33 new species to the British list (Fisher, 1966). They differentiated between Grey and Yellow Wagtail, Chiffchaff and Wood Warbler, Redpoll and Twite, and Marsh and Coal Tit. Ray's notes were sufficiently accurate to identify the pelicans presented to Charles II by the Russian Ambassador as Dalmatian, not White, Pelicans (Fisher, 1966). Ray accurately described the Solan Geese seen on Bass Rock in 1661 as having four toes webbed together, discriminated the juvenile black, white-speckled, plumage from that of the adults, and remarked that the flesh smelt and tasted of the herrings and mackerel on which they fed. Surprisingly, observing Gannets fishing off Cornwall the next summer, but failing to obtain a specimen, he did not realize that it was the same species, and seems to have confused them (juveniles, presumably) with skuas (Gurney, 1921). At the end of the century, the celebrated visit of Martin Martin to St Kilda in 1697 added Fulmar to the British list, as well as prompting the description of the likely breeding season of the Great Auk.

The bird list produced by Ray seems to have formed the basis for the scientific listing produced by Linnaeus in successive editions of *Systema Naturae*, of which the 10th, dating to 1758, is taken as the starting point for scientific zoological nomenclature. By that time, suggests Fisher, the British list had grown to 214 species. Perhaps bird-watching, too, started around this time, with the remarkable success of the perennial favourite, *The Natural History of Selbourne* (White, 1789). Gilbert White (1720–93) travelled widely through southern England before settling in Selborne, having relatives and friends from Devon to Sussex and north to Rutland, but it was his local observations in Selborne (as now spelt) and surroundings that provided most of the material for his letters to Daines Barrington and Thomas Pennant. His discrimination, by song, of Willow Warbler, Chiffchaff and Wood Warbler, and of Lesser Whitethroat, his attempts to investigate the migration or hibernation of Swallows, his description of the old female Raven sitting tight on her nest in the giant oak while it was felled, and his descriptions of the numbers of birds, as well as their habitats, around Selborne, have fuelled generations of inquisitive bird-watchers. During his lifetime, others sorted out Sedge from Reed Warbler, Tree from House Sparrow, Tree and Rock Pipits from Meadow Pipit, and Great from Lesser Black-backed Gull.

By the end of White's century, another 40 species had been added to the British list, bringing it to 240. Few breeding birds remained to be recognized. George Montagu separated Cirl Bunting from Yellowhammer in 1800, the harrier that was bestowed with his name from Hen Harrier in 1802, and recognized the Roseate Tern. He also added Gull-billed Tern, Little Crake, and Little Gull to the British list. Leach's Petrel was described in 1817 by Veillot, recognized and added to the British list the next year, 1818, by William Bullock who collected a specimen at St Kilda. The following year, Naumann finally resolved the separation of Arctic and Common Terns, Yarrel sorted out Bewick's Swan in 1824, Blyth separated Reed from Marsh Warbler in 1871, and Marsh and Willow Tits were finally discriminated in 1897 (Fisher, 1966). By 1900, the British list stood at 380 species, and the combination of guns and collecting, writing, and communication, had resulted in most counties having their own published avifaunas. Most of the species added in the twentieth century have been strays from distant places, a tribute to the sharp eyes and ears, or mist-nets, of new generations of bird-watchers, but rather insignificant ornithologically. The current list recognizes about 573 species, but of these, about 230 only are breeding species and 43 are regular migrants or winterers. Some 300 are vagrants from far afield, including 105 from North America, unlikely ever to breed (BOU, 1998). Having said that, some expected and a few new and unexpected breeding species have also been added. Many of the expected breeders have been former breeders, returning after lessening of persecution or responding to habitat restoration (Crane, Avocet, Little Egret, Black-tailed Godwit, White-tailed Eagle, Osprey, Goshawk, Savi's Warbler). Others have at least given some indication of their likely colonization by turning up as migrants, or as isolated singing males, well before achieving breeding status (Little Ringed Plover, Bluethroat, Cetti's Warbler, Scarlet Rosefinch, Serin). The arrival of Collared Doves in Norfolk in 1954 seems unremarkable now, but caused considerable interest then. The north-westwards spread of the Collared Dove had been documented in European circles, but had hardly been appreciated or anticipated in Britain. Perhaps even more remarkable have been the attempts of the North American Spotted Sandpiper and Pied-billed Grebe to nest, and the successful recent attempt by Pectoral Sandpipers, emphasizing that in zoology, even the unlikely and unexpected can happen.

Birds lost and gained

Perhaps the most ornithologically significant feature of Elizabeth I's reign was the passage in 1566 of *The Act for the Preservation of Grayne* (8 Elizabeth c. 15). This gave both the legal duty to persecute many species (of mammals as well as birds), and the requirement for church wardens to make payment for their heads. The rates of payment were specified, and probably give some indication of the rarity or otherwise of the species listed. As these payments had to be listed, this Act also precipitated the churchwardens' accounts, documenting early persecution, and may well have led on to the persecution later conducted by gamekeepers. Ecologically, the listings do not make good sense. Although 'Crowes Chawghes Pyes or Rookes' were three heads a penny, and Stares (Starlings) 12 a penny, the predators of these potential agricultural pests were also listed. Every 'Martyn Hawkes Fursekytte Moldkytte Busarde Scharp [Carmerant] or Ryngtayle' was worth two pence, every Iron or

Ospray four pence, every 'Woodwall Pye Jaye Raven or Kyte' one penny, as was 'the bird which is called the Kyngs Fyssher'. The payment of a penny for every 'Bulfynche or other Byrde that devoureth the blowth (flower) of fruite' makes more sense. Oddly, Sparrows, which one might expect to be serious raiders of grain, in store or in the fields, were not mentioned, although rats and mice were certainly listed. Commenting on this strange omission, Lovegrove (2007) suggests that House Sparrows did not become really abundant, and pests, until late in the seventeenth century. The typical legal conventions that omit any punctuation make it difficult in places to decide which species are involved. Choughs (Chawghes) were then Jackdaws (not to be confused with Cornish Choughs), while the Martyn Hawkes were presumably Hobbies (what else would hunt martins?), Hen Harriers were probably both the Ringtail (females or young) and the Furze Kites (males). The identity of Moldkyttes is uncertain (Mole-kites? perhaps Buzzards, known to eat large numbers of Moles), but Woodwalls were woodpeckers. Were Irons (Ernes, perhaps intended for male Eagles, cf. Harting, 1864) really as rare, or common, as Ospreys? And why should Pyes (Magpies) appear twice, and at different rates? Although rural Kites and Ravens were to be persecuted, a later section of the same Act excluded payment for any killed in or within 2 miles of any town or city, suggesting that their role as public hygiene assistants was still appreciated. And persecution of corvids was not to extend to disturbance of nesting Hawks, paupers Swans, Herons, Egrets or Shovelers (= Spoonbills). These were either wanted for falconry or food, the paupers' Swans presumably being other than royal swans.

While it seems likely that this Act was the legal spur to the increasing persecution of birds of prey over the following centuries, and consequent losses of bird species from Britain, several related changes were also essential components – fenland drainage, inclosure, and deforestation, all largely the consequence of the drive to improve agricultural production. The rise of the sporting estate, and therefore of gamekeeping as a profession, followed from these.

In Chapter 3 (Table 3.3) it was estimated that fenland, represented by sedge pollen, might have occupied 17,851 km^2 of Mesolithic Britain; for Chapter 4, a different approach, assessing the maps in Darby & Versey (1975), suggested that about 3,164 km^2 of fenland occupied much of East Anglia in Norman times, and indicated a figure for the whole of England of 8,427 km^2, in other words about half the Mesolithic figure. However, the line of coastal settlements in Domesday Lincolnshire, running along the siltlands from Swine, Mere Haven, and Saltfleet in the north to Skegness in the south, and then a line parallel to the coast of the Wash, from Wrangle south to Spalding and then across east to Long Sutton, implies at least a coastal bank of long-standing. It probably reflects an early, largely unnoticed, drainage by the Romans that Rackham (1986) mentions. They dug a long drainage ditch, the Car Dyke, running for 140 km from Cambridge to Lincolnshire, to capture the small rivers running in from the higher ground to the west, and perhaps also to allow an easier route for inland shipping (Darby, 1983). Much of their drainage work seems to have fallen into disrepair, but renewed assaults on the Fenland were undertaken in the early Mediaeval period, when a substantial new sea wall was constructed most of the way round the Wash, perhaps a rebuilding of an earlier Roman earthwork, in what Rackham calls the Second Draining. As Darby (1983) documents, it involved smaller encroachments on the floodlands, both seawards on to the salt marshes and inland on to the fenland, that increased the wealth of the siltlands by five to 10 times between 1086 and 1334. Then in the Third Draining, the Old (1637) and New

(1651) Bedford Rivers were dug to divert the Great Ouse and Cam towards the Little Ouse and King's Lynn. By draining much of the southern peat fen, this allowed more intensive agriculture to displace the fen life, and ecology, that depended on fishing for eels, harvesting water birds, and summer grazing on the water meadows. With the loss of fens and meadows, the last remnants of the large colonies of Herons, Egrets, and other birds that had been supplied to royal and noble banquets for 400 years were also lost.

The same agricultural pressure led to the enclosure of common lands and the end of the open strip fields of Mediaeval agriculture. Although popularly associated with the eighteenth century, and the protests of John Clare, much of this change had already occurred during the previous centuries. Complaints about inclosure are traced throughout the fifteenth, sixteenth, and seventh centuries, but, as shown by Darby (1976), most of England outside the Midland belt (the Planned Countryside as mapped by Rackham (1986)) was already largely enclosed by 1700. The Parliamentary Enclosures extended this process to that Midland belt. Some 2,837 Acts passed between 1700 and 1903, 82% of them between 1760 and 1820, transformed both the agriculture and appearance of these counties. The land had to be mapped, its ownership established and reallocated, and new roadways had to be imposed; this might have taken 6–7 years in each case (Darby, 1976). The new ownerships had to be 'well and sufficiently hedged and ditched'. From a largely arable landscape, much was transformed into pasture; from open unbroken 'champain' countryside, the new roads, ditches and especially hedges produced instead a more intimate, fragmented landscape. This must have had a major negative impact of birds of truly open landscapes, such as Great Bustard and Stone Curlew. The undoubted positive impact on hedgerow species was poor compensation, and much less notable or noticed. Lovegrove (2007) argues that the more intensive agriculture was what led to the House Sparrow becoming much more numerous, and a serious agricultural pest; he also documents how intensively it was persecuted, and speculates that as many as 100 million were killed between 1700 and 1930.

Deforestation in this period was a less dramatic change, and its extent is contentious. Undoubtedly the industrial pressures from mining (pit props), metal smelting and glass-making (charcoal) and tanning (oak bark) were increasing as the population itself and these industries expanded (Darby 1976). There were complaints that insufficient timber remained to support shipbuilding, and powerful advocacy of increased planting, in remaining Royal forests as on private estates, followed from Parliamentary enquiries, especially that of John Evelyn in 1664. On the other hand, Rackham (1986) points out that coppiced woodlands regrow, that the iron masters of the Weald sustained their woods and their activities for several centuries before turning to coke, and that the Weald is still as well-wooded as anywhere in England. He argues that prices for ship-building timber changed little relative to other costs, though the Admiralty, which complained loudest, was particularly frugal. Agricultural pressure, rather than overuse of timber, seems to have been again the main cause of lost woodland. Rackham points out that Norfolk, never well-wooded but at the centre of agricultural improvement, lost 75% of its medieval woods between 1600 and 1790, suffering worse than any other county. It does though seem that a combination of Evelyn's proselytizing and the emphasis anyway on elaboration of estates and estate forestry led to increasing extents of woodland during the eighteenth century, and that the low point of woodland cover in England dates to about 1700. In Highland Scotland, both the final decline of woodland and the subsequent fashion of replanting woods seem to have been phased perhaps 50 years later, so that minimum tree cover occurred about 1750, and extensive tree planting started in the subsequent half-century (Smout, 2003).

Estate woodlands and gamekeeping go together. In lowland Britain, some of the main game, Pheasants, Wood Pigeons, and Woodcock, are associated with woodland. Partridges and Brown Hares need farmland, but appreciate the shelter of hedgerows as nest sites, as cover, and as alternative feeding areas. In Scotland, Red Deer need the shelter and forage of valley woodland in winter, even if they can survive on open moorland in summer. With land owned and enclosed, protecting its wild animals became a viable prospect. Legally, wild animals in Britain belonged, and belong, to no-one while alive, though to the landowner when dead. This was largely overcome by redefining some species as game. Even now, the Wildlife & Countryside Act does not recognize Pheasants, Partridge, or Red Grouse as birds, protected or pest: they are game. The Game Act of 1671 allowed game to be taken only by landowners with a rental income exceeding £100 annually. Thus the richest 16,000 landowners had the right to kill game, but the other 300,000 did not (Tapper, 1992). They were able to employ gamekeepers to protect their estates and the game within them, and the profession seems to have begun about 1800; it was legalized by the Game Act of 1831 (Allen, 1978). At their maximum in 1911, there were 23,000 gamekeepers, but now, after two World Wars and enormous social changes, only about 2,500 (Tapper, 1992). Intensive shooting, justifying the maintenance of sporting estates, required also technical advances. The replacement of the flintlock by percussion caps in the 1830s, the improved, standardized round lead shot created by pouring molten lead down shot towers, and the arrival from France of the first breech-loading shotguns in 1861 all contributed (Allen 1978; Tapper, 1992).

Harvesting and protecting game shared the benefits of improved technology with collecting specimens for scientific and artistic purposes. Developments in taxidermy were also necessary. Again, the French seem to have been well in advance of the British in this art. In Paris, Compte de Réaumur apparently had a collection of stuffed birds mounted in life-like poses around 1750, but a sensible way of preserving bird skins, using salt, pepper, and alum, was not published in Britain, and then anonymously, until 1763 (Allen, 1978). The use of arsenic as an effective preserver of skins was discovered in France around 1750, apparently, but only reached Britain around 1800 (Morris, 1993). The earliest surviving example in Britain of a successful mount seems to be the Duchess of Richmond's African Grey Parrot, preserved in 1702 (Morris, 1981), and the Hawfinch perhaps preserved by Gilbert White in 1791 has also survived (Morris, 1990). However, these were only partly prepared, eviscerated but not skinned out. The practice of skinning out the specimen, and removing the brains through the back of the skull, only became general after about 1820, following further transference of French knowledge. Taxidermy reached its maximum activity as a profession around 1900, supplying both the common-place, including mounted domestic pets, and the rarities treasured by competitive collectors, as well as representative collections of specimens for public museums (Morris, 1993). Museum study skins, skinned out but made into round skins, not formally mounted, equally depended for their successful preservation on the use of arsenical preservatives, so the earliest surviving scientific collections likewise date to that same period in the early nineteenth century. Along with collecting skins went collecting eggs, both for private and scientific collections. Allen (1978) reports John Martyn of London trying to preserve eggs by boiling them, but then reporting in 1724 that they could be blown out. So the earliest surviving egg collections are somewhat older than skin collections. A few of Willughby's eggs collected in the 1660s apparently survive at Wollaston Hall (Gurney, 1921).

These various eighteenth to nineteenth century changes impacted on different birds to different degrees. It has become fashionable to blame collectors and collecting, even taxidermists, for the loss of many species, and to blame deliberate persecution by gamekeepers for the loss of many others. Undoubtedly, they did indeed play a part, but in many cases only by completing the losses already inflicted by habitat change. It is worth examining some of the better cases in more detail, for there are obvious lessons for modern conservation science. If habitat loss was the main cause, then only reversal of that change will be effective. The loss of the Crane and Capercaillie may well have been exacerbated by hunting, but habitat loss was clearly a major factor in their declines (see Chapter 6). If persecution is, or was, the main cause for a species' decline, protective legislation, duly enforced, is crucial. White-tailed Eagles and Goshawks clearly did not lose their habitat in the nineteenth century, but did suffer heavily from persecution.

Great Bustard

The fate of the Great Bustard is a striking case in point. The last flocks in Norfolk and Suffolk were mercilessly persecuted, and the gamekeeper George Turner of Wretham and Thetford in Norfolk was accused by Stevenson (Stevenson, 1870) of exterminating them. True, Turner did a great deal to exterminate the last herd, setting batteries of large-bore wildfowling guns on baited feeding stations, and managed to kill seven with a single shot on one occasion in 1812. On the other hand, by this date, Great Bustards seem to have been reduced to just the two droves, the 'Thetford drove' around Icklingham in Suffolk and the 'Swaffham drove' in Norfolk. The decline of these two is well documented, and owes much to persecution, by Turner and others, but it seems to mark the terminal phase of a long decline.

It seems that the Great Bustard was, in fact, never either widespread or common in Britain (*contra* Osborne, 2005). There are currently only five archaeological records, of which three are from Late Glacial or (probably early) Mesolithic times, when the landscape was still open (Chapter 3). There is then one Roman record, from Fishbourne Palace (Eastham, 1971), that might indicate importation for food rather than native status (though there are rumours that this might in fact be a misidentified Crane: both belong to the Ralliformes). There appears to be no Anglo-Saxon name for Great Bustard, and none of the illustrated manuscripts or other sources suggests an early presence in Britain (Yapp, 1981b, 1982a). There is then one late Mediaeval archaeological record, from about 1520, at Baynard's Castle, London (Bramwell, 1975a), which appears to substantiate the culinary interest evinced in the species, for instance the nine records, between 1520 and 1548, of bustards being brought to the le Strange household at Hunstanton Hall, Norfolk (Gurney, 1921). Mostly these were singletons, and four at least had been killed with crossbows. The exception concerns two young bustards brought in on the 25 July 1537, and clearly indicates that they bred locally at that time.

The recent and detailed review of available eighteenth to nineteenth century information by Waters & Waters (2005) can be interpreted as confirming that it was never very abundant, despite being very conspicuous when present. The only record of possible breeding in Scotland is the early suggestion by Boece, in about 1526, that it bred in the Merse, the coastal lowlands of East Lothian and Berwickshire. Gurney (1921) quotes Muffet, writing in 1595, as having seen 'half a dozen of them lie in a wheatfield, fatting themselves...with ease....', and this is probably a reference to Salisbury Plain near Wilton, Wiltshire where Muffet lived. Gilbert White referred to them only four times: in Letter VII to Barrington, dated 8 October 1770 'there are bustards on the wide downs near Brighthelmstone' (Brighton); 13 February 1770 'saw bustards on Salisbury Plain (between Charlton and Fyfield): they resemble fallow deer at a distance'; 15 December 1773 'Some bustards are bred in the parish of Findon' (which is on the South Downs 15 km west of Brighton); 16 November 1787 'the carter told us that about 12 years ago he had seen a flock of 18 bustards at one time on that farm (between Fyfield and Winton, near Andover, Hampshire) and once since only two'. The fact that he recorded each so assiduously is an indication of their scarcity across these southern counties at that time. There seem to be no later records from Sussex. Other records from Wiltshire (Thomas, 2000) include the early descriptions by John Aubrey of them as abundant in the seventeenth century, particularly around Stonehenge and Lavington; a flock of about 25 near Winterslow Hut in the 1760s; three at Tilshead in 1788; several near Chittern Bam in 1785–6; two in 1800 near Tilshead; one near Upavon, 1801; singles near Amesbury, supposedly the last in Wiltshire, in 1803 and 1804, and less defined records at about the same time from four other sites. The last in Berkshire was reputedly in 1802, on the Lambourn Downs (Jones, 1966). They were still present during Gilbert White's lifetime in Yorkshire, with two killed and sold at Wansford in 1720, and another single bird killed at Rudstone on the Wolds in October 1792. In 1808, 11 were killed near Sledmere, at Borrow, as were another three of five seen at Flixton Wold in 1811. The only egg known from Yorkshire was taken at North Dalton in 1810. Birds lingered in the northern Wolds, in the Flixton–Hunmanby–Reighton area, and the last Yorkshire record was perhaps in 1835 around Foxholes, a little to the west (Nelson, 1907). They disappeared from the Lincolnshire Wolds at about the same time – as did the small flocks in Norfolk and Suffolk.

The Suffolk drove numbered 40 in 1812, but with persecution by Turner and others, declined to 24, then 15, finally to just two females; a female seen on Icklingham Heath in 1832 and a nest found that summer on the borders of Thetford Warren were perhaps the last records (Stevenson, 1870). In Norfolk, 19 were reported together near Westacre in 1819 and 27 in 1820; these declined to just three females by 1833, when they appear to have moved to Great Massingham Heath, and the five eggs (two c/2, one c/1) laid were taken in part on the assumption that, with no male seen, they were in any case sterile. The last two were probably those shot in 1838, one at Dersingham near Castle Riding and the other at Lexham near Swaffham. One possibly lingered to 1843 or 1845, but these were the last of the British native population (Stevenson, 1870; Stevenson & Southwell, 1890).

One of the major contemporary mysteries surrounding the Great Bustards in early nineteenth century England was their whereabouts in autumn–winter. Did they migrate out of Britain, to return to their breeding grounds in spring? (Waters & Waters (2005) review the discussions.) All the evidence from modern studies is that Great Bustards are not migratory, although they are great wanderers, especially in severe winter weather. The fact that they apparently vanished in autumn, and that their wintering grounds were uncertain, implies to me that they were in fact thinly distributed, and able to hide themselves in unexpected sites or habitats.

Great Bustards do of course still occur occasionally in Britain as migrants or vagrants, especially during severe winters in their homelands, for instance in 1871, more recently after the 1962/63 winter (one dead in Norfolk under powerlines, 28 March 1963) and again in 1981 (one to three birds in Kent in December 1981). Most remarkable was the flock of five to nine individuals seen in Suffolk from 16 January to 7 March 1987. However, as they succumb to agricultural changes across Europe, the chances of further sightings become less (Osborne, 2005). The serious reintroduction programme begun in 2004, using juveniles imported from Russia, deserves to succeed. Whether it does so will depend in part on why it became extinct in the nineteenth century. The records suggest only a thinly scattered population occurring in only a few favoured localities – Salisbury Plain, the Suffolk Brecklands, and the Norfolk, Lincolnshire, and Yorkshire Wolds – far from the abundant or widespread bird that has sometimes been supposed (Figure 7.1). It is likely that recruitment was only ever just sufficient to maintain the population, and any persecution was likely to be catastrophic. With very

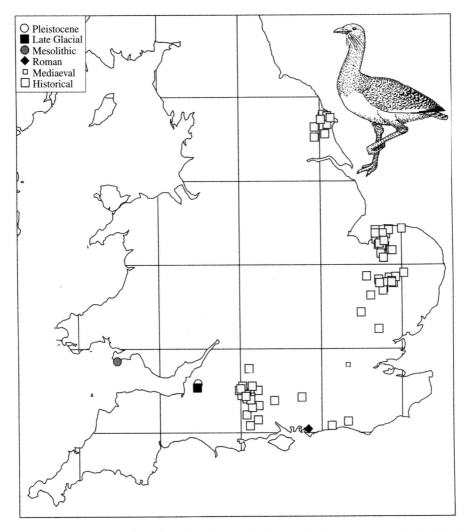

Fig. 7.1 Historical distribution of the Great Bustard in Britain. Archaeological records listed in the text; historical breeding distribution (1750–1850) from county avifaunas.

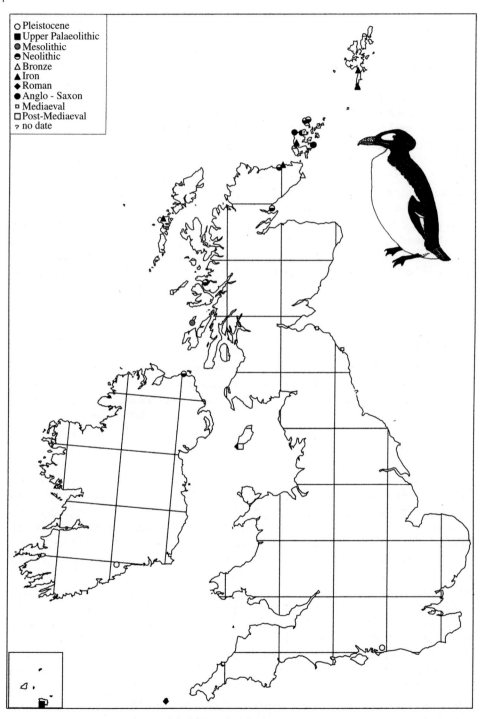

Fig. 7.2 Archaeological distribution of the Great Auk in Britain (data in Table 7.3).

long-lived birds only laying clutches of two, recruitment could never be high. Shrubb (2003) argues that habitat loss, the hedging and tree planting on its preferred open habitat, was critical. This is what is said too in near-contemporary accounts: in Berkshire, ploughing of the downland grass during the Napoleonic wars was regarded as the final cause of their loss there (Jones, 1966), and more determined hand-weeding of the crops was considered harmful there and in Norfolk (Stevenson, 1870). The change from rye, in which they usually nested, to wheat, which was more carefully cultivated, was considered especially harmful in Norfolk; more determined weeding, by hand or horse-hoe, meant that nests were invariably found. The planting of hedges and tree-breaks on the Breckland also altered their habitat. As a number of the early records refer to its eating cereals, especially wheat, the change following enclosure to more pasture may also have been highly detrimental. However, Salisbury Plain is still remarkably treeless, and is of course the target area for the current reintroduction (Osborne, 2005). Modern cereal production may be too intense, with insufficient spilt grain for the adults or insects for the chicks, and the densities of modern roads, indeed the shear human pressures, are far greater. Fortunately, interest in wilderness and wild birds is also far greater. Persecution clearly was a major factor in its final decline. Almost all the later accounts refer to bustards being shot, and to very few escaping that fate. With much more goodwill, and the legislative support that it needs, at least that sad litany should not be repeated.

Great Auk

At least there are still Great Bustards in Iberia and southern Russia to be used in reintroduction projects. The saddest, most complete, irrevocable loss, the one Palaearctic bird to have become extinct in the last 400 years, the Great Auk or Garefowl was historically only a tenuous British breeding bird. Historical records of possible breeding sites in the British Isles seem to be only three (Fuller, 1999). It evidently bred on St Kilda, as recorded by Martin Martin in 1698; it spent, according to what he was told by the islanders, only about 6 weeks on land, arriving at the start of May and leaving by mid-June (mid-May to late June in the modern calendar). It is estimated that incubation lasted perhaps 39 days, and the chicks, like Razorbills and Guillemots, left after a very short period, perhaps only 9 days (Fuller, 1999). Although some 75 fully grown (mostly adult, at least two juvenile) skins are preserved, there are no preserved chicks. It probably continued to breed on St Kilda throughout the next century, as one was captured in 1821, and the last killed in 1840. A pair turned up off Papa Westray, Orkney, in 1816, and one of them was finally killed (it is preserved still in the British Museum (Natural History) at Tring). It is surmised that the neighbouring islet of Holm of Papa Westray might have been a breeding site, Papa Westray itself being unsuitable. Lastly, tenuous evidence suggests that they bred around the south of the Isle of Man, perhaps on the Calf, on the basis of an old drawing, dated about 1652, which reports them to inhabit the area. As Table 7.3 shows, this is supported by archaeological evidence from Man, as is the probable breeding at Papa Westray, for the species has a much longer and extensive subfossil record than most other British birds. Given its remarkable swimming and diving ability, it seems likely that these records do in fact record breeding attempts, for only when it came on land for its short breeding period was it vulnerable to hunting, unless, of course, it was sick or perhaps storm-driven. In time, records spread from the early record from Boxgrove (see Chapter 3) to the latest from the Isle of Man, this last matching reasonably well what the historical records show about the time of extinction. The

Table 7.3 Archaeological records of Great Auk *Pinguinus impennis* from the British Isles (cf. Fig. 7.2).

Site	Grid Ref	Date	Citation
Boxgrove	SU 91 08	Cromerian	Harrison & Stewart (1999)
Jersey – St Brelade's Bay	WV6147	Late Pleistocene	Bell (1922); Tyrberg (1998)
Ballynamintra Cave, Ireland	X 10 95	Late Pleist-Holocene	Bell (1915, 1922); Tyrberg (1998)
Oronsay, Caistell-nan-Gilean	NR 35 88	Mesolithic	Grieve (1882)
Risga, Argyll	NM 61 60	Mesolithic	MacDonald (1921); Lacaille (1954)
Embo, Sutherland	NH 82 92	Neolithic	Clarke (1965); Henshall & Ritchie, (1995)
Oronsay	NR 35 88	Neolithic	Henderson-Bishop, (1913); Bramwell in Mellars (1987)
Knowe of Ramsay, Rousay	HY 40 28	Neolithic	Davidson & Henshall (1989)
Knap of Howar, Papa Westray,	HY 48 51	Neolithic	Bramwell (1983c)
Toft's Ness, Sanday, Orkney	HY 76 47	Neolithic	Serjeantson (2001)
Reay – Cnoc Stanger	NC 95 65	Neolithic	Finlay (1996)
Links of Noltland, Orkney	HY 42 49	Neolithic	Armour-Chelu (1988)
Toft's Ness, Sanday, Orkney	HY 76 47	L. Neolithic	Serjeantson (2001)
Toft's Ness, Sanday, Orkney	HY 76 47	E. Bronze Age	Serjeantson (2001)
Elsay Broch, Scotland	ND 38 52	Bronze Age	Harrison (1980a)
Scalloway	HU 40 39	E. Iron Age	O'Sullivan (1998)
Toft's Ness, Sanday, Orkney	HY 76 47	E. Iron Age	Serjeantson (2001)
Old Scatness Broch, Shetland	HU 390111	Iron Age	Nicholson (2003)
Sollas, North Uist	NF 81 74	Iron Age	Finlay (1991)
Crosskirk Broch	ND 02 70	E. Iron Age	MacCartney (1984)
Cnip, Lewis	NB 09 36	Iron Age	Serjeantson (2001)
Deerness – Skail	HY 58 06	Iron Age	Allison (1997b)
Howe, Orkney	HY 27 10	Iron Age	Bramwell (1994)
The Udal XI-XII, N. Uist	NF 78 82	L. Iron Age	Serjeantson (2001)
Pool 6, Sanday	HY 61 37	L. Iron Age	Serjeantson (2001)
Jarlshof, Shetland	HU 39 09	Iron Age	Platt (1933a, 1956)
Halangy Down, Scilly	SV 91 12	Roman	Locker (1999)
Perwick Bay, IOM	SC 20 67	AD 90	Garrad (1972)
Buckquoy	HY 36 27	Norse	Bramwell (1977b)
Pool 7, Sanday	HY 61 37	Norse	Serjeantson (2001)
Newark Bay, Orkney	ND 56 04	Norse	Serjeantson (2001)
Deerness – Skail	HY 58 06	Viking	Allison (1997b)
Brough of Birsay	HY 23 28	Viking	Allison (1989)
Lindisfarne	NU 13 41	Medieval	O'Sullivan & Young (1995)
Dunbar – Castle Park	NT 66 79	Medieval	Smith (2000)
Pool 8, Sanday	HY 61 37	Medieval	Serjeantson (2001)
Castletown, Isle of Man	SC 26 67	17th C	Fisher (1996)
Rosapenna, Donegal	C 10 38	Holocene	Bell (1922)
Whitepark Bay, Co Antrim	D 02 45	Post-Glacial	D'Arcy (1999)
Rousay, Orkney	HY 40 30	Post-Glacial	Bramwell (1960a)
Cleadon Hill, Whitburn	NZ 39 64		Jackson (1953)

species is particularly well represented from Neolithic and Iron Age Britain, occurring in most sites where seabirds were important, and suggesting that it must have been a regular, possibly even abundant, breeding bird in earlier times. Geographically, most sites are, not surprisingly, from the Scottish isles (Figure 7.2), and at Howe, bones of juveniles were found, confirming that as another early breeding site (Bramwell, 1994). It was found in several layers there, implying a long-lived colony. At Knap of Howar on Papa Westray, at least four adults and a juvenile were recorded, again confirming a breeding site nearby (Bramwell, 1983c). Oronsay too seems to have yielded numerous remains (Grigson & Mellars, 1987; Grigson pers. comm.),

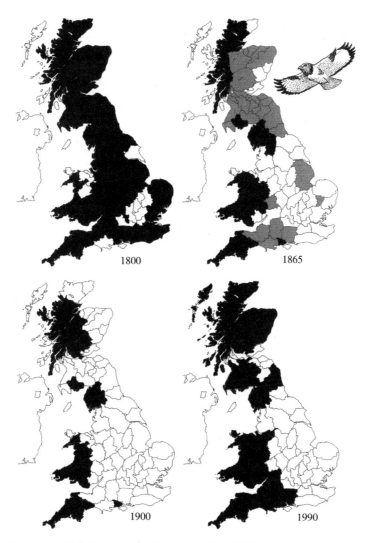

Fig. 7.3 Declining breeding distribution of Buzzard 1800–1900 (based on Moore 1957); grey shading indicates scarce, declining or uncertain by 1865. Recent (1990) distribution based on Gibbons *et al.* (1993).

though most sites have only a few bones. Analysing these records, Serjeantson (2001) points out that Great Auk bones decline from 14% or more of the bird bones in some of the Neolithic and Iron Age sites to no more than 2–3% in Norse times.

Capercaillie

The once widespread distribution of the Capercaillie is quite well documented in the palaeontological and archaeological record (see Chapter 6, Table 6.5), with 27 occurrences, though the possibility of confusion, in either direction, with other gamebirds (Turkey, Peacock, Pheasant) has to be acknowledged. In fact, the Capercaillie is not only much the largest tetraonid, but as such morphologically readily distinguished from the phasianids, given reasonably well preserved bones. Turkey and Peacock are much larger, Pheasant about the same size as female Capercaillie. The archaeological record confirms its occurrence in northern England at least to Mediaeval times, and its wide occurrence in Ireland until then. Willoughby remarks on its presence still in Ireland, and its absence in England, around 1650. It seems as though the deforestation of both countries in the sixteenth to seventeenth century led to their extermination. In Wales, there is neither much of an archaeological nor much of a historical record, though Pennant (1778) implies they were still present, and he thought them still present in County Tipperary in 1760 (Pennant, 1776). In Scotland, it was still well known, though evidently declining. It seems to have been exterminated in Inverness-shire in the 1770s, and the last were shot in Aberdeenshire in 1785 (Holloway, 1996). They had, however, already been given some legal protection as early as 1621, implying that they were already scarce and declining. This decline matches all too well the decline of woodland in this period, and it seems likely that any persecution was a minor, additional, cause of decline. It also parallels the near extinction then of three other woodland species in Scotland: Red Deer, Roe Deer, and Red Squirrel (summarized by Yalden, 1999). The fact that Capercaillie were reintroduced successfully in 1837–38, by which time the new conifer plantations established during the mid eighteenth century were well-grown woodland, restoring their habitat, tends to confirm that this is a case where habitat loss, not persecution, was the main cause of extinction. Red Squirrels also recovered in this period, following major reintroductions in 1772, 1790, and 1844. Likewise, Roe Deer spread into the southern Highlands in 1828, and crossed into the Southern Uplands by 1840–45. These changes all reflect the increasing availability of woodland.

Raptors

Collectively, the birds of prey suffered more than any from persecution, and habitat loss can be essentially discounted as a factor in their nineteenth century decline. Conceivably Marsh Harriers suffered more than other raptors from loss of their special habitat, and perhaps deforestation caused a decline in wasps and hornets, so also of Honey Buzzards. For most species, enclosure and afforestation provided if anything more habitat during the eighteenth to nineteenth centuries, yet most species declined most rapidly during the nineteenth century. County bird books as well as the records of collectors and taxidermists document this. N.W. Moore (1957) mapped the declining range of the Buzzard: still present and widespread in 1800, it had gone from most of lowland England by 1865, was scarce in eastern and lowland Scotland, and remained well established as a breeding bird only in south-west England, Wales and the Marches, north-west England, and north-west Scotland (Figure 7.3). By 1915, it survived only where it had been numerous in 1865. The New Forest, where it was actively protected by the enlightened chief forester (Deputy Surveyor of the New Forest), the Honourable Gerald Lascelles, provided an interesting and instructive outlier (Tubbs, 1974). It is possible to map the similar declines of two other formerly widespread species, the Red Kite and (as an honorary raptor) the Raven, using the last reports of breeding reported in county avifaunas, *Victoria County Histories* and other regional works as summarized in an unpublished project report by Steven Bond (1988). Ravens apparently bred in every county in 1800 (Figure 7.4), and

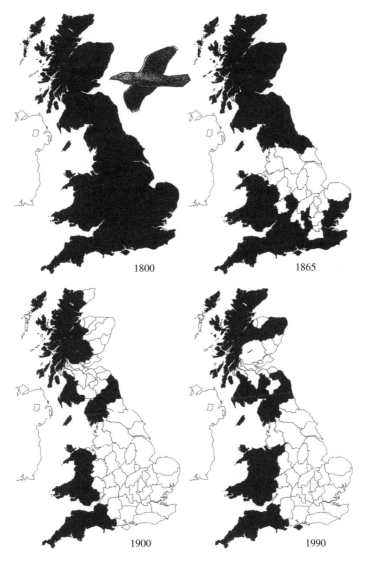

Fig. 7.4 Declining breeding distribution of Raven 1800–1900 (after Bond 1988, Holloway 1996). Recent (1990) distribution based on Gibbons *et al*. (1993).

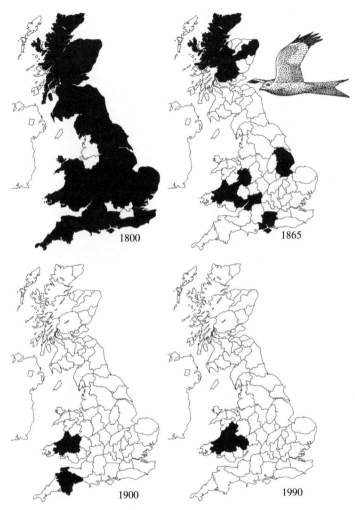

Fig. 7.5 Declining breeding distribution of Red Kite 1800–1900 (after Bond 1988, Holloway 1996). Recent (1990) distribution based on Gibbons *et al.* (1993). Negligible spread led to reintroductions to S England and Scotland which started just after this; the 2007 range is much more extensive.

Red Kites in all but five (London; Middlesex, Surrey; Monmouth, Glamorgan) (Figure 7.5). By 1865, Ravens had gone from nine English and nine Scottish counties, while Red Kites had similarly gone from eight English and nine Scottish counties, but only half the counties involved had lost both Kites and Ravens. This is perhaps a comment on the different vulnerabilities, and different nesting habits, of the two species, for, by 1900, while the Red Kite survived in only five Welsh counties, the Raven still persisted in 48 counties (all those in Wales, most of the Scottish and 15 of the 40 English ones (Figure 7.4). The Goshawk suffered even more severely: like the Osprey and White-tailed Eagle, it was driven to extinction. By 1800, it was already absent from most of southern England, and Wales, a sad decline from the time when it was carefully husbanded for hawking. Even in Scotland, it was patchily distributed.

Table 7.4 Persecution of raptors and other species by gamekeepers. Data from 5 Deeside parishes (Braemar, Crathie, Glenmuick, Tulloch, Glengarden), 1776–86; Langwell and Sandside estates, Sutherland, 1819–26; and the Duchess of Sutherland's estate, 1831–34 from Ritchie (1920). Glen Garry 1837–40 from Pearsall (1950). R = Ravens only. More recent data, for 1912–1969, from specimens submitted for taxidermy in Inverness (McGhie 1999).

	Deeside 1776–86	Sutherland 1819–26	Sutherland 1831–34	Glen Garry 1837–40	Inverness 1912–69
Eagles	70	295	171	42	
Hawks and Kites	2520	1115	1055	1379	
Ravens and Crows	1347	1962	936R	1906	
Cull per year	394	482	721	1109	
Golden Eagle				15	161
White-tailed Eagle				27	–
Osprey				18	2
Goshawk				63	–
Red Kite				275	–
Hen Harrier				63	11
Marsh Harrier				5	1
Montagu's Harrier				–	1
Peregrine				98	131
Hobby				11	1
Gyr Falcon				6	7
?Red-footed Falcon				7	–
Merlin				78	30
Kestrel				462	43
Common Buzzard				385	108
Rough-legged Buzzard				371	8
Honey Buzzard				3	1
Raven				475	?
Hooded Crow				1431	?
Short-eared Owl				71	15
Long-eared Owl				35	49
Tawny Owl				3	81
Barn Owl				–	52
Snowy Owl			–	–	3

By 1900, it possibly just survived as a breeding bird in Gloucestershire, but was gone by 1914 (Bond, 1988).The persecution documented by Ticehurst (1920), Ritchie (1920), and Pearsall (1950) has been well quoted. Ticehurst documented the early start of this persecution in Kent, concentrated in the period 1676–90, when 432 Red Kites were killed in Tenterden parish in just 14 years. For Scotland, the later start and increased rate of killing through the nineteenth century (Table 7.4) supports the judgement given above that this was most serious in the nineteenth century. By 1837–40, not only had the rate of killing increased, but the largest (and most vulnerable) species, the eagles, were already thinned. The more detailed breakdown of species killed in Glen Garry (Pearsall, 1950) is also remarkable. Not only were the smaller, more common (and less relevant to gamekeeping) species such as Kestrel and owls being severely culled by that time, but such rarities as Gyr Falcon, Hobby, and Honey Buzzard were also being shot.

The 'Orange-footed Falcons', suspected to be Red-footed Falcons, are also remarkable, as are the numbers of Red Kite and Rough-legged Buzzard being killed. A similar detailed breakdown of birds of prey submitted in the twentieth century to the main taxidermists in Inverness, Macpherson's, is given by McGhie (1999). The comparison is instructive. Evidently, the most impressive species, Golden Eagle and Peregrine, were most often submitted, out of all proportion to their relative abundance, but the wide range of species bears strong comparison with the nineteenth century toll. White-tailed Eagles and Red Kites were no longer available, and the number of Rough-legged Buzzards is far lower. Were they really more numerous in the 1830s, or were Common Buzzards being misidentified then? Either way, not only were the resident species being harried to extinction, but the level of persecution was high enough to ensure that even rare winter visitors were killed.

Tapper (1992) has shown that gamekeeper density was higher than 0.8 keepers per 1000 ha over most of lowland Britain in 1911, when the peak of 23,056 gamekeepers was enumerated in the national census. Only four counties (Caithness, Sutherland, Cardiganshire, Carmarthenshire) had a lower density than 0.4 keepers/1000 ha; note that the last two of these were among the counties where Red Kite (and Polecat) just survived total extinction in Britain. By the 1951 census, there were only 4,391 keepers, and only four counties (Hampshire, Berkshire, Bedford, Norfolk) had densities higher than 0.4/1000 ha. It is considered that a well-keepered estate needs 2.5 keepers/1000 ha, so in 1911 the whole of Norfolk was well keepered. Another way of regarding the same figures is to assume that the 23 thousand keepers in 1911 looked after 400 ha each; 92,000 km^2 of Britain would have been well keepered, or about 40% of the whole land area. That gamekeepers had indeed depressed predator numbers so severely was also evident from the fact that, with reduced numbers of keepers both during and after the two World Wars, predator numbers slowly but steadily recovered. The more thorough recent enumeration of the persecution of raptors over the last four centuries by Lovegrove (2007) serves only to strengthen this brief summary.

Conclusions

The period from about 1600 onwards marks a rapid transition from a patchy historical record of our avifauna documented by archaeological remains to one reliant on documentary evidence. As a consequence, it becomes increasingly more detailed, but also increasingly well known and well discussed by more competent ornithologists. It is also the period when the increasing Human population of Great Britain had an increasingly direct impact on birds, for better or worse. We were only 5 million in 1600, 6.75 million in 1700, and 10.25 million in 1800, but we numbered 37.5 million by 1900 and 55 million by 2000 (McEvedy & Jones, 1978). The number of known bird species documented as occurring in Great Britain increased from about 150 in 1600 to 380 in 1900, and now stands at over 570, though the number of breeding species is under half of this number. Losses and gains in the last 150 years or so have played a large part in the recent history of our avifauna. The interaction between the increasingly large number of people and the changing bird fauna is a major topic for Chapter 8.

8
Now and hereafter

Birds in the twentieth century

The history of birds in the twentieth century has been thoroughly documented by better authors than us; each species has its own history, just as it has its own ecology, and there is no point in attempting a detailed review of such a well-travelled road. On the other hand, it does seem worthwhile to comment on some broader trends, (1) because they have resulted in our present fauna, and our responses to it, and (2) because our perceptions of what has happened in the recent past colour our perceptions of what should happen in the future, in so far as we can control it. Food supplies are the major constraint on the size of any animal population, our own included. For birds of many species, the changes in farming to ensure our own food supply, well reviewed by Shrubb (2003), have been the main determinants of their food supplies too. The loss of horses as working animals meant the loss of oats as a crop, and also allowed the change from a mixed rotational farming system to one in which the country is polarized into an arable east and pastoral west (Tapper, 1992; Shrubb, 2003). The pressures (particularly subsidies) to increase numbers of sheep in the uplands, wheat in the eastern lowlands and cattle, therefore silage, in the western lowlands have seen larger fields, fewer hedges, fewer weeds, fewer insects, and the loss of fallow land in winter. It is not surprising that seed-eating farmland birds – buntings, finches, sparrows in particular – have declined, nor that waders and Yellow Wagtails, nesting in wet meadows that have been increasingly drained and turned to cereal or silage, have also declined. For upland birds, the impact of overgrazing by sheep is less well documented, but the loss of heather moorland and the consequent decline of Red Grouse numbers has been substantial (Hudson, 1992; Fuller & Gough, 1999). It is still not clear whether declines of such species as Ring Ouzel can be related, nor whether upland Meadow Pipits and their Cuckoo parasites have declined as a consequence. Moorland birds have faced a second pressure during much of the twentieth century, from the increase the amount of woodland in Britain, in a vain attempt to make us self-sufficient in timber and paper pulp. The consequence has been the planting of large areas under alien conifers, particularly Sitka Spruce, which now accounts for 26% of the total woodland area (http://www.forestry.gov.uk). This has had enormous benefits for some forest birds, notably Siskin, Crossbill, and Coal Tit. Furthermore, the young stages of plantations, with sheep and deer excluded, have been excellent but short-lived habitat for a diversity of species such as Short-eared and Long-eared Owl, Hen Harrier, Whinchat, and Black Grouse. Much of this was fuelled by an anomalous tax system that encouraged the planting of trees on totally unsuitable land, because the costs of establishing the woodlands were tax deductible, and tax would not be paid until the trees were harvested. Many never were. The end of the tax system in 1988 by the then Chancellor Nigel Lawson quickly ended this misuse of land, but not before substantial damage had been done to the Flow Country in northern Scotland (Marren, 2002). Stroud *et al.* (1987) estimated that 19% of the Golden Plovers and 17% of the Dunlins and Greenshanks nesting in the area, totalling 1,833 pairs of waders, had been lost to afforestation.

Increasing recreational use of the countryside has also been a problematic change for birds. On the one hand, it has encouraged many more people to take an interest in birds, and other wildlife, and easier access has made it possible to carry out better surveys. On the other hand, ground-nesting birds in particular are very susceptible to disturbance, to trampling, and, especially, to the dogs that accompany about 1 in 22 people. It seems likely that the Kentish Plover was an early victim of these pressures. It only ever had a rather marginal distribution in south-east England, and on just the sandy beaches that were most popular with early post-war holiday makers. The Ringed Plover is another beach-nesting species whose range appears already to be circumscribed in the south by the levels of recreational use (Pienkowski, 1984). Some beach-nesting species nest in dense colonies, and can be wardened to protect them at the critical times. Even so, Little Terns have had a fragile record in recent years, and some colonies have been lost. Others, like the plovers, that nest in a dispersed manner, relying on camouflage to protect nests and young, are impossible to protect this way. In the uplands, increasing numbers of hikers threaten some populations of nesting waders and others. Along the well-walked Pennine Way, a long-distance path that receives up to 500 visitors a day on peak summer weekends, Golden Plovers retreated about 150 m away from either side of the path, and about 25% of their nesting habitat was lost. Fortunately, surfacing the path with stone slabs has reduced the level of disturbance, and the birds have reclaimed much of the lost habitat (Finney *et al.*, 2005). The Nightjar is another ground-nesting species that seems vulnerable. Its favoured habitat, the heathlands of southern England, are similarly popular with dog walkers and others, and the effects of too many people are signalled by the heathlands that Nightjars avoid (Liley & Clarke, 2003). Dartford Warblers, which share that habitat, breed well enough in gorse-dominated territories, but breed up to 30 days later in much-disturbed heather-dominated territories (Murison *et al.*, 2007).

Changing attitudes

The persecution that attended both predators in general and rarities in particular during the nineteenth century was rapidly replaced by a more tolerant attitude to wildlife in general, and birds in particular. It is hard to know the forces driving this change. Undoubtedly the social upheaval caused by the 1914–18 war played a significant and critical part, if only by removing so many gamekeepers from immediate antagonism; the peak count of 23,036 in the 1911 census was reduced to about 1,400 in 1921 (Tapper, 1992). However, the beginnings of a reaction to the killing of so much spectacular wildlife can be dated much earlier. There were, for instance, forlorn efforts to preserve the last Great Bustards in the 1840s, and when a male appeared in 1876 at Hockwold, attempts were made both to protect it and release a mate for it (Stevenson & Southwold, 1890). Similarly, attempts were made by the Grants, the estate landowners, to protect the Loch an Eilein Ospreys from raiding egg thieves as early as the 1880s (Lambert, 2001). The beginnings of what is now the RSPB date to the 1890s and attempts by three remarkable ladies, notably Winifred Dallas-Yorke, later Duchess of Portland, to protect plume-bearing birds from the voracious trade in elegant feathers. It is hard to believe that the Great Crested Grebe, now numbering some 6,000 pairs and common on most lowland waters in the British Isles, was reduced to about 32 pairs, mostly on a few strictly protected Cheshire meres, by the 1860s. It was shot to provide 'grebe furs', the dense breast feathers, for muffs carried by fashionable ladies. Breeding egrets were shot in their hundreds and imported to provide the elegant head plumes for ladies' hats. Formed as the Society for the Protection of Birds in 1891 to campaign against such excesses, and becoming the Royal Society in 1904, the RSPB, the largest bird charity in Europe, now boasts a membership of over 1 million, has a staff of over 1,000 and owns a network of 140 bird reserves totalling 111,500 ha (Marren, 2002). Even larger is the National Trust, with a membership of over 3 million. Formed in 1893, particularly to look after the heritage of houses and gardens, it early became the owner too of important wildlife sites such as Wicken Fen, Blakeney Point, and Box Hill. It has very special legal status, under which it has benefited from special taxation arrangements that allow it to accept, on behalf of the nation, estates and land in lieu of their owners paying death duties to the government. It has thus become the owner of large areas of coastal, heath, and upland habitat, which, while not all overtly bird reserves, are in fact important for many species. Its byelaws proscribe damaging any plant or animal life, birds included, and its wider estate is increasingly in fact managed as a series of large nature reserves.

The conservation of birds, and other wildlife, became a formal governmental responsibility after the 1939–45 war with the creation of the Nature Conservancy in 1949. Charged with both owning as well as managing national nature reserves and with advising government more generally on wildlife conservation, the original Nature Conservancy also carried out the ecological research necessary to understand how to manage the reserves for wildlife. Subsequent politically motivated changes to its organization saw the research arm split off in 1973 into the Institute of Terrestrial Ecology ITE, (now part of the Centre for Ecology and Hydrology, CEH), leaving it as the Nature Conservancy Council (NCC), then split it into country agencies in 1989 (English Nature, Scottish Natural Heritage, Countryside Commission for Wales). It is not at all clear that bird conservation has benefited from these changes. More important was the passage of the 1954 Protection of Birds Act, subsequently included in the 1981 Wildlife and Countryside Act, which finally gave virtually all birds, their eggs and nests legal protection. Special protection (higher fines) were given to most of the rarest species (those with fewer than 100 pairs, plus a few very conspicuously persecuted species, including Barn Owl, Kingfisher, and Peregrine), while a few pest species were excluded and game species were allowed protection during their breeding seasons but allowed to be shot in specified hunting seasons. This is remembered by those old enough as the formal end of the boyhood hobby of bird nesting, but the protection given to rarer birds by their inclusion is undoubtedly one of the successes of the Act, even though prosecutions of egg collectors remain one of the major signs of infringement. Illegal persecution of raptors in game-rearing areas remains the other. The government agencies have been poor at enforcing these laws, but the RSPB has been more diligent, thanks to its own crime investigation team.

RSPB members care much for birds, though some may know little about them. The 1000+ professional staff of the RSPB, however, include the best group of research ornithologists in the country, perhaps in the world. They do though have friendly rivals in the British Trust for

Ornithology (BTO), a much smaller charity, with a membership of about 11,000 and a staff of 400, but whose members and staff are all good field ornithologists. These are the volunteers who go out routinely to count birds at all times of the year, check nest-boxes and fill in nest record cards, ring birds to see where they go and how long they live, and take part in various atlas schemes. Increasingly, the BTO and RSPB cooperate to conduct surveys and analyse their results. The periodic counts of nesting seabirds, for instance, see ornithologists from all possible bodies, and none, collaborating to achieve maximum coverage. It is one of the remarkable successes of the British conservation scene that so many volunteers are willing and able to contribute to the increasingly complex surveys conceived by the professional ornithologists. As a consequence, we have two national atlases of breeding birds (Sharrock, 1976; Gibbons et al., 1993), one of wintering birds (Lack, 1986) and a migration atlas summarizing the ringing recoveries (Wernham et al., 2002). Survey work for a combined breeding and wintering atlas, intended to last from 2007 to 2011, is getting under way as we write. We also have, for instance, a continuous run of annual counts of Heron nests that started in 1927, five counts of breeding Peregrines at decadal intervals, and annual indices of the numbers of commoner breeding birds that go back to 1962. A third, more specialized, society contributes to regular counts of wintering coastal and wetland birds, especially ducks, geese, and waders. Formed by the visionary efforts of Sir Peter Scott, what started as the Severn Wildlife Trust, and is now the Wildfowl and Wetlands Trust, owns 12 major wetland bird reserves (including not only the original one at Slimbridge, but also Martin Mere near Southport, Arundel, Welney near Peterborough, and most recently the London Wetland Centre near Hammersmith Bridge). Ducks, geese, and swans are its main concern, but wetland conservation in general, and therefore counts of birds, waders as well as wildfowl, sees it collaborating with BTO and RSPB members and staff to achieve a complete coverage of all estuaries, lakes, and other wetlands at least once a month every winter in the Wetland Bird Survey (WeBS).

It requires a certain dedication to go out to count birds routinely, once a winter month, on the nearby lake or estuary, for these winter wetland bird counts. It can require about 3 years' training to become a self-confident and fully licensed solo bird-ringer. It requires no less dedication, of a different sort, to find nests and record their contents every 3 or 4 days to follow through their success (or failure). Even more dedication is required when the winter survey results for a particular square or lake are likely to be nil, but that has to be proved, not assumed. Yet the surveyors for the Breeding Bird Survey (BBS), monitoring common breeding birds, cover 2,250 one-kilometre squares, twice each breeding season. About 2,000 ringers ring some 850,000 birds each year, and 11,000 recoveries were reported in 2003 (Clark et al., 2005). In 2004, over 31,000 Nest Record Cards, documenting the nesting attempts of 170 species, were submitted to the BTO by about 750 active nest finders (Nest Record News 21, June 2005). Some 3,400 wetland counting areas are surveyed each winter month, usually on a predetermined date, by at least that number of WeBS counters. The 1993 breeding atlas involved 92,346 hours of timed visits that yielded over 320,000 records, and a further 230,000 records from untimed visits (Gibbons et al., 1993). The sheer scale of the volunteer inputs into these projects is remarkable. Nor should one overlook the efforts of the contributors to the Garden Bird Survey (Garden Bird Watch, GBW). Some 12,600 participants report the species present in their gardens each week, and moreover pay for the privilege of contributing. Most ornithologists prefer to count birds in the countryside, and urban birds have consequently been neglected, despite the best efforts of a few dedicated urban ornithologists who have, for instance, mapped the birds of Inner London and the Royal Parks. In compiling Table 8.6 (see later), the most uncertain estimate concerns the population of Feral Pigeons in the British Isles, yet these are regarded as an important pest species, and also important prey for the urban Peregrine population. An interesting antidote is provided by a questionnaire survey of these GBW contributors, asking them how many nests of which species they recorded on their properties in 2000. It suggests that some suburban and urban bird populations have been severely underestimated: the Swift population, for instance, might be as many as 395 thousand pairs, rather than the 80 thousand estimated in Gibbons *et al.* (1993) (Bland *et al.*, 2004).

This cooperative ornithology between an amateur network of field recorders and the professionals who collaborate with them, conceiving and designing surveys, interpreting the data collected and writing the papers and books based upon them, is the envy of the scientific world. It would cost an enormous amount to provide such services through an entirely professional organization, and it would be very difficult even then to provide the breadth of coverage provided by the enormous amateur network. Greenwood & Carter (2003) estimated that the amateur input to monitoring was about 1.5 million man-hours a year, compared with about 13,000 professional man-days (equivalent to 104,000 man-hours of notional 8-hour days – but it is unlikely that professional bird researchers have the luxury of notional days, either). Implicitly, amateurs contribute about 14 times as much field-work time as the professionals. Another estimate, made by Eaton *et al.* (2006), is that four main monitoring surveys (Breeding Birds, Waterways Birds, Wetland Birds, Swans and Geese) involved 6,020 volunteer surveyors contributing 74,160 man-hours; ringers, nest-recorders, garden bird surveys, and specialist surveys would be additional to these. At a notional £25 per hour this represents a £1.8 million contribution to the research budget. There is one cause for concern, one that afflicts many amateur societies. This is the ageing force on which it is based. Many younger (and some older) birdwatchers (are they ornithologists?) seem more concerned with chasing rarities than the routine monitoring work that all these surveys need. Their skills in bird identification are undoubted, but there is a risk that what they are indulging is not much more than stamp-collecting, with the added problem that their zeal to see a rarity can lead to its harrying, even its death (the tragic case of a Sora Rail that was reputedly trampled to death by the twitchers trying to flush it cannot be the only such event). Greatly improved transport now makes it possible for a twitcher to be on the Scilly Isles one weekend and on Fair Isle the next, if the greatly improved telephone and internet communications advise that the requisite rarity is there to see. The hope, expectation, and some experience, is that twitchers, or many of them, will in fact develop beyond chasing rarities to apply their skills to the surveys, and will become more than adequate replacements for the present older generations.

So now much effort goes into bird conservation, and into the routine monitoring needed to detect its successes, and failures. One undoubted success in the late twentieth century was the recovery of raptors first from the persecution that they suffered in Victorian times, and then from the serious problems caused by organochlorine pesticides from 1957 to the mid-1960s. It is hard now to recall just how scarce the Sparrowhawk had become, entirely absent from much of south-eastern Britain, by 1965. And the Peregrine, having recovered by 1955 to about 550 pairs, after wartime persecution because of its affects on carrier pigeons, then plummeted to about 350 pairs, most of which were failing to breed. Fortunately the efforts of Derek Ratcliffe, as Chief Scientist for the Nature Conservancy, aided by others such as

Norman Moore at ITE Monks Wood, first documented the decline and the breeding failures, then demonstrated the effects of eggshell thinning by the organochlorine pesticides involved, and finally managed to get them withdrawn from use, initially voluntarily and eventually by law. When he wrote the first edition of his monograph on the Peregrine (Ratcliffe, 1980), he thought the British Isles might hold enough space for about 750 pairs, at maximum. It is then truly remarkable that the most recent (2002) census has documented about 1,700 pairs.

The balance of the bird fauna now

Accepting the species mapped by Gibbons *et al.* (1993), combined with the analysis by Brown & Grice (2005), there are 208 bird species which regularly breed in the British Isles, and another 12 species bred irregularly and/or in very small numbers during 1988–91 (13 if the definition of British Isles is stretched to include the Channel Isles, so also Short-toed Treecreeper). There are another 32 that winter regularly but do not breed. At least one more species, the Little Egret, has established itself as a regular breeder since 1996, when it first bred in England; it had reached about 150 pairs by 2002. In 1997, it also started to breed in Ireland. Several other species have also added themselves to the list of at least occasional breeding birds (e.g. Muscovy Duck, Spoonbill, Eagle Owl, Hoopoe, Bee-eater, Bluethroat) (Brown & Grice, table 2.9). The abundance of some species has genuinely changed since the estimates in Gibbons *et al.* (1993), and for others, better population estimates are now available from better censuses or methodology. We have resisted the temptation to update the figures in favour of staying with a comprehensive and widely used set of data, one which, moreover, covers the British Isles (updates such as that by Baker *et al.* (2006) cover GB or the UK, but not the whole archipelago). It is illuminating to examine the overall abundance of British birds, and their abundance relative to each other, using three different criteria – numerical abundance, ubiquity, and biomass.

The 1993 figures suggest that about 167 million individuals contributed a spring biomass of about 22,988 metric tonnes of wild birds of the 220 species that breed (or bred during 1988–91) in the British Isles (a complete listing is in Table 8.6). Numerically, the top 30 species include 26 passerines and only four non-passerines, two of them seabirds (Guillemot, Fulmar) in 28th and 29th places (Table 8.1). The Wren is much the most numerous, at nearly 20 million birds, followed by Chaffinch, Blackbird, and Robin which each contribute over 10 million. The Wood Pigeon, in seventh place and Pheasant, in 13th, are the other two non-passerines. Ecologically, the list is interesting, too. Of the 30 species, 20 are woodland or woodland-edge species, needing trees at least for nesting if not as feeding habitat as well. Two are birds of open country (Meadow Pipit, Skylark), four are hedgerow or scrub species and two are farmland birds (House Sparrow, Swallow), while there are only the two seabirds. Does this reflect our present countryside (not that well wooded!) or the historical background from which our fauna is derived? Clearly, with woodland covering only about 10% of the countryside, it more probably represents the woodland heritage of 8,000 years ago (see Chapter 3) than the present availability of habitats.

The table of ubiquity too (Table 8.2) is dominated by passerines, most of them the same ones that are the most numerous species. There are, however, some interesting changes of rank. Several very widespread species are thinly spread across these islands. Thus the Pied Wagtail, in fourth place, Mallard in 15th, Kestrel in 19th, and Coal Tit in 23rd, are all much

Table 8.1 Top 30 wild breeding birds in the British Isles listed in decreasing order of abundance (after Table 9 of Gibbons *et al.* 1993; means of their ranges are used where necessary, numbers of territories multiplied by 2, numbers of breeding males or females doubled, assuming (dubiously) an even sex ratio. Individual masses come from various sources, mostly Hickling (1983) and HBWP: means of male and female masses are used for dimorphic species). The consequent biomass column does not match at all well, though the species appearing in the ubiquity column are a better match. Note that passerines dominate this table.

Species	N. individuals	Mass (kg)	Biomass (kg)	Hectads	Ubiquity
Wren	19,800,000	0.010	196,020	3748	0.9715
Chaffinch	15,000,000	0.020	300,000	3564	0.9238
Blackbird	12,400,000	0.095	1,178,000	3654	0.9471
Robin	12,200,000	0.019	235,460	3610	0.9357
House Sparrow	9,400,000	0.027	253,800	3440	0.8917
Blue Tit	8,800,000	0.012	101,200	3424	0.8875
Wood Pigeon	6,420,000	0.524	3,364,080	3469	0.8992
Willow Warbler	6,260,000	0.009	53,836	3539	0.9173
Hedge Sparrow	5,620,000	0.021	119,706	3472	0.8999
Meadow Pipit	5,600,000	0.020	112,000	3497	0.9064
Skylark	5,140,000	0.038	195,320	3669	0.9510
Great Tit	4,040,000	0.019	76,760	3340	0.8657
Pheasant	3,100,000	1.131	4,061,000	3123	0.8095
Starling	2,920,000	0.082	239,440	3591	0.9308
Yellowhammer	2,800,000	0.027	74,200	2814	0.7294
Song Thrush	2,760,000	0.060	165,600	3581	0.9282
Rook	2,750,000	0.488	1,342,000	3156	0.8180
Chiffchaff	1,860,000	0.008	14,880	2949	0.7644
Magpie	1,820,000	0.237	431,340	2932	0.7600
Goldcrest	1,720,000	0.006	9,804	3189	0.8266
Swallow	1,640,000	0.019	31,160	3622	0.9388
Carrion Crow	1,620,000	0.570	923,400	2653	0.6877
Whitethroat	1,560,000	0.016	25,428	2824	0.7320
Greenfinch	1,380,000	0.028	38,364	3150	0.8165
Linnet	1,300,000	0.015	19,890	3065	0.7945
Blackcap	1,240,000	0.018	21,700	2417	0.6265
Jackdaw	1,200,000	0.246	295,200	3290	0.8528
Guillemot	1,200,000	1.002	1,202,400	274	0.0710
Fulmar	1,142,000	0.808	922,736	716	0.1856
Reed Bunting	1,100,000	0.018	20,130	3023	0.7836

more widespread than their abundance suggests. Most extreme from this viewpoint is the Grey Heron, nationally not a numerous species, but very widely distributed. Conversely, the Guillemot and Fulmar, which rank well in the abundance scale, are far from ubiquitous, and absent from this table.

The ranking by biomass looks very different (Table 8.3). Over half (17 of 30) of the top species are non-passerines; moreover, many of them are large seabirds with very limited ranges, and they do not appear in either the abundance or ubiquity tables. The Guillemot, in fourth place and the Fulmar, in eighth, which do appear in the table of abundant species, are joined here by Gannet (sixth), Kittiwake, Herring Gull, Puffin, Manx Shearwater, and Shag. Most importantly, the top species in Table 8.3 is the introduced Pheasant, and the

Table 8.2 Top 30 wild breeding birds in the British Isles listed in decreasing order of ubiquity (after Table 9 of Gibbons *et al.* 1993; means of their population estimates are used where appropriate. Individual masses come from various sources, mostly Hickling (1983) and HBWP). Again, 25 of the 30 are passerines.

Species	N. individuals	Mass (kg)	Biomass (kg)	Hectads	Ubiquity
Wren	19,800,000	0.010	196,020	3748	0.9715
Skylark	5,140,000	0.038	195,320	3669	0.9510
Blackbird	12,400,000	0.095	1,178,000	3654	0.9471
Pied Wagtail	860,000	0.022	18,920	3640	0.9435
Swallow	9,400,000	0.019	178,600	3622	0.9388
Robin	12,200,000	0.019	235,460	3610	0.9357
Starling	2,920,000	0.082	239,440	3591	0.9308
Song Thrush	2,760,000	0.060	165,600	3581	0.9282
Chaffinch	15,000,000	0.020	300,000	3564	0.9238
Willow Warbler	6,260,000	0.009	53,836	3539	0.9173
Meadow Pipit	5,600,000	0.020	112,000	3497	0.9064
Hedge Sparrow	5,620,000	0.021	119,706	3472	0.8999
Wood Pigeon	6,420,000	0.524	3,364,080	3469	0.8992
House Sparrow	9,400,000	0.027	253,800	3440	0.8917
Mallard	246,000	1.785	439,110	3437	0.8909
Blue Tit	8,800,000	0.012	101,200	3424	0.8875
Great Tit	4,040,000	0.019	76,760	3340	0.8657
Kestrel	120,000	0.202	24,240	3298	0.8548
Jackdaw	1,200,000	0.246	295,200	3290	0.8528
Mistle Thrush	640,000	0.130	83,200	3264	0.8460
House Martin	960,000	0.018	17,280	3217	0.8339
Goldcrest	1,720,000	0.006	9,804	3189	0.8266
Coal Tit	176,000	0.009	1,602	3170	0.8217
Rook	2,750,000	0.488	1,342,000	3156	0.8180
Greenfinch	1,380,000	0.028	38,364	3150	0.8165
Cuckoo	347,000	0.114	39,558	3136	0.8129
Grey Heron	27,900	1.361	37,972	3129	0.8110
Pheasant	3,100,000	1.131	4,061,000	3123	0.8095
Spotted Flycatcher	310,000	0.015	4,650	3117	0.8079
Linnet	1,300,000	0.015	19,890	3065	0.7945

equally alien Canada Goose appears at number 23. The placement of the Mute Swan and Feral Pigeon in this table might reflect their semi-domestic status in times past and present. However, none of these comes close to rivalling the contribution made by our most abundant bird. If we assign the modest individual mass of 2 kg to each Domestic Fowl (probably near enough for egg-laying breeds such as Warrens and Black Rocks, which make up some of the domestic population, but surely a very modest mass for the broilers and roasting fowl, which are about 75% of the June flock – see Chapter 5), the 155 million birds must contribute a biomass of at least 310,000 metric tonnes, outweighing by 13 times all the wild birds put together. This disproportion is not quite so great as for mammals, where the domestic ungulates contribute a biomass some 21.5 times greater than all the wild mammals, alien and native (Yalden, 2003). Neither is the contribution of alien birds so disproportionate. Alien wild mammals exceed the native species in biomass, but the strictly alien birds are only 17%

Table 8.3 Top 30 wild breeding birds in the British Isles listed in decreasing order of biomass (kg) (after Table 9 of Gibbons *et al.* 1993; means of their abundance values used where appropriate. Individual masses come from various sources, mostly Hickling (1983) and HBWP). From this perspective, non-passerines are much more important, reflecting their greater impact on the ecology of the British Isles.

Species	N. individuals	Mass	Biomass	Hectads	Ubiquity
Pheasant	3,100,000	1.131	4,061,000	3123	0.8095
Wood Pigeon	6,420,000	0.524	3,364,080	3469	0.8991
Rook	2,750,000	0.488	1,342,000	3156	0.8180
Guillemot	1,200,000	1.002	1,202,400	274	0.0710
Blackbird	12,400,000	0.095	1,178,000	3654	0.9471
Gannet	373,000	3.010	1,122,730	24	0.0062
Carrion Crow	1,620,000	0.570	923,400	2653	0.6877
Fulmar	1,142,000	0.808	922,736	716	0.1856
Hooded Crow	900,000	0.570	513,000	1657	0.4294
Mute Swan	45,900	10.750	493,425	2141	0.5550
Mallard	246,000	1.785	439,110	3437	0.8909
Magpie	1,820,000	0.237	431,340	2932	0.7560
Kittiwake	1,087,200	0.387	420,746	315	0.0816
Herring Gull	411,400	0.951	391,241	904	0.2343
Puffin	941,000	0.395	371,695	181	0.0469
Red Grouse	560,000	0.651	364,560	1086	0.2815
Chaffinch	15,000,000	0.020	300,000	3564	0.9238
Jackdaw	1,200,000	0.246	295,200	3290	0.8528
House Sparrow	9,400,000	0.027	253,800	3440	0.8917
Manx Shearwater	550,000	0.453	249,150	38	0.0099
Starling	2,920,000	0.082	239,440	3591	0.9308
Robin	12,200,000	0.019	235,460	3610	0.9357
Canada Goose	60,140	3.780	227,329	1215	0.3149
Stock Dove	540,000	0.400	216,000	2190	0.5677
Wren	19,800,000	0.010	196,020	3748	0.9715
Skylark	5,140,000	0.038	195,320	3669	0.9510
Moorhen	630,000	0.299	188,370	2758	0.7149
Shag	95,000	1.814	172,330	520	0.1348
Song Thrush	2,760,000	0.060	165,600	3581	0.9282
Rock Dove/Feral Pigeon	400,000	0.400	160,000	2443	0.6332

of the total biomass of wild birds in the British Isles (most of that being the aforementioned Pheasants). Why the balance of bird and mammal faunas should be so different is an intriguing zoological puzzle. It belongs with a related puzzle, that birds are, size for size, much less abundant than mammals (Greenwood *et al.*, 1996). Demonstrating this puzzle, as well as exploring it, requires some elaboration.

Larger animals, birds or mammals, are naturally scarcer than small ones. How much scarcer? There are theoretical arguments suggesting that abundance should perhaps scale at either –2/3 or –3/4; in algebraic terms, as $Mass^{-0.67}$ or $Mass^{-0.75}$. These relationships derive, simply, from surface/volume relationships (2/3) or from the relationship between size and metabolic rate (3/4). Plotted on a log-log plot (Figure 8.1), the result should be a straight line declining with a 2/3 or 3/4 slope. Empirically, the terrestrial mammals (but bats

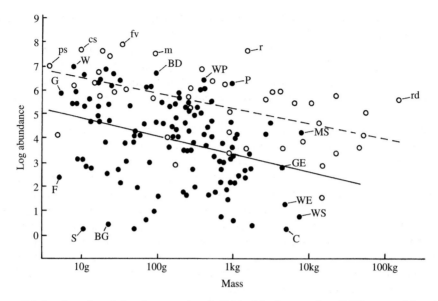

Figure 8.1 Log-log plot of abundance against individual body mass for all GB terrestrial resident birds (solid dots) and terrestrial mammals (circles). The slopes do not differ, but mammals average about 45 times more abundant than birds. Outliers identified: BD Blackbird; BG Brambling; C Crane; F Firecrest; G Goldcrest; GE Golden Eagle; MS Mute Swan; P Pheasant; S Serin; W Wren; WE White-tailed Eagle; WP Wood Pigeon; WS Whooper Swan. cs common shrew; fv field vole; m mole; ps pigmy shrew; r rabbit; rd red deer.

excluded this time, as well as domestic species) produce an equation that log (Abundance) – log (Mass)$^{-0.62}$, while for resident birds log (Abundance) – log (Mass)$^{-0.79}$. However, mammals are on average 45 times more abundant than resident birds at the same body masses (Greenwood et al., 1996). This esoteric argument is best illustrated by some examples. Such a very abundant wild bird as the Chaffinch, with 15 million inhabitants, is much less numerous than the similar-sized Field Vole with 75 million. At the other end of the scale, even the very abundant Pheasant, with about 3.1 million birds, does not match the abundance of the rather larger Rabbit, estimated at 37.5 million. None of the wild mammals exists as only a few hundred animals (as, for instance, the rather stable 430 or so pairs of Golden Eagle, let alone the one to three pairs of Crane). Given that there are about 200 breeding bird species, but only about 60 terrestrial mammals (including bats but not seals or whales), one explanation for this might be that each bird species occupies a smaller 'ecological space' than each mammal. This argument implies that birds as a group should contribute as many animals to the landscape as mammals, but that is clearly not true, nor is there any obvious reason to expect that it should be.

As flight is the obvious characteristic of birds, and one moreover shared with bats, which seem to scale as though they were birds, it is tempting to suggest that perhaps this very energetic method of travel requires so much energy that birds and bats are scarcer than terrestrial mammals. However, the very high metabolic rates of shrews mean that they use much more energy over a year than bats, which hibernate in winter and enter torpor in summer. Two ecologically equivalent insectvorous birds, Robins and House Martins, which might be expected to represent the extremes – a non-migratory bird that uses a lot of short flights (very expensive) and stays here through the cold of winter (also very expensive) and a migrant that spends much of its life on the wing (very expensive) but avoids the cold weather, use very comparable amounts of energy with shrews (Table 8.4).

Birds, as a group, exploit scattered resources such as seeds and berries, insects and worms, or other birds, mammals, and fish. Few are herbivores – the large gut needed to digest herbage, using bacteria and protozoa to assist the digestion of bulky plant leaves, does not fit so well inside a body designed for flight. Geese and grouse are the two groups that are most clearly herbivorous in the British avifauna. Many mammals, rodents, lagomorphs, and ungulates, are by contrast herbivores, with relatively large guts, though others are granivores, insectivores, and carnivores like most birds. The fact that birds can fly might enable them to exploit such scattered resources more efficiently than can terrestrial mammals. Conversely, terrestrial mammals might cope better eating abundant bulky foods. This hypothesis is difficult to test, but there are some obvious contradictions to it. If, for instance, birds such as finches can exploit well the scattered resources (in space and time) represented by seeding plants, how do Wood Mice, feeding on a very similar food supply, manage to be so much more abundant (estimated at 38.5 million in Great Britain) than Chaffinches (10.8 million) or Greenfinches (1.06 million)? One possibly illuminating case is the comparison by Summers of the feeding ecologies of Red Squirrels and Crossbills, both feeding on seeds extracted from cones of Scots Pine in the Caledonian pine forests. Red Squirrels preferred to eat the seeds from the larger cones born by trees within the continuous areas of pine forest. Crossbills can exploit the scattered trees away from the main forested areas; their cones were more numerous but smaller, so accessible to the smaller vertebrates' bills. Nationally, the much larger Red Squirrels are thought to number 160,000 (Harris *et al.*, 1995), far more than the 1,500 Scottish Crossbills, though perhaps comparable with total numbers of Crossbills after a major irruption (Knox, in Gibbons *et al.*, 1993). One should also expect carnivorous mammals and birds, both exploiting scattered and elusive prey, to fit on about the same regression. On that basis, a 200 g bird predator, such as the Kestrel, should be at least as numerous as the similar-sized Stoat; instead they are believed to number about 100,000 and 460,000 respectively. Likewise, the Common Shrew, numbering about 41.7 million, far outnumbers its avian insectivorous equivalent the Wren, at about 19.8 million, though both weigh about 8–10 g.

The smaller contribution of alien birds than alien mammals is surely a reflection of the larger bird fauna (fewer open niches to invade) and the poorer balance of the mammal fauna. Not only was the number of mammal species able to invade Great Britain limited by the opening of the English Channel – North Sea connection, but the mammal fauna has been hit proportionately much harder by extinctions since then; loss of Lynx, Wolf, Brown Bear, Root Vole, Beaver, Elk, Wild Boar, and Beaver makes much more of an ecological gap in the mammal fauna than does loss of Kentish Plover, Black Tern, Hazel Hen, Pygmy Cormorant, Great Auk, Spoonbill, Eagle Owl, White Stork, Great Bustard, and Dalmatian Pelican from the bird fauna (Greenwood *et al.*, 1996; Yalden, 1999). While none of the introduced mammals obviously fits the niche left by any of the extinctions, lack of the

Table 8.4 Notional annual energy expenditures of 4 insectivorous vertebrates. These species were selected because the detailed studies were available – daily activity and energy budgets had been studied, the latter mostly using doubly-labelled water. The comparison is a little distorted, because Pipistrelles, at 5 g, are so much lighter that flight is less costly for them than for the much larger birds (both around 20 g); a 20 g bat would have a flight cost of about 6,700 kJ per year (Speakman & Thomas 2003). Flight is 3×more expensive for Robins than House Martins because they use more short flights, involving many take-offs and landings.

Robin *Erithacus rubecula*		
Summer days, hopping	667.1 J/hr × 182 d × 14 hr	1699.8 kJ
Summer nights, rest	1,293.6 J/hr × 182 d × 8 hr	1883.5 kJ
Summer, hopping	1,980.0 J/hr × 182 d × 1 hr	360.3 kJ
Winter days, rest	1,519.7 J/hr × 183 d × 6 hr	1668.6 kJ
Winter nights, rest	1,180.4 J/hr × 183 d × 16 hr	3456.2 kJ
Winter, hopping	2,499.5 J/hr × 183 d × 1 hr	457.4 kJ
Flight	25,600 J/hr × 365 d × 1hr	9344.0 kJ
Annual Total		**18,869.8 kJ**
House Martin *Delichon urbica*		
Brood-rearing	3,233.2 J/hr × 85 d × 24 hr	274.8 kJ
Perching	1,596.1 J/hr × 280 d × 6 hr	2,681.4 kJ
Flying	7,449.5 J/hr × 280 d × 6 hr	12,515.1 kJ
Roosting	806.2 J/hr × 280 d × 12 hr	2,708.8 kJ
Annual Total		**18,180.1 kJ**
Common Shrew *Sorex araneus*		
Active, Summer, Juvenile	2,133.8 J/hr × 120 d × 13 hr	3,328.7 kJ
Resting, Summer, Juvenile	1,268.6 J/hr × 120 d × 11 hr	1,674.6 kJ
Active, Winter, Juvenile	2,169.5 J/hr × 120 d × 13 hr	3,384.4 kJ
Resting, Winter, Juvenile	1,210.8 J/hr × 120 d × 11 hr	1,598.3 kJ
Active, spring, Adult	3,176.6 J/hr × 125 d × 13 hr	5,162.0 kJ
Resting, Spring, Adult	1,829.1 J/hr × 125 d × 11 hr	2,515.0 kJ
Annual Total		**17,663.0 kJ**
Common Pipistrelle *Pipistrellus pipistrellus*		
Torpor, Winter	21.6 J/hr × 183 d × 23.6 hr	93.3 kJ
Arousal, Winter	5,400 J/hr × 183 d × 0.3 hr	296.5 kJ
Active, Pre-flight	1,980.0 J/hr × 182 d × 1 hr	360.4 kJ
Active, Post-flight	1,288.8 J/hr × 182 d × 1 hr	235.9 kJ
Flight	4,032 J/hr × 182 d × 4 hr	2,953.3 kJ
Torpor/Rest, Summer	216 J/hr × 182 d × 18 hr	707.6 kJ
Annual Total		**4,629.0 kJ**

Compiled from:
Robin: Tatner & Bryant (1986).
House Martin: Bryant & Westerterp (1980).
Common Shrew: Genoud (1985).
(with temperatures set at 15, 5, 10 C and weights set at 8, 7, 11 g, respectively)
Pipistrelle: Speakman & Racey (1991) and John Speakman pers. comm.

Table 8.5 Population sizes, biomass and ubiquity for introduced (alien) bird species in Britain (after Table 9 of Gibbons *et al.* 1993; means of their abundance values used where appropriate. Individual masses come from various sources, mostly Hickling (1983) and HBWP). Restored or returning native species (White-tailed Eagle, Goshawk, Capercaillie, Greylag Goose and others) discounted.

Species	Number	Mass (kg)	Biomass (kg)	Occurrence	Ubiquity
Pheasant	3,100,000	1.131	4,061,000	3123	0.8099
Red-legged Partridge	180,000	0.484	87,120	1226	0.3178
Canada Goose	60,140	3.780	227,329	1215	0.3149
Little Owl	18,000	0.168	3,024	1228	0.3183
Mandarin	7,000	0.570	3,990	219	0.0568
Ring-necked Parakeet	6,000	0.122	732	63	0.0163
Golden Pheasant	1,500	0.700	1,050	47	0.0122
Ruddy Duck	1,180	0.560	6,618	300	0.0778
Egyptian Goose	775	1.863	1,444	87	0.0226
Lady Amherst's Pheasant	150	0.800	120	9	0.0023
Red-crested Pochard	100	1.157	116	13	0.0034
Totals	3,374,845		4,395,543		

larger predators has certainly made it easier for larger herbivores (Rabbit, Brown Hare, Fallow Deer, Sika, Muntjac) to establish themselves. The ecological positions of the birds that have been introduced (Table 8.5) is notable. The two numerous gamebirds are well adapted to the farmland landscape that has evolved over the last few thousand years. Two more ornamental gamebirds and three ornamental waterfowl are well adapted to the parkland and new forestry landscapes of southern Britain; moreover, the Canada Goose had the good fortune to establish itself at a time when the native Greylag Goose was much reduced. The Ruddy Duck, as a member of a different tribe, the stifftails Oxyurini, is presumably sufficiently different from other British diving ducks to have found a free niche. Its diet perhaps is biased towards eating submerged plants, rather than animals, in contrast with its commonest potential competitor, the Tufted Duck. More surprising, perhaps, is the success of the Little Owl, which as a hole-nesting, somewhat insectivorous, diurnal predator seems likely to have been in direct competition with the well-established Kestrel. However, the latter's usual method of hunting small mammals in flight may have left enough ecological space for a small hedgerow, perch-hunting, rival. Moreover, Kestrels thrive in the uplands as well as the lowlands, whereas the Little Owl, perhaps reflecting its Mediterranean origins, is a lowland bird. The equally surprising success of the Ring-necked Parakeet probably reflects the gap in suburbia for a large hole-nesting exploiter of the increasingly abundant bird feeders. The severe winter seems to have eliminated the Greater Manchester population in about 1986, and it remains to be seen if a severe winter in southern England will curtail its expansion there; winters over the period of its rapid expansion have been unusually mild.

There are additional contributions of aliens that should be noted. Most Pheasants and Red-legged Partridges shot in this country now are reared in captivity and released in autumn, prior to the shooting season, rather than bred in the wild. This has reached levels that have become controversial. As many as 20 million Pheasants, reaching local densities of 350/km², have been released in recent autumns, and about 45% are shot in their first winter (Tapper, 1992). In the 1980s, about 800,000 hybrid *Alectoris* partridges (mostly Red-leg × Chukar) were also released. Combined, these would contribute something like 22,384,000 kg to the autumn biomass of birds in the countryside, or five times more than the breeding biomass of alien birds estimated in Table 8.5. It is also a figure close to the total spring biomass of breeding birds in the British Isles (Table 8.6), though by spring the majority of them will be long dead and eaten. The biomass contributed by cage birds is uncertain, and most do not appear as wild birds. Budgerigars have been imported at least since 1840, and Canaries much earlier. The popularity of these, and other cage birds, is hard to quantify, but the possibility that a rare species observed in the wild derives in fact from a cage continues to perplex those responsible for the 'British List'. The same problem can even occur in archaeological circumstances. The parrot recovered from post-Mediaeval Norwich, about the size of an African Grey, was surely a pet of some sort (Albarella *et al.*, 1997); so was the Pygmy Cormorant at Abingdon another one?

The bird fauna in the future

It is fashionable to suppose that global warming is inevitable, proceeding, and already affecting our wildlife. Some migrants are returning earlier, and many species are nesting earlier, by between 4 days (Starling) and 17 days (Magpie) (Mead, 2000). It is currently topical to attempt to evaluate what changes this will bring in our bird fauna. The RSPB has recently (2005) issued a press release suggesting a list of 10 likely newcomers. They are largely southern species that either already occur in the British Isles occasionally as migrants or vagrants, even, rarely, breeding, or were once regular breeders, now lost. Bee-eaters nested in Sussex in 1955, and then in Durham, very publicly and spectacularly, in 2002. Another pair, attempting to breed in Herefordshire in 2005, suggest that this might become more regular if summers really do become warmer, though their success may depend on the continued abundance of bees, which seems currently uncertain. Its equally colourful southern companion the Hoopoe has also bred sporadically, on 42 occasions in England by Brown & Grice's (2005) listing. Over 100 are recorded most years, usually in spring having apparently overshot their usual breeding grounds on migration, so the possibility for colonization is evident. On the other hand, the four pairs that nested in 1977, perhaps following the exceptionally warm summers of 1975 and 1976, have not been matched since, and Brown & Grice remark that it seems no more likely to establish itself than at any other time in the last 200 years. The Black-winged Stilt has bred successfully twice, in Nottinghamshire in 1945 and then in Norfolk in 1987, but has also nested unsuccessfully on at least three other occasions (Cambridgeshire 1983, Cheshire 1993, Lancashire 2006). Given that its habitat and distribution closely resemble those of its near relative the Avocet, which has staged such a successful comeback, a more determined colonization is certainly possible. The Wryneck and Red-backed Shrike, two insectivores thought lost because of poorer summers and fewer insects, might well return if a combination of warmer summers and agricultural changes recreate the food supply that they seem to have lost. Wrynecks feed extensively on ants, especially the Black Ant *Lasius niger*, but also on various other insects, and nest sites are in crevices and

Table 8.6 A complete list of breeding birds of the British Isles (after Table 9 of Gibbons *et al.* 1993), in order of their biomass contribution. Means of their population estimates (individuals) are used where appropriate. Individual masses come from various sources, including Hickling and HBWP. Ubiquity is the proportion of the total of the 3858 hectads (10 x 10 km squares) in which each species was recorded.

Species	N. individuals	Mass (kg)	Biomass (kg)	Hectads	Ubiquity
Pheasant	3,100,000	1.131	3,506,100	3123	0.8095
Wood Pigeon	6,420,000	0.524	3,364,080	3469	0.8992
Rook	2,750,000	0.488	1,342,000	3156	0.8180
Guillemot	1,200,000	1.002	1,202,400	274	0.0710
Blackbird	12,400,000	0.095	1,178,000	3654	0.9471
Gannet	373,000	3.010	1,122,730	24	0.0062
Carrion Crow	1,620,000	0.570	923,400	2653	0.6877
Fulmar	1,142,000	0.808	922,736	716	0.1856
Hooded Crow	900,000	0.570	513,000	1657	0.4295
Mute Swan	45,900	10.750	493,425	2141	0.5550
Mallard	246,000	1.785	439,110	3437	0.8909
Magpie	1,820,000	0.237	431,340	2932	0.7600
Kittiwake	1,087,200	0.387	420,746	315	0.0816
Herring Gull	411,400	0.951	391,241	904	0.2343
Puffin	941,000	0.395	371,695	181	0.0469
Red Grouse	560,000	0.651	364,560	1086	0.2815
Chaffinch	15,000,000	0.020	300,000	3564	0.9238
Jackdaw	1,200,000	0.246	295,200	3290	0.8528
House Sparrow	9,400,000	0.027	253,800	3440	0.8917
Manx Shearwater	550,000	0.453	249,150	38	0.0099
Starling	2,920,000	0.082	239,440	3591	0.9308
Robin	12,200,000	0.019	235,460	3610	0.9357
Canada Goose	60,140	3.780	227,329	1215	0.3149
Stock Dove	540,000	0.400	216,000	2190	0.5677
Wren	19,800,000	0.010	196,020	3748	0.9715
Skylark	5,140,000	0.038	195,320	3669	0.9510
Moorhen	630,000	0.299	188,370	2758	0.7149
Shag	95,000	1.814	172,330	520	0.1348
Song Thrush	2,760,000	0.060	165,600	3581	0.9282
Rock Dove/Feral Pigeon	400,000	0.400	160,000	2443	0.6332
Eider	64,600	2.229	143,993	533	0.1382
Lesser Black-backed Gull	177,400	0.765	135,711	525	0.1361
Hedge Sparrow	5,620,000	0.021	119,706	3472	0.8999
Razorbill	182,000	0.620	112,840	301	0.0780
Meadow Pipit	5,600,000	0.020	112,000	3497	0.9064
Black-headed Gull	402,400	0.276	111,062	816	0.2115
Grey Partridge	291,000	0.374	108,834	1665	0.4316
Lapwing	466,000	0.228	106,248	2833	0.7343
Blue Tit	8,800,000	0.012	101,200	3424	0.8875
Collared Dove	460,000	0.196	90,160	2783	0.7214
Great Black-backed Gull	47,000	1.854	87,138	636	0.1649
Red-legged Partridge	180,000	0.484	87,120	1226	0.3178
Mistle Thrush	640,000	0.130	83,200	3264	0.8460
Greylag Goose	22,700	3.465	78,656	741	0.1921
Great Tit	4,040,000	0.019	76,760	3340	0.8657
Yellowhammer	2,800,000	0.027	74,200	2814	0.7294

Table 8.6 (*Continued*)

Species	N. individuals	Mass (kg)	Biomass (kg)	Hectads	Ubiquity
Curlew	95,000	0.725	68,875	2564	0.6646
Common Gull	143,200	0.411	58,855	664	0.1721
Shelduck	48,850	1.152	56,275	1142	0.2960
Jay	340,000	0.161	54,740	1986	0.5148
Cormorant	23,400	2.319	54,265	272	0.0705
Willow Warbler	6,260,000	0.009	53836	3539	0.9173
Oystercatcher	83,000	0.519	43,077	1979	0.5130
Cuckoo	347,000	0.114	39,558	3136	0.8129
Greenfinch	1,380,000	0.028	38,364	3150	0.8165
Grey Heron	27,900	1.361	37,972	3129	0.8110
Coot	54,600	0.668	36,473	1963	0.5088
Swallow	1,640,000	0.019	31,160	3622	0.9388
Black Grouse	25,000	1.202	30,050	432	0.1120
Buzzard	33,000	0.875	28,875	1637	0.4243
Whitethroat	1,560,000	0.016	25,428	2824	0.7320
Raven	21,000	1.200	25,200	1823	0.4725
Kestrel	120,000	0.202	24,240	3298	0.8548
Great Skua	15,800	1.415	22,357	97	0.0251
Tawny Owl	40,000	0.545	21,800	2054	0.5324
Turtle Dove	150,000	0.145	21,750	969	0.2512
Blackcap	1,240,000	0.018	21,700	2417	0.6265
Reed Bunting	1,100,000	0.018	20,130	3023	0.7836
Linnet	1,300,000	0.015	19,890	3065	0.7945
Pied Wagtail	860,000	0.022	18,920	3640	0.9435
Corn Bunting	380,000	0.046	17480	932	0.2416
House Martin	960,000	0.018	17,280	3217	0.8339
Sparrowhawk	86,000	0.199	17,114	2845	0.7374
Black Guillemot	40,000	0.413	16,520	473	0.1226
Chiffchaff	1,860,000	0.008	14,880	2949	0.7644
Woodcock	46,250	0.316	14,615	1383	0.3585
Tufted Duck	18,750	0.698	13,088	1739	0.4508
Bullfinch	580,000	0.022	12,644	3010	0.7802
Great Crested Grebe	12,150	1.036	12,587	1117	0.2895
Redshank	73,600	0.159	11,702	1686	0.4370
Ptarmigan	20,000	0.535	10,700	173	0.0448
Golden Plover	46,000	0.229	10,534	814	0.2110
Siskin	720,000	0.015	10,440	1442	0.3738
Goldcrest	1,720,000	0.006	9,804	3189	0.8266
Snipe	80,000	0.119	9,520	2447	0.6343
Sandwich Tern	36,800	0.242	8,906	81	0.0210
Arctic Tern	93,000	0.094	8,742	375	0.0972
Goldfinch	550,000	0.016	8,580	2972	0.7703
Goosander	5,400	1.500	8,100	676	0.1752
Sedge Warbler	720,000	0.011	8,064	2571	0.6664
Storm Petrel	320,000	0.025	8,064	68	0.0176
Swift	200,000	0.039	7,800	2971	0.7701
Garden Warbler	400,480	0.018	7,169	1933	0.5010
Sand Martin	527,000	0.014	7,115	2160	0.5599
Redstart	420,000	0.015	6,090	1338	0.3468
Red-breasted Merganser	5,700	1.063	6,059	841	0.2180
Capercaillie	2,000	2.900	5,800	66	0.0171

Species	N. individuals	Mass (kg)	Biomass (kg)	Hectads	Ubiquity
Nuthatch	260,000	0.022	5,720	1270	0.3292
Green Woodpecker	30,000	0.189	5,670	1555	0.4031
Redpoll	460,000	0.012	5,290	2292	0.5941
Tree Sparrow	238,000	0.022	5,236	1476	0.3826
Leach's Petrel	110,000	0.045	4,895	11	0.0029
Little Grebe	24,000	0.201	4,824	1613	0.4181
Red-throated Diver	2,700	1.780	4,806	389	0.1008
Spotted Flycatcher	310,000	0.015	4,650	3117	0.8079
Great Spotted Woodpecker	55,000	0.082	4,488	1962	0.5086
Treecreeper	490,000	0.009	4,410	2687	0.6965
Long-tailed Tit	500,000	0.008	4,100	2660	0.6895
Mandarin	7,000	0.570	3,990	219	0.0568
Common Tern	32,000	0.116	3,712	535	0.1387
Golden Eagle	840	4.383	3,682	408	0.1058
Wheatear	134,000	0.026	3,484	2175	0.5638
Tree Pipit	140,000	0.022	3,080	1529	0.3963
Little Owl	18,000	0.168	3,024	1228	0.3183
Arctic Skua	6,700	0.443	2,968	113	0.0293
Barn Owl	10,300	0.270	2,781	1304	0.3380
Long-eared Owl	9,400	0.280	2,632	676	0.1752
Peregrine	2,930	0.898	2,630	1338	0.3468
Dipper	34,750	0.065	2,259	1738	0.4505
Rock Pipit	93,000	0.024	2,232	927	0.2403
Grey Wagtail	112,000	0.019	2,128	2796	0.7247
Twite	137,000	0.015	2,110	711	0.1843
Common Sandpiper	36,600	0.057	2,075	1737	0.4502
Lesser Whitethroat	160,000	0.012	1,872	1279	0.3315
Ring Ouzel	17,040	0.109	1,857	573	0.1485
Shoveler	2,700	0.678	1,831	500	0.1296
Yellow Wagtail	100,000	0.017	1,700	1050	0.2722
Reed Warbler	120,090	0.014	1,681	812	0.2105
Coal Tit	176,000	0.009	1,602	3170	0.8217
Eqyptian Goose	775	1.863	1,444	87	0.0226
Short-eared Owl	4,500	0.305	1,373	690	0.1788
Gadwall	1,600	0.800	1,280	382	0.0990
Marsh Tit	120,000	0.011	1,272	1133	0.2937
Barnacle Goose	650	1.786	1,161	45	0.0117
Golden Pheasant	1,500	0.700	1,050	47	0.0122
Pied Flycatcher	75,000	0.014	1,013	735	0.1905
Dunlin	18,650	0.047	884	638	0.1654
Stonechat	56,250	0.016	872	1611	0.4176
Teal	2,590	0.323	837	1335	0.3460
Black-throated Diver	300	2.740	822	201	0.0521
Whinchat	45,750	0.016	750	1528	0.3961
Chough	2,290	0.324	742	256	0.0664
Ring-necked Parakeet	6,000	0.122	732	63	0.0163
Pochard	860	0.828	712	500	0.1296
Hen Harrier	1,620	0.427	692	621	0.1610
Ruddy Duck	1,180	0.560	661	300	0.0778

Table 8.6 (*Continued*)

Species	N. individuals	Mass (kg)	Biomass (kg)	Hectads	Ubiquity
Wigeon	800	0.700	560	385	0.0998
Quail	5,380	0.099	533	838	0.2172
Greenshank	2,700	0.192	518	244	0.0632
Hawfinch	9,500	0.054	513	315	0.0816
Willow Tit	50,000	0.010	510	1100	0.2851
Kingfisher	12,200	0.039	476	1531	0.3968
Nightjar	6,060	0.078	473	285	0.0739
Water Rail	3,900	0.120	466	597	0.1547
Grasshopper Warbler	32,000	0.013	426	1598	0.4142
Corncrake	2,980	0.141	420	407	0.1055
Whimbrel	930	0.449	418	83	0.0215
Goshawk	400	1.004	402	237	0.0614
Common Scoter	340	1.079	367	67	0.0174
Wood Warbler	34,460	0.010	341	1298	0.3364
Little Tern	5,640	0.057	321	146	0.0378
Hobby	1,400	0.211	295	628	0.1628
Ringed Plover	4,180	0.071	295	1274	0.3302
Avocet	900	0.295	266	28	0.0073
Merlin	1,440	0.181	261	851	0.2206
Nightingale	11,000	0.022	243	457	0.1185
Osprey	144	1.528	220	170	0.0441
Dotterel	1,790	0.106	190	99	0.0257
Red Kite	184	1.016	187	85	0.0220
Lesser Spotted Woodpecker	9,000	0.020	178	792	0.2053
Goldeneye	190	0.895	170	186	0.0482
Stone Curlew	310	0.459	142	54	0.0140
Lady Amherst's Pheasant	150	0.800	120	9	0.0023
Red-crested Pochard	100	1.157	116	13	0.0034
Crossbill	2,500	0.046	115	919	0.2382
Roseate Tern	980	0.110	108	30	0.0078
White-tailed Eagle	22	4.792	105	9	0.0023
Marsh Harrier	190	0.403	77	121	0.0314
Whooper Swan	8	9.350	75	51	0.0132
Little Ringed Plover	1,895	0.039	74	422	0.1094
Garganey	160	0.359	57	146	0.0378
Pintail	72	0.790	57	94	0.0244
Slavonian Grebe	120	0.375	45	24	0.0062
Bittern	32	1.231	39	13	0.0034
Honey Buzzard	60	0.626	38	27	0.0070
Black-tailed Godwit	72	0.305	22	68	0.0176
Dartford Warbler	1,900	0.011	21	50	0.0130
Woodlark	700	0.029	20	73	0.0189
Crested Tit	1,800	0.011	20	51	0.0132
Black-necked Grebe	54	0.281	15	35	0.0091
Cetti's Warbler	900	0.014	13	89	0.0231
Bearded Tit	800	0.016	13	63	0.0163
Cirl Bunting	458	0.024	11	32	0.0083
Crane	2	5.440	11	2	0.0005
Redwing	120	0.065	8	140	0.0363
Scaup	6	1.146	7	21	0.0054

Species	N. individuals	Mass (kg)	Biomass (kg)	Hectads	Ubiquity
Mediterranean Gull	22	0.300	7	7	0.0018
Snow Bunting	170	0.034	6	42	0.0109
Fieldfare	50	0.112	6	104	0.0270
Golden Oriole	80	0.069	5	45	0.0117
Red-necked Grebe	6	0.819	5	9	0.0023
Snowy Owl	2	1.762	4	4	0.0010
Black Redstart	200	0.017	3	103	0.0267
Spotted Crake	30	0.078	2	27	0.0070
Firecrest	330	0.006	2	99	0.0257
Ruff	10	0.156	2	42	0.0109
Red-necked Phalarope	42	0.035	1	10	0.0026
Wood Sandpiper	12	0.062	1	8	0.0021
Savi's Warbler	30	0.016	0	29	0.0075
Wryneck	10	0.032	0	6	0.0016
Marsh Warbler	24	0.012	0	15	0.0039
Purple Sandpiper	4	0.065	0	3	0.0008
Temminck's Stint	6	0.026	0	3	0.0008
Parrot Crossbill	2	0.052	0	2	0.0005
Brambling	4	0.024	0	13	0.0034
Red-backed Shrike	2	0.030	0	15	0.0039
Serin	4	0.012	0	10	0.0026
Scarlet Rosefinch	2	0.023	0	5	0.0013
Totals	166,718,680		22,987,788		

holes in old trees, especially in orchards (HBWP). Loss of old orchards and old grasslands, with numerous anthills, might have progressed too far. Red-backed Shrikes, by contrast, need an abundance of larger insects, often taken in flight but sometimes from the ground, such as bumble-bees, dragonflies, and dung beetles. Bumble-bees are certainly much less abundant, and many rarer species have disappeared completely from central England (Williams, 1982); we know that most farmland ponds, the homes of dragonflies, have been lost; dung beetles have been reduced by the use of ivermectin pesticides applied to livestock, though we do not know enough about their populations to appreciate the severity of any decline. Changes in farmland might restore their food supplies, but climatic change alone is unlikely to suffice. The combination of improved wetlands and warmer climates has already seen Little Egrets succeed in colonizing southern Britain. Cattle Egrets and Great White Egrets are two more, ecologically related, species that are already expanding their western European ranges, and occurring as vagrants more frequently. Purple Herons, Night Herons, and Little Bitterns, not on the RSPB list of 10 but already nesting in Holland, might be more likely early colonists of managed wetlands, especially as wetland conservation in the Netherlands seems already to be producing a surplus of colonists for us; however, the Little Bittern seems to be a declining species in Europe at present. Evidence of ringing recoveries shows that both the rapidly increasing inland Cormorant population in southern Britain and the more regular Spoonbills come from there, and the Spoonbill is another returning species that has already nested, though unsuccessfully, in East Anglia in 1997, 1998, and 1999, and then successfully in Cheshire in 1999 (Brown & Grice, 2005). Two songbirds suggested as colonists by the RSPB list are Serin, which has also already nested spasmodically in southern England, and Fan-tailed Warbler, also known as Zitting Cisticola. This is a small warbler that sings a persistent, irritating, zit-zit-zit note in a switch-back song flight, and which is the sole European representative of an enormous genus of African warblers, *Cisticola*. It was confined to Mediterranean areas during the nineteenth century, but spread northwards through France in the 1920s, until the severe 1939–40 winter reduced it to its Mediterranean source. It spread again, at least as far as the Channel coast, through the post-1945 period, until another severe winter in 1985–6 again reduced it to its heartland (HBWP). It is now spreading again, and if mild winters persist could easily cross the Channel to colonize tussocky grasslands in southern England.

A further group of possible newcomers, not acknowledged by the RSPB list but a worrying diminution of international biodiversity, are the alien species. They have been largely accepted whenever they have seemed likely to settle as breeding birds, but should not be. We should have a western Palaearctic bird fauna, and polluting it with alien species diminishes biodiversity world-wide. After all, a large part of the excitement of travelling overseas is to see the characteristic fauna of those areas. No British ornithologist travels to the USA, or New Zealand, in the hope of seeing House Sparrows or Starlings. Conversely, American ornithologists don't come to Britain to see Canada Geese or Ruddy Ducks. Worse, of course, is the fact that these aliens may well displace our native species (Greylag Geese should be our native species), or start to migrate and hybridize with endangered Palaearctic species (i.e. Ruddy Ducks with White-headed Ducks). The worst prospect of all is that we end up with a 'slum avifauna' of the ubiquitous followers of Humans, including Canada Goose, Mallard, Pheasant, House Sparrow, Starling, Blackbird, and Chaffinch, world-wide. This would match a 'slum mammal fauna' of Rabbit, Brown or Black Rat, House Mouse, Feral Cat, Pig, Fallow Deer, Goat, and Sheep. The most likely additions seem to be the waterfowl, many of which are kept in, and have escaped from, wildfowl collections (Bar-headed Goose, Muscovy Duck, Snow Goose?), though additional gamebirds (Guinea-fowl, Black Francolin, Peacock?) and parrots cannot be dismissed. There was a colony of feral Budgerigars on the Isles of Scilly from 1972 to 1977. The addition of Mandarin, and even Ring-necked Parrakeet, to our fauna is seen as a benign event that adds an attractive bird at little cost. Perhaps. As Kear (2003) points to the shortage of nest holes generally in British woodlands, ousting of Barn Owls, Tawny Owls, Jackdaws, Stock Doves and others might not be so trivial. We have managed to eliminate Muskrat and Coypu from the British Isles. Most ornithologists would cheer if American Mink followed them into local oblivion. So why do they object so strongly to efforts at removing Ruddy Ducks?

The converse issue is which lost birds might be helped to return. Capercaillies were returned as early as 1837–38, and the Goshawk, lost as a breeding bird in the 1880s, returned by way of an unofficial release scheme, using mostly Norwegian birds that would otherwise have been culled, in the 1960s–70s. White-tailed Eagles were the subject of an official reintroduction, after a couple of false starts, begun in 1975 by NCC (and now continued by RSPB and SNH) (Love, 1983). Migrant species have had a better chance of returning naturally, as Ospreys and Avocets have done, and as Wrynecks and Red-backed Shrikes might still do. What other species does the archaeological record suggest might be returned? The most obvious suggestion is the Eagle Owl, which, like the Goshawk, is already 'leaking' from captive stock, and has bred in Yorkshire fairly regularly, albeit as no more than one pair. Strangely omitted completely from Brown & Grice's (2005) impressive compendium on the birds of England, Mead (2000) remarks that had they colonized naturally they would have been fully protected, but that most ornithologists think the escaped birds should be returned to captivity. They certainly do not take that attitude to Goshawks, Capercaillies, or Sea Eagles! The Great Bustard is another obvious candidate, and subject of an active reintroduction campaign at present (Osborne, 2005). What about Dalmatian Pelican and Pygmy Cormorant? Their current ranges are so far away, and the quality of our remaining wetlands so poor, that there seems no chance of them ever returning as breeding birds. On the other hand, the improving quality and quantity of grazing marshes in the Ouse Washes has seen Black-tailed Godwits return naturally, and the reintroduction there of breeding Corncrakes is being attempted. Cranes have returned of their own volition to the Broads, and have rapidly discovered other improving wetlands, for instance at Lakenheath. So wetlands are improving, and strenuous attempts by RSPB and others to increase the area of reedbed for Bitterns, as well as for Marsh Harriers and Bearded Tits, might yet see such unlikely events. Cranes, at least, should be much more widespread, as they were in Saxon times, if we can get enough wetlands returned to good ecological condition. Kentish Plovers ought to be able to return, if appropriate sandy beaches in quiet enough areas can be created. Could we create enough mature woodland in southern Britain to tempt in (or back?) Black Woodpeckers, or was Kear (2003) right in supposing that their ant food is too scarce?

If warming is really a prospect, then there will also be losses, of northern or montane species. Some very rare and occasional breeders (Purple Sandpiper, Temminck's Stint, Wood Sandpiper, Bluethroat, Redwing, Brambling) are obviously vulnerable. The three established specialist montane birds, Ptarmigan, Dotterel, and Snow Bunting, are also susceptible. The last of these, with a slender toehold of only 50 or so breeding pairs, and confined to the highest mountains, is clearly the most vulnerable of the three. Its habit of exploiting late snow patches for the insects knocked out by the cold must add to its vulnerability. Moreover, it has already had a rather chequered history, with few or none nesting in the warmer interlude from 1920 to 1940, though as a species nesting thinly and at high altitude, it has also always been difficult to census properly. Its compatriot on the high tops, the Dotterel, has equally had a poorly counted but apparently chequered history, perhaps declining to as few as 50 known pairs in the period 1930–50 (Nethersole-Thompson, 1973). However, signs of a recovery were recorded during the 1950s, and Nethersole-Thompson thought that there were about 75 pairs then. By the time of the first breeding atlas, numbers were thought to be about 100 pairs, with a few breeding records from northern England and even a record, the first ever, from north Wales (Sharrock, 1976). At about that time, the first ever breeding record from Ireland was also recorded (in 1975), but that has not been repeated. Subsequent surveys have recognized even more pairs in Scotland, breeding at higher densities (up to nine pairs/km^2) on well established hills but also spreading to other hills, so that by the time of the second breeding atlas at least 840 and perhaps 950 pairs were thought to be present (Thompson & Whitfield in Gibbons et al., 1993). There is no doubt that some of this increase reflects targeted studies of this species and better coverage of its likely range, the consequence of more, and more mobile, surveyors. However, despite these factors, numbers breeding in northern England have not increased, and no further breeding has been reported from Wales; the contrast with Scotland confirms that change there has included a real increase. It is very likely that heavy grazing by sheep has reduced the suitability of the habitat for this species in England and Wales, replacing moss-rich (*Racomitrium*) heath by overgrazed grassland which supports fewer craneflies. The higher Scottish mountains, which remain snow covered in most winters, are not suitable for winter grazing by sheep, and the Dotterel, which return quite late, in mid May, may have to wait for the snow to clear. Global warming ought to affect this species severely, and the increase in numbers through the 1970s and 1980s is thought to coincide, perhaps correlate with, a number of severe winters with prolonged late snow-lie. In that case, the mild winters of the 1990s should have started a decline in numbers, but this has not been reported. Ptarmigan, which also breeds on the higher mountains, above about 700 m in the central Highlands, disappeared from its English sites, in the Lake District, by 1800, and from the Southern Uplands by 1900. Within Highland Scotland, it is more abundant and widespread than its two montane colleagues, with a population roughly estimated at 10,000 pairs in the 1993 breeding atlas (Watson & Rae in Gibbons *et al.*, 1993), spread over 173 hectads. The population seemed to have declined little in range since the previous atlas, though local declines have been noted. Most conspicuously, the birds that used to breed near the Ptarmigan Restaurant at the top of the Cairn Gorm ski-lift have gone, exterminated by the crows and other predators of eggs and chicks that have followed the increasing numbers of tourists to the high tops, attracted by the picnic scraps (Watson & Moss, 2004). Where the study area had 10 territories in 1968 and 1969, numbers thereafter declined, and none bred from 1978 to 1995. A few immigrant adults have attempted to colonize the area each winter, but have generally been killed on the abundant wires and fences. It seems less likely that Ptarmigan would be completely lost as breeding birds from the British Isles; the montane heath on which they nest and, as herbivores, on which they feed, is less likely to be lost than the *Racomitrium* heaths and their insect communities that support the smaller numbers of Dotterel and Snow Bunting. The fact that they are sufficiently tolerant to have survived as a relict since the last glaciation in the Alps and Pyrenees, whereas Red Grouse (Willow Ptarmigan) did not, also suggests a robust species, tolerant of considerable climatic change!

Other species that could also be vulnerable to climate change are those dependent on moist conditions in summer and autumn, particularly those relying on earthworms for their food. There have been frequent concerns in drier summers about the water supply (for humans) in south-east England. Part of the decline in numbers of Song Thrushes over the past 30 years, one of the species whose decline has been most severe, must be blamed on drier soils as a result not only of poorer rainfall (one possible consequence of global warming) but also through agricultural drainage. The lack of food supplies for the fledglings, from damp ditches and field margins, has been identified as one of its biggest problems (Gruar *et al.*, 2003). Other species of damp pasture, such as Lapwing, Curlew, Snipe, and Yellow Wagtail, could also be affected. It is unlikely that these would be lost completely as British breeding birds, given their wide ranges further south in Europe, but some retreat from southern England seems very plausible. However, as they all occur southwards into the Mediterranean region, perhaps they are more tolerant than these pessimistic comments suggest. One group that certainly does not occur so far south, and is already showing worrying signs of breeding failure that might be related to global warming, are the seabirds for which British coasts are so important. Species such as Fulmar, Gannet, Kittiwake, Arctic Tern, Puffin, Black Guillemot, and Razorbill have northern ranges that reach their southern limits in the British Isles or, just a little further south, in Brittany and the Channel Isles. It is evident already that many of the fish on which they rely are also northern species, that inhabit the cooler waters to the north; scarcity of Cod has dominated human concern, but Herring, Greater Sand-eel and Sand-eel are also northern species, and, when abundant, the major food supply for many of these birds. While overfishing by Humans has been the main cause of their decline, it is evident that rising sea temperatures have also affected breeding and recruitment of Cod, and must be expected to affect other northern fish too. The spectacular and alarming breeding failures of Kittiwakes and terns in some of the southern, especially North Sea, colonies in some recent years, including 2005, seem likely to be the consequence, though, again, it is not clear how much human overfishing, rather than temperature change, is to blame. Other northern species that might be vulnerable, because they occur now only in the very north of the British Isles and in relatively small numbers, include Common Scoter, Red-necked Phalarope, Arctic Skua, Great Skua, and Whimbrel. However, an analysis of the range changes between the two breeding bird atlases (Sharrock, 1976; Gibbons *et al.*, 1993) showed that while southern species had on average expanded northwards by 18–19 km, northern species had not retreated northwards from the southern edges of their ranges (Thomas & Lennon, 1999). Possibly they are less sensitive to warming than the southern species; alternatively, the topography of Britain (higher, therefore cooler, in the north) is ameliorating the effects of any global warming.

The future of predators

Two more possible colonists on the RSPB list of 10 species are raptors, the Black Kite and Booted Eagle. Both are reasonably common and widespread in western Europe, and the Black Kite, as a migratory species, already turns up in the British Isles with some regularity. Though there were only four records in England between 1866 and 1958, it now occurs annually, with as many as 31 in 1994 (Brown & Grice, 2005). This reflects its increasing range and abundance in France (HBWP). The Booted Eagle is a less likely incomer, but it too breeds quite widely in France. However, the status of raptors in general is a contentious issue, or perhaps a series of contentious issues, which merits some discussion. Raptors as a group have been a major conservation success story in the latter half of the twentieth century. After severe persecution in the nineteenth century, persisting into the first half of the twentieth (see Chapter 7), all species were much scarcer than the habitat or food supply would allow, and six of the 15 species became extinct as British breeding birds during this period (Honey Buzzard, extinct 1900–11; White-tailed Eagle, 1916–75; Marsh Harrier, 1898–11; Montagu's Harrier, 1974–75; Goshawk, 1883–1950; Osprey, 1916–54) (Galbraith *et al.*, 2000). Two more survived only as small regional populations: Red Kites, perhaps only two pairs in Wales at various times between 1900 and 1935; Hen Harriers only on Orkney from 1920 to 1940, and reduced to 30–50 pairs. The other seven species also showed reduced ranges and populations at various times. For some, the nadir was not in the face of persecution in the period from 1900 to 1950, but from organochlorine pesticides in the late 1950s and 1960s. DDT, an insecticide invented during the 1939–45 war as a way of controlling the incidence of malaria and typhus, by killing their mosquito and louse vectors, became commercially available during the 1950s. Among other uses, pigeon fanciers used it to control feather lice, but it rapidly achieved agricultural use to control insect pests of cereal crops. It is often forgotten that the chemical found much favour because it could be used in very low doses, at which it was effective against insects but had no affect on vertebrates. What was not fully appreciated, when it was tested experimentally on laboratory animals for these effects, was that it would be very persistent in the ecosystem. Its persistence was, if anything, a virtue, because it remained toxic to the target insects. Unfortunately, the low doses that killed insects persisted when those insects were eaten by insectivorous songbirds, and again when they were eaten by such raptors as Sparrowhawks or Hobbies, and the concentrations therefore multiplied up the food chain. The four bird-eating specialists, Sparrowhawk, Hobby, Merlin, and Peregrine, suffered worse than the mammal-eating Kestrel and Buzzard; birds in south-eastern lowland, agricultural areas suffered worse than birds in north-western uplands. High levels killed the adults directly; worse, lower levels turned out to be hormone mimics for birds that affected the secretion of calcium-rich eggshells, so that breeding success was badly affected even when doses were not sufficient to kill. This was best documented for wild birds in Peregrines, but later was also shown experimentally using American Kestrels (Ratcliffe, 1980). At the population level, the consequence was minima for the Peregrines in 1962–63, for Sparrowhawks at about the same time, but, somewhat puzzlingly, for Merlins about 1980. The Hobby, as a migrant to southern Britain, seems to have been more affected by weather than pesticides, and was reduced to about 50–70 pairs, confined to the sunniest areas of southern England, during 1900–50 (Galbraith *et al.*, 2000). Whether released from previous constraints by withdrawal of pesticides or by warmer summers, its numbers have increased markedly in the last decade, and it is now thought to number 500–900 pairs, and to breed as far north as southern Scotland. This increase is matched by equally spectacular recoveries of Osprey, Sparrowhawk, Buzzard, Marsh Harrier, and Peregrine. Goshawks, surreptitiously, and White-tailed Eagles, officially, have been reintroduced, while the Welsh Red Kite population has finally climbed away from the dire straits in which it languished for so many years, and has been joined by a very successful reintroduced population in England and another, more constrained one, in Scotland. However, the Hen Harrier has not increased so well, particularly not in England, suffering from the same problem that afflicts the Scottish Red Kites – continued illegal persecution. The very success of Peregrines, recolonizing not only the coastal and montane cliffs but also urban sites, and of Sparrowhawks also recolonizing suburbs and city centres, has led to calls for their culling.

There are four separate problems: predation on grouse from grouse moors; predation on other gamebirds; predation of homing pigeons; predation of garden birds. Each has been well studied and is well understood by ornithologists, but their results are either not accepted or not acceptable to the interested parties. The UK Raptor Working Group produced excellent summaries of the first three (Galbraith *et al.*, 2000), and many have considered various aspects of the last.

The first is perhaps most acute, because gamekeepers and others concerned with grouse moor management are generally well equipped to ignore the legal protection of raptors. The breeding success of Hen Harriers was only 0.8 fledglings per breeding female per year on grouse moors, compared with 2.4 on other moorland, and the survival rate of females was poorer. Other raptors (Buzzard, Red Kite, Golden Eagle) have also been shown to suffer poorer breeding success and survival in grouse-shooting areas. Prosecutions, usually by the

RSPB, have confirmed some cases of direct and deliberate persecution; other cases have involved the illegal setting of poison baits in the open, perhaps aimed at Foxes and Crows, but killing also protected species of carrion-feeding birds. Unfortunately, detailed studies have also shown that, so far as Hen Harriers are concerned, game interests have some reason for genuine concern (Redpath & Thirgood, 1997). At Langholm Estate in the Borders, Hen Harriers were protected, and increased from two breeding females in1992 to 20 in 1997. In 1995–96, with eight and 12 female harriers nesting, about 30% of the adult Red Grouse were killed by raptors between October and March, and, worse, Hen Harriers fed about 37% of the grouse chicks to their own young during the summer. Although the size of the breeding grouse population was not seriously affected, the numbers of grouse available for shooting in the autumn was reduced by 50%. With further increases in harriers, there were eventually so few August grouse available that shooting was suspended, and the gamekeepers lost their livelihoods. Paradoxically, numbers of Hen Harriers are not related to numbers of grouse – grouse chicks are, as it were, a bonus food for Hen Harriers provisioning their chicks. Rather, the number of Hen Harriers is determined by the density of Field Voles and Meadow Pipits in spring. These in turn depend on grass, not heather, so a well managed grouse moor (which requires gamekeepers), predominantly covered in heather of various ages, will contain little grassland, few voles or pipits, and support few Hen Harriers. What converts heather moorland into grassland is excessive sheep grazing; on Langholm, heather had declined by 48% between 1948 and 1988. Sheep have been subsidized, but grouse moors are taxed. If we want to see well managed moors, with numerous Red Grouse and other moorland birds, including a modest number of Hen Harriers, the economic balance has to be changed. The payments for heather management under recent ESA (Environmentally Sensitive Area) schemes in England are a step in the right direction. It is unreasonable to expect game interests to sustain heavy losses from the predators that the rest of us want to see with no financial compensation. Subsidies for fledgling Hen Harriers produced would be no less reasonable than subsidies for supporting excessive numbers of sheep.

Predation of other gamebirds is a less critical issue now that most Pheasants are reared for release, rather than depending on successful breeding in the wild, though it is necessary to ensure that the rearing pens are safe from marauding Tawny Owls, Buzzards, and Red Kites (as well as Foxes, Pine Martens, and other mammals). In the past, nesting lowland gamebirds, mostly Pheasants and Grey Partridges, were vulnerable to mammalian predators and corvids that hunted along the hedgerows for eggs and nesting females. Grey Partridges have declined largely as a result of changes in farming. Killing of Foxes and corvids may be needed to ensure their adequate breeding success (Tapper *et al.*, 1999), but this is legal; farm-land raptors play no significant part in regulating their numbers, and illegal killing of them is quite unjustified. One problem might be the impact of Goshawks, newly recolonizing after illicit reintroduction, on Black Grouse, in serious decline everywhere. It is well documented that Black Grouse are vulnerable to Goshawk, as well as other predators. There is no suggestion that the decline of Black Grouse has been driven by predation: rather it seems that the main problem is the loss of bilberry, and the insect food it supports, because of heavy sheep and deer browsing, combined perhaps with the intensive farming and forestry that has 'tidied up' its moorland edge habitat. On the other hand, there is evidence that a low density Black Grouse population may suffer from a 'predator trap', in which even small predator populations, supported by other more abundant prey, can kill enough of a rarer prey to prevent it recovering, particularly when that rarer prey is particularly favoured by or vulnerable to the predator. The classic example of this is the Tree Pipits predated by Sparrowhawks in Holland, as studied by Tinbergen (1946). Tree Pipits are relatively scarce, and could not support a Sparrowhawk population; woodland Sparrowhawks are usually sustained by the much more abundant Chaffinches, Blue Tits, and Great Tits. Because Tree Pipits sing from exposed twigs, or parachute slowly down to their song posts, they are extremely vulnerable to Sparrowhawks, and were exterminated from Tinbergen's study site. Black Grouse might, in low numbers, be similarly vulnerable to Goshawks.

The problem of Peregrine predation on Homing Pigeons is also a contentious and difficult one. It was requests from pigeon racers for a cull of Peregrines because they were losing too many birds that first drew attention, paradoxically, to how few Peregrines were left in 1962, and how poorly they were then breeding. At that time, it is clear that Peregrine predation played at most an insignificant part in pigeon losses, and other factors, such as weather, were the principal cause. This raises a cautious scepticism from ornithologists and conservationists now, when the undoubtedly far more numerous Peregrines are blamed for considerable losses of pigeons. There is no doubt that Peregrines do kill racing pigeons, and recoveries of pigeon rings are a routine matter for ringers and others monitoring the success of Peregrine eyries. The scale of these losses, in relation to losses caused by weather (including magnetic storms), inexperienced birds losing themselves, and other accidents, is less clear. The UK Raptor Working Group reported that some 2.25 million adults and another 2.5 million young pigeons are raced each year, an earlier (April–July) season for adults and a later (July–September) one for young birds; cumulatively about 9.2 million bird-days raced annually. It is thought that about 52% are lost to all causes, but that only 7.5% of all racing pigeons are lost to raptors (Galbraith *et al.*, 2000). It seems likely that the considerable increase in Peregrine numbers in the Lake District owes much to the heavy passage of vulnerable racing pigeons at just the time when the Peregrines are feeding their growing young. This suggests two ways in which losses of racing pigeons could be minimized without having to kill Peregrines: change the routes on which pigeons are raced, to avoid the densest populations (in particular, flying them down the Eden valley, between the Lake District and Pennines, seems foolhardy), and alter the timing of the races, particularly those involving training of young inexperienced birds, to avoid this critical time in the Peregrine breeding season. It might even cause the number of occupied Peregrine eyries to decline if this prey source were to be reduced at a critical time. Meantime, Peregrines have begun to move into cities, where the large populations of Feral Pigeons, many descended from lost racers and some 3.6% still carrying their racing rings, provide a prey base that few would object to being culled. Urban Peregrines, and Sparrowhawks, do pose a second threat to racing pigeons; training flights can provide another predictable and vulnerable source of food. Mitigating such losses is less easy, though varying the times of training flights would make the raptors' food supply less predictable.

The impact of Sparrowhawks and other predators, including Magpies, on their nests and young, is regularly cited by correspondents to newspapers and others as the major cause of the decline of common songbirds. This is invariably joined with emphatic suggestions that these predators should be culled. This is a much researched topic, on which the scientific evidence is very clear but apparently uncomprehended, perhaps incomprehensible, to the suburban majority. There is no doubt that some songbirds, notably, in a garden context, Song

Thrushes and House Sparrows but also such once common species as Yellowhammer, have declined. If this were due to raptor predation, it should show itself as reduced adult survival rates or poorer nesting success, but neither of these is apparent. Moreover, many species that are the main prey of Sparrowhawks, including Robins, Blackbirds, Blue Tits, and Great Tits, have not declined over the period in which Sparrowhawk numbers have recovered. The most informative study is from Wytham Woods, Oxford, where most Blue and Great Tits nest in boxes, and have been well studied continuously for over 50 years. Thus their populations have been studied while Sparrowhawks were still common, when they were absent due to organochlorine poisoning (1964–70) and since their return (1971–84).The population of Great Tits averaged 188 pairs and 206 pairs in the 320 ha wood during these two periods, not a significant difference; adult survival rates were not significantly different either (Perrins & Greer, 1980; Newton, 1986). A study of the population sizes of a wider range of 20 songbirds, before, during and after the Sparrowhawk's absence from south-east England, similarly shows no greater abundance while the predator was absent nor a decline since it has returned (Galbraith *et al.*, 2000). While they were absent, far more songbirds died in winter of starvation; now they have returned, more are killed by predators, but the overall populations are little affected. That is not to say that the return of the Sparrowhawks had no effect on the Wytham Great Tits. They now keep themselves slimmer in mid winter by about 1 g, so that they are more agile and able to avoid their predators (Gosler *et al.*, 1995). One reason for the impact of Sparrowhawks on human perceptions has been the fact that they were absent for so long; in the London Area, Sparrowhawks were reported from only 50 tetrads in the 1968–72 breeding bird atlas, but from 579 in 1988–94 (67% of the total 859 tetrads; Hewlett, 2002). Their return as regular predators of garden birds, even in urban areas, has come at a time when feeding birds in gardens, therefore making them good hunting sites for Sparrowhawks, has reached a peak of popularity. Much the same is true for Magpies, which have also increased about threefold since the 1960s. From a first appearance in Central London in 1971 (single pairs in Hyde Park and Regent's Park), they now breed in nearly every square, park, and cemetery (Hewlett, 2002). As a consequence of their abundance and familiarity, their depredations on songbirds are very noticeable, and resented. A study of Blackbird nests in Manchester parks suggested that only 7% were successful to rearing young, and that most eggs were taken by Magpies (dummy plasticine eggs revealed their beak marks) (Groom, 1993). On that basis, Blackbirds should be extinct as urban birds, but of course they are as common as ever. The problem is that we have made parks excellent habitat for Magpies (lawns for hunting insects, tall trees for safe nesting) and very bad nesting habitat for Blackbirds (no hedges or thickets); fortunately, Blackbirds in gardens nest much more successfully than those in parks (but are much harder to study) so support the park populations. Just as Hen Harrier populations on grouse moors are supported by numbers of Meadow Pipits and Field Voles, urban Magpies are supported by the large numbers of grassland insects (Tatner, 1983). Chicks are a minor but welcome supplementary food for both moorland and urban predators at critical times in their breeding cycles.

In a world where concern and support for birds has never been stronger, and where declines of many birds, especially of farmland, has been so strongly documented, the success stories tend to get overlooked. We now have more species of breeding raptor, and far more individual raptors, than at any time in the twentieth century. Leslie Brown (1976), in his superbly written and provocative synthesis, thought we had 13 breeding species, possibly

14 (the Goshawk was then an uncertain breeder) totalling about 188,500 birds in Great Britain. By 1993, there were certainly 15 species and about 205,100 birds (Gibbons *et al.*, 1993). More recent estimates suggest an increase to 221,600 birds, despite some reductions in apparent populations due to better surveys (Galbraith *et al.*, 2000); all raptor species are thought to be more numerous and widespread than formerly. The total number of breeding bird species in the British Isles, about 220 species, is also as high as it has ever been, and the few certain recent losses in the last 50 or 60 years (Kentish Plover, perhaps Wryneck, Red-backed Shrike) have certainly been exceeded by returnees (Savi's Warbler, Avocet, Bittern, Marsh Harrier, Spoonbill, Little Egret) and new colonists (Collared Dove, Little Ringed Plover, Firecrest, Cetti's Warbler).

It is a great time to be an ornithologist in Great Britain, Ireland, and Man.

Appendix
An annotated historical list of British birds

This list attempts to summarise the historical record of each species, list all the regular breeding and wintering birds, and provide a list of the scientific names of birds mentioned in the text. For many smaller birds there is no useful historical record, and they are simply listed.

The cautions offered in Chapter 1, about having to accept published identifications at face value and the difficulties of identifying closely related species, for example, in *Anser*, *Anas*, *Tringa*, *Turdus*, particularly apply here. The numbers of records quoted are those in our database, about 9,000 in all as of December 2007, covering as comprehensive a record as we could accumulate, from middle Pleistocene (Cromerian) onwards. We loosely refer to this as the archaeological record, though many sites are strictly not archaeological sites (do not contain Human remains or artefacts). Dating and the periods we quote are discussed in Chapter 1. Where possible, we refer back to earlier tables.

Mute Swan *Cygnus olor*
Once supposed to be a Mediaeval introduction, but in fact a native species with a good archaeological record, 59 sites from Late Glacial and Mesolithic onwards (Table 4.3).

Bewick's Swan *Cygnus columbianus*
Thinly recorded in the archaeological record, 13 reports, several from Late Pleistocene cave sites, and then from Iron Age Meare and Danebury, Roman York and Longthorpe, and Early Christian Lagore.

Whooper Swan *Cygnus cygnus*
Better recorded, from 33 sites, including Cromerian Boxgrove, Mesolithic Gough's Cave, 3 Neolithic, 1 Bronze Age, 3 Iron Age, 6 Roman and 10 later sites. Mostly N sites, but, for example, Iron Age Meare, Roman Silchester, suggest not very different from modern wintering range.

Bean Goose *Anser fabalis*
Only claimed from six definite records: Ipswichian Bacon Hole, two Late Glacial caves, Bronze Age Elsay Broch, Roman Towcester and Early Christian Lagore; also three possible (Bean/Greylag) records.

Pink-footed Goose *Anser brachyrhynchus*
With 23 records and another 4 uncertain Pink-footed/White-fronted Geese, reasonably well recorded. Identified at Devensian Pinhole and Late Glacial Robin Hood's Cave, also

St Brelade's Bay, and at Neolithic Rousay, Dunagoil, Iron Age Bu, Harston Mill, Roman York, London Wall, Saxon Flixborough, Mediaeval Perth, Beverley, Northampton, Kings Lynn.

White-fronted Goose *Anser albifrons*
The 23 records span Wolstonian Swanscombe, Ipswichian Ilford, Devensian Pinhole, Robin Hood and Langwith Caves, to Mesolithic Port Eynon, Iron Age Howe, Meare, Saxon West Stow, N Elmhan, York (Coppergate), Mediaeval Dyserth Castle. In Ireland, at several poorly dated cave sites (Alice, Keshcorran, Catacomb, Castletownroche, Newhall) and from late Christian Lagore, Mediaeval Valencia.

Lesser White-fronted Goose *Anser erythropus*
Reported from Late Glacial Soldier's Hole.

Greylag Goose *Anser anser*
With 67 records, well reported, from Cromerian West Runton, Boxgrove, Ipswichian Kirkdale Cavern, numerous Late Glacial, Mesolithic and later records (summarised with other wild geese in Table 5.1).

Domestic Goose *Anser anser domesticus*
Certainly kept by the Romans, arguable whether they had domestic geese in Britain; certainly common from Anglo-Saxon times onwards, the common table bird at feasts through Anglo-Saxon to Mediaeval times, until supplanted by Turkey (see Table 5.1).

Canada Goose *Branta canadensis*
Introduced from Canada around 1660, present St James Park 1665; establishing in the wild by early nineteenth century. Now second only to Pheasant in biomass of introduced wild species (see Table 8.5).

Barnacle Goose *Branta leucopsis*
Well recorded, for a bird with a limited modern distribution and status: 53 records, ranging from the Wolstonian (at Swanscombe) and Late Glacial caves (Inchnadamph, Robin Hood, Port Eynon) to Mediaeval; many records northern (Orkney S to York), but also, for example, Iron Age Meare, Mediaeval Oxford. In Ireland at Shandon Cave and Lagore, where commoner even than Domestic Fowl. Identifications mostly OK, because much larger than Brent, and *Branta* has distinctions from *Anser*, though difficulties indicated by additional seven uncertain Barnacle/White-front and four Barnacle/Brent. Molecular confirmation of identity at Flixborough.

Brent Goose *Branta bernicla*
Distinctively smaller than Barnacle, much larger and morphologically distinct from Shelduck. Present at 29 sites, from Ipswichian Tornewton Cave, Late Glacial Pinhole, Robin Hood's Cave and St Brelade's Bay, Jersey; then Mesolithic Star Carr, Iron Age Bu, Howe, through to Roman, Anglo-Saxon, Mediaeval and later sites. Many coastal (Lindisfarne, Caldicot, Flixborough) matching present distribution, but also inland at, for example, York, Lincoln, presumably traded for food. In Ireland, present at Lagore, also in later Clonmacnoise, Carrickfergus, Dublin.

Red-breasted Goose *Branta ruficollis*
Reported twice, from Ipswichian Grays and from post-Mediaeval Newcastle (Mansion House); now rare winter visitor.

Ruddy Shelduck *Tadorna ferruginea*
Recorded from four caves, Ipswichian Tornewton, and Devensian/Late Devensian Pinhole, Robin Hood and Ossiferous Fissure C8 at Creswell. Now a rare vagrant from SE Europe.

Shelduck *Tadorna tadorna*
Recorded 29 times, from Wolstonian and Ipswichian Tornewton Cave, Late Glacial Neale's Cave, Torbryan Cave and Kent's Cavern. From Mesolithic Port Eynon, Neolithic Oronsay, Links of Noltland, Iron Age Meare, Skaill through Roman (Lincoln, Portchester), Saxon/Norse (N Elmham Park, Buckquoy, Jarlshof) to Mediaeval times (Lincoln, Thetford, Abingdon, Oxford).

Mandarin *Aix galericulata*
Claimed from Cromerian West Runton. Modern populations of this Chinese species derive from releases/escapes in 1930s and later.

Wigeon *Anas penelope*
Recorded 75 times, so presumably formerly as numerous in winter as now. Earliest Cromerian West Runton, then Boxgrove, Ipswichian, Tornewton Cave. In Late Glacial caves (e.g. Walton, Soldier's Hole, Pinhole, Kirkdale), Mesolithic Inchnadamph. At 6 Iron Age, 13 Roman, 9 Saxon/Norse, 20 Mediaeval or later sites. In Ireland, at three cave sites (uncertainly dated), Mesolithic Mount Sandel and Christian Lagore.

Gadwall *Anas strepera*
Reported from 20 sites, Ipswichian Waterhall Farm, then from Mesolithic Demen's Dale through 4 Iron Age and 4 Roman sites to 6 Mediaeval sites (e.g. Oxford, King's Lynn, Coventry, Beverley).

Teal *Anas crecca*
With 173 records, well documented because distinctively smaller, though two more "Teal/Garganey" records add a caution over id. From Cromerian West Runton, post-Cromerian Boxgrove, Devensian Pinhole, Tornewton, Late Glacial Robin Hood's, Cat Hole, then Mesolithic Demen's Dale, Inchnadamph, Neolithic Point of Cott, Dowel Cave, but mostly Roman (45) and Mediaeval/post-Mediaeval (78) records.

Mallard *Anas platyrhynchos*
With 251 records, as abundant in the archaeological record as the modern landscape. Recorded in post Cromerian Boxgrove, Westbury, then Devensian Chelm's Combe, Late Glacial Merlin's Cave, Pinhole. Numerous in Roman and Mediaeval sites (see Table 5.1).

Domestic Duck *Anas platyrhynchos domesticus*
Difficult to distinguish from its wild ancestor. Romans certainly had domestic ducks, argued whether they brought them to GB. Numerical analysis (Table 5.1) suggests they did, uncertain if known to Saxons, alternative view that not domesticated in GB till early Mediaeval period.

Pintail *Anas acuta*
Of 19 records, earliest are Mesolithic Star Carr, Demen's Dale. Later sites include Neolithic Mount Pleasant, Bronze Age Caldicot, Iron Age Meare, Glastonbury, Howe, Roman Caerwent, Barnsley Park, Saxon Westminster Abbey, Portchester and Mediaeval Barnard Castle, Kings Lynn. Also, in Ireland, at Newhall Cave (undated), Keshcorran Cave, Ballinderry and Lagore crannogs.

Garganey *Anas querquedula*
Marginally larger wing bones, and some morphological details, sometimes allow differentiation from very similar Teal. Recorded at Cromerian Boxgrove, Devensian Pinhole, Mesolithic Demen's Dale, and at a range of Roman -Mediaeval sites. An additional three Garganey/Teal records attest to that obvious uncertainty.

Shoveler *Anas clypeata*
With 20 records, from Hoxnian Swanscombe, Devensian Chelme's Combe, Kent's Cavern, also Ightham Fissure (date?), then well scattered through Mesolithic Demen's Dale, Iron Age Meare, Scatness, Roman York, Caeseromagnus, Saxon Thetford to Mediaeval Carisbrooke and Baynard's Castles.

Red-crested Pochard *Netta rufina*
Twice reported from Cromerian (West Runton, Ostend), also from Iron Age Glastonbury.

Pochard *Aythya ferina*
Reported 25 times, from Cromerian Ostend, West Runton, then Mesolithic Demen's Dale, Gough's Cave; Iron Age Meare, Glastonbury, Howe; Roman Fishbourne, Barnsley Park, Longthorpe, Verulamium; 12 later sites, and 2 undated Irish caves (Newhall, Catacomb).

Tufted Duck *Aythya fuligula*
With 34 records, the most common diving duck, as in the present fauna. Known from Cromerian West Runton, Devensian Pinhole, Late Glacial Walthamstow, Gough's Old Cave, Kent's Cavern; then Mesolithic Inchnadamph, Demen's Dale. No Neolithic records, but at Iron Age Meare, Glastonbury, Danebury, Roman Godmanchester, Colchester, Lincoln, 12 later records.

Scaup *Aythya marila*
Recorded eight times, but unusually more Irish than British records: Castlepook and Keshcorran Caves (dates uncertain), Ballinderry crannog (both Bronze Age and Early Christian), Lagore (Late Christian); also Iron Age Meare, Glastonbury and post-Mediaeval Peel (Man).

Eider *Somateria mollissima*
Of 20 records, 18 from N Scotland, mostly Orkney and Shetland, Mesolithic Inchnadamph to Norse Buckquoy: only two further S (Mediaeval Hartlepool, undated Whitrig Bog, Wigton). Suggests that never widespread, but surprising that apparently no Late Glacial records from S.

Long-tailed Duck *Clangula hyemalis*
Only six records, from the Norwich Crag at Southwold, then Late Glacial Inchnadamph, Mesolithic Port Eynon, Roman Wroxeter, Longthorpe and Postmediaeval Peel.

Common Scoter *Melanitta nigra*
For a relatively scarce modern duck, well represented, 23 records, from Cromerian Mundesley and Wolstonian Swanscombe, Late Glacial Inchnadamph and Merlin's Cave, then from Mesolithic Star Carr to Mediaeval Lindisfarne. Three Mediaeval "scoter sp." (Hartlepool, King's Lynn, Oxford) are probably also this species.

Velvet Scoter *Melanitta fusca*
Only seven or eight records, from Mesolithic Port Eynon Caves, Inchnadamph, Neolithic Links of Noltland, Papa Westray, Iron Age Howe, Viking Jarlshof, post-Mediaeval Lindisfarne.

Goldeneye *Bucephala clangula*
An osteologically distinctive species with 24 records, from Cromerian West Runton, Boxgrove, through Late Glacial Pinhole, Robin Hood's Cave and Mesolithic Demen's Dale, Thatcham to Mediaeval York, Beverley and Norwich. Found on S sites (Iron Age Meare, Glastonbury) as well as N sites (Iron Age Howe, Norse Buckquoy), so probably wintered further S, bred further N, just as now.

Smew *Mergellus albellus*
Recorded 15 times, from Cromerian West Runton, Ipswichian Crayford, Late Glacial Merlin's cave, Chudleigh Fissure, Mesolithic Port Eynon, Bronze Age Burwell Fen, Iron Age Meare, Glastonbury, Howe, Saxon Lincoln to Mediaeval Leicester, Beverley. Only Irish record Keshcorran cave, date uncertain.

Red-breasted Merganser *Mergus serrator*
The 16 records range from Cromerian West Runton, Hoxnian Swanscombe, Late Glacial Merlin's Cave through Mesolithic (Star Carr, Risga), Neolithic (Point of Cott, Oronsay), Iron Age (Meare, Glastonbury, Howe) to Mediaeval Baynard's Castle, later Lindisfarne. In Ireland, at Lagore.

Goosander *Mergus merganser*
Has 15 records, from Wolstonian Tornewton, mid-Devensian Pinhole, Late Glacial Robin Hood's, Gough's Old Caves, Mesolithic Demen's Dale and a scatter in time and space from Iron Age Howe and Meare to Mediaeval Baynard's Castle, Post-Mediaeval Peel.

Red Grouse *Lagopus lagopus scotica*
Earliest post-Cromerian Westbury; with 54 records, present in most (20) Devensian/Late Glacial caves; well distributed in N and W, as far S as Iron Age Danebury, Meare. Few Roman (Corbridge, Great Staughton), Saxon (Ipswich) or Mediaeval (York (Bedern), Castle Sween, Freswick Castle) records. Numerous records from Neolithic to Norse Orkney (Bu, Skaill, Isbister, Howe, Buckquoy, Quanterness); in Ireland, from Castlepook, Ballynamintra and Alice caves (undated), Mesolithic Mount Sandel.

Ptarmigan *Lagopus muta*
With 29 records, widely distributed in space, but not time. Earliest record, post-Cromerian Westbury. Numerous in Devensian/Late Glacial period, as far S as St Brelade's Bay (Jersey), and in various caves in Devon, Somerset, Derby. Latest records Mesolithic (Demen's Dale, Gough's Cave, Inchnadamph) and Neolithic Elbolton Cave, Yorkshire. Known from two caves in Ireland (Shandon, Ballynamintra) but dating uncertain.

Capercaillie *Tetrao urogallus*
Quite well represented archaeologically, with 27 records; from Wolstonian Swanscombe, then Late Glacial Kent's and Kirkdale Caverns; present in Mesolithic Wetton Mill, Dowel Cave, Neolithic Fox Hole, through to Mediaeval York, Leicester; in Ireland, from Mesolithic Mount Sandel, Anglo-Norman Trim Castle to Mediaeval Dublin, Waterford, Wexford and later Carrickfergus, Galway (see Table 6.6).

Black Grouse *Tetrao tetrix*
Well represented archaeologically, by 63 records; in time, spread from Late Glacial caves (Gough's Old Cave, Ossom's Soldier's Hole, Pin Hole) to Mediaeval; geographically, many N records, but as far S as Iron Age Meare. One dubious record from Ireland, Ballynamintra Cave.

Hazel Hen *Bonasia bonasus*
Five records: Post-Cromerian Westbury and then four Late Glacial records which suggest that was native to S Britain for a short period (see Table 3.1).

Grey Partridge *Perdix perdix*
With 126 records, one of the best recorded wild birds: a common bird of open ground throughout its long history (Table 3.6), and popular prey for both Humans and other predators. Known from post-Cromerian Boxgrove. Frequent in Glacial and Late Glacial cave sites. Fewer Mesolithic-Iron Age records, common from Roman times onwards. Seems a good indicator of more open conditions, from earlier tundra to farmland when it became more available.

Red-legged Partridge *Alectoris rufa*
Claimed from Late Glacial St Brelade's Bay, Jersey, perhaps just within its native range, and from Roman Fishbourne. Latter a likely import for food, was illustrated in Roman mosaics. Modern populations date from introductions for hunting, especially to Suffolk, from 1790 onwards.

Quail *Coturnix coturnix*
Recorded 23 times, from Devensian Torbryan, Late Glacial Chudleigh Fissure, Merlin's Cave; then Neolithic Quanterness, Iron Age Bu, Roman Frocester, Great Staughton, Maxey, York and 10 later records. From Ireland, reported from Newhall and Castlepook caves (undated), Trim and Armoy.

Pheasant *Phasianus colchicus*
If some dubious early records are ignored, known from a few Roman sites, but only numerous from Late Saxon through Mediaeval and later times. Some 58 records (see Table 5.3).

Domestic Fowl *Gallus domesticus*
Domesticated in China by 7000 bp, reached Britain in Late Iron Age about 100 BC, common and widespread ever since; our most common bird, by a wide margin (see Table 5.1).

Guineafowl *Numida meleagris*
Known to the Romans in N Africa, but no evidence in the British archaeological record.

Peacock *Pavo cristatus*
An Indian bird, known to the Romans; a scatter of early British records from Roman (Portchester, Great Staughton) and Anglo-Norman (Thetford, Faccombe Nettleton, York) times, rather more in Mediaeval and later times, 35 in all (see Table 5.1).

Turkey *Meleagris gallopavo*
Nearly all of 47 records late or post-Mediaeval, as expected from known historical date of introduction from America around 1530 (see Table 5.1). Just five earlier records: "hoodwinks" or misidentified?

Red-throated Diver *Gavia stellata*
Recorded from 12 sites, earliest Cromerian Mundesley, then Mesolithic Star Carr; Iron Age Meare, Scalloway, Viking Jarlshof, Skaill. Also in Ireland, at Newhall and Shandon caves (undated), Mesolithic Mount Sandel, Christian Lagore, later Dublin (Wood's Quay).

Black-throated Diver *Gavia arctica*
Recorded from six sites: mid Devensian Pinhole Cave, Mesolithic Port Eynon Cave, then Neolithic Papa Westray, Roman Brancaster, Mediaeval Baynard's Castle, Exeter; also one possible record, of Black/Red-throated Diver from Viking Brough of Birsay.

Great Northern Diver *Gavia immer*
With 18 records from Neolithic Papa Westray and Links of Noltland onwards, surprisingly well represented. While 10 records are from N sites (mostly Orkney), also recorded from Iron Age Meare, Roman Halangy Down, Saxon Southampton and three layers at Porchester. One Irish record, undated Catacomb Cave.

Dabchick (Little Grebe) *Tachybaptus ruficollis*
Only nine records: Mesolithic Star Carr, Neolithic Papa Westray, Bronze Age Caldicot, Iron Age Meare, Glastonbury, later London Wall, and Norwich (Castle Mall), plus two undated Irish records (Newhall and Catacomb Caves).

Great Crested Grebe *Podiceps cristatus*
A thin archaeological record, including four from Irish cave sites, with dubious dates, then Mesolithic Star Carr, Iron Age Meare, Christian Lagore, Mediaeval Poole and Beverley, and later Lindisfarne. Its heavy exploitation in Victorian times is not matched by evidence of heavy earlier killing.

Red-necked Grebe *Podiceps grisegena*
Reported from Neolithic Embo, Sutherland.

Slavonian Grebe *Podiceps auritus*
Reported from Iron Age Howe and tenth to eleventh century Jarlshof.

Fulmar *Fulmarus glacialis*
Historically only occurred on St Kilda in the nineteenth century, then spread widely round British coasts during the twentieth century. However, 20 records span from Mesolithic Morton through to numerous N sites in Neolithic and Iron Age, especially in Orkney. Perhaps nearly exterminated by Human overhunting (see Table 4.2).

Cory's Shearwater *Calonectris diomedea*
A single record, from Ipswichian Bacon Hole; a southern/Mediterranean species which appears regularly offshore in SW Britain. Was this an early example of a wrecked bird, or an indication of its presence here in warmer times?

Manx Shearwater *Puffinus puffinus*
With 39 records, from Late Glacial Potter's Cave, numerous N sites, for example, Neolithic Papa Westray, Links of Noltland, Bronze Age Nornour, Elsay Broch, Iron Age Crosskirk, Scatness, Howe. At several high status Mediaeval sites (Iona, Castletown, Hartlepool, Newcastle, Launceston and Dudley Castles). Three "shearwater sp." (Neolithic Westray, Papa Westray; Iron Age Glastonbury) probably belong here.

Sooty Shearwater *Puffinus griseus*
Reported from Happaway Cave (date uncertain) and Iron Age Howe; since substantially larger than Manx (which also recorded at Howe) id safe enough, but significance of this S Atlantic breeder unclear. Regular now off British coasts in late summer, so storm-driven?

Storm Petrel *Hydrobates pelagicus*
Only recorded from Bronze Age Jarlshof.

Leach's Petrel *Oceanodroma leucorrhoa*
Two records, from Neolithic Quanterness and Norse Jarlshof.

Fea's Petrel? *Pterodroma* cf. *feae*
A small *Pterodroma* recorded from three sites in the Scottish isles (The Udal, N Uist; Kilellan Farm, Islay; Brettaness, Rousay) is closest to *P feae*. All Scottish Iron Age, but probably contemporary with Anglo-Saxon England. Mixed with the bones of other seabirds that had been eaten, so presumably also food (see Chapter 6).

Gannet *Morus bassanus*
A distinctive species, on size and morphology, recorded 55 times; mostly from N/island sites, as expected; from Late Glacial Paviland, Mesolithic Morton, Risga and Port Eynon Cave through to Mediaeval sites (see Table 4.1); but records from, for example, Mediaeval Launceston and Okehampton Castles, Hereford, indicate some trading inland.

Cormorant *Phalocrocorax carbo*
Well represented, 81 records, from Hoxnian Swanscombe, Mesolithic Risga, Morton; common in northern and coastal sites, but also inland, for example, Iron Age Meare, Glastonbury, Roman Stonea. Also five uncertain Cormorant/Shag records, all coastal.

Shag *Phalocrocorax aristotelis*
Also well represented, 54 records, though only one early (Devensian Kent's Cavern); then Mesolithic Risga, Morton, Port Eynon Cave, Neolithic Point of Cott, Links of Noltland, Rousay, etc. Forty-two later sites. Mostly coastal, from Shetland (Jarlshof) to Guernsey (Le Dehus) but, for example, Roman Stonea, Mediaeval Stafford Castle suggest some trading inland for food.

Pygmy Cormorant *Phalacrocorax pygmaeus*
A single record, a pair of distinctive metacarpi from fifteenth to sixteenth century Abingdon (Table 4.4). Now no nearer than Balkans. Traded, wild or a pet?

Dalmatian Pelican *Pelecanus crispus*
Bred in the fenlands of East Anglia and Somerset in Bronze and Iron Age times, where known from 10 records (see Table 4.4). Now no nearer than Balkans.

Bittern *Botaurus stellaris*
Recorded from about 21 sites; well-dated records from Mesolithic Star Carr, Neolithic Rousay, and through Bronze Age to Mediaeval times (see Table 4.4). Surprisingly, no Irish archaeological records, despite known historical occurrence.

Night Heron *Nycticorax nycticorax*
Only two records, Roman London Wall and Postmediaeval Royal Navy Victualling Yard (Table 4.4); historical documentation suggests that frequently eaten (and known as Brewes; imported or native?) in late Mediaeval period.

Little Egret *Egretta garzetta*
A single record, probably of Roman date, from London Wall; historical documentation suggests that frequently eaten in late Mediaeval period. Barely recorded before 1950, but increasingly regular from then, started breeding England 1996, Ireland 1997.

Heron *Ardea cinerea*
Distinctive size and morphology, plus targeted hunting, give this species a good record, from 84 sites. Scarce in caves (Devensian Walton, Ossiferous Fissure C8, Pinhole), apparently no Mesolithic or Neolithic records, but more common from Bronze Age (Nornour, Caldicot, Jarlshof) and later times. Numerous Mediaeval-Post Mediaeval records from high-status sites (e.g. Stafford, Baynard's, Hertford and Okehampton Castles, Faccombe Netherton), presumably indicating hawking (see Table 6.4).

White Stork *Ciconia ciconia*
Recorded 10 times, from Wolstonian Tornewton Cave, Mid-Late Devensian Pinhole, Robin Hood Caves, Bronze Age Nornour, Jarlshof, Iron Age Harston Mill, Dragonby, Roman Silchester, Saxon London (Westminster Abbey). Only Mediaeval record, Oxford (St Ebbes), but three records of "stork sp.", from Leicester (Little Lane), Beaurepaire and Poole probably belong here.

Black Stork *Ciconia nigra*
Two possible records of this solitary woodland species, from Devensian Tornewton Cave and from Lynx Cave, Clwyd, dated by C^{14} to 2,945 bp; identification is difficult, however, but these would be interesting indicators of former wooded conditions.

Spoonbill *Platalea leucorodia*
Only two records, Mediaeval Castle Rising Castle and Southampton (Cuckoo Lane).

Honey Buzzard *Pernis apivorus*
Scarce at best as a breeding bird. Ought to have been more numerous in wooded Mesolithic Britain, but no archaeological record. Overlooked, or never common? (see Table 3.5).

White-tailed Eagle *Haliaeetus albicilla*
Well recorded, much more widespread and numerous than Golden Eagle; 58 records, from Wolstonian Tornewton Cave, Devensian Soldier's Hole, Walton Cave, Walthamstow,

Mesolithic Skipsea, Port Eynon. Numerous records from N and W islands (Isbister, Links of Noltland, Iona), but as far S as Iron Age Meare, Dragonby, Roman Leicester. In Ireland, from Neolithic Lough Gur, Dublin (Dalkey Island), Christian Lagore, Mediaeval Waterford, Dublin (Woods Quay) (see Tables 6.3 and 6.7).

Griffon Vulture *Gyps fulvus*

Monk (Black) Vulture *Aegypius monachus*

Bearded Vulture (Lammergeier) *Gypaetus barbatus*

Bonelli's/Booted Eagle *Hieraaetus fasciatus/pennatus*
One record of one or other of these similar-sized relatives from Ightham Fissures, of uncertain date.

Red Kite *Milvus milvus*
Well recorded, 71 sites; first from Ipswichian Bacon Hole. Absent from Glacial/Late Glacial/Mesolithic sites. Present from Neolithic Durrington Walls, Iron Age Meare, Glastonbury, Danebury, Howe, onwards. At 14 Roman, 8 Saxon, 40 Mediaeval/post-Mediaeval sites (see Table 6.3); evidently responded to the development of open farmland and towns – a sign of scavenging? In Ireland, present at four Mediaeval sites in Dublin, also Dundrum and post-Mediaeval Roscrea.

Black Kite *Milvus nigra*
No archaeological evidence of this smaller species, now a scarce but regular migrant.

Marsh Harrier *Circus aeruginosus*
Only 15 records, all relatively late: Iron Age Meare, Glastonbury, Harston Mill, Saxon Flixborough, London (Westminster Abbey), Mediaeval Beverley, Portchester, Faccombe Netherton. In Ireland, from Neolithic to Mediaeval Lough Gur, Ballinderry and Dublin (Table 4.4).

Hen Harrier *Circus cyaneus*
Only seven records, from Iron Age Gussage All Saints, Saxon Ossom's Eyrie, West Stow and Ipswich, Mediaeval Royal Naval Victualling Yard and twice from Dublin.

Montagu's Harrier *Circus pygargus*
Reported from Iron Age Meare. Appreciably smaller than Hen Harrier, particularly in hind limb, so probably well identified.

Goshawk *Accipiter gentilis*
Mostly Saxon – Mediaeval records (23/44 records), but with a scatter back to Late Glacial (Pinhole and Robin Hood's Caves), Mesolithic (Mount Sandel), Neolithic, Iron Age and Roman (see Table 6.3).

Sparrowhawk *Accipiter nisus*
Scarce in early periods (Late Glacial Soldier's Hole, Mesolithic Port Eynon Cave), no Neolithic-Iron Age records; present Roman Barton Court, Boreham, Colchester, then 38/45 records from Saxon or later date (see Table 6.3).

Buzzard *Buteo buteo*
Well recorded, from 111 archaeological sites. Earliest possibly Walton, otherwise scarce/absent in Late Glacial cave sites; present Mesolithic Star Carr, Wetton Mill, Neolithic Rousay, Westray, Papa Westray, Links of Noltland; more numerous and widespread in Iron Age, Roman and later sites (see Tables 3.5 and 6.3).

Rough-legged Buzzard *Buteo lagopus*
Just four records, from Devensian Pinhole Cave, Ightham Fissures (date uncertain), Iron Age Howe and Neolithic Links of Noltland. Wing bones average larger than Common Buzzard, hind limbs have shorter tarsometatarsus but longer femur, so plausibly identifiable from good material.

Golden Eagle *Aquila chrysaetos*
Only 15 records, 5 of them Late Glacial cave sites (Robin Hood's, Pinhole, Gough's Old, Cat Hole, Aveline's Hole) and also post-Roman Ossom's Eyrie Cave. Records from Iron Age Howe, Christian and Mediaeval Iona, Mediaeval Stafford Castle hint at a wider range, but all records are essentially from upland/northern sites.

Osprey *Pandion haliaetus*
Reported only seven times, from Late Glacial Pinhole and Robin Hood's Caves, Iron Age Meare, Pictish Buckquoy, two sites in Mediaeval Dublin, and Post-Mediaeval Exeter.

Kestrel *Falco tinnunculus*
With 45 records, one of the better known raptors. Given its nesting habits, perhaps unsurprising that present in many cave sites, from Ipswichian and Wolstonian Tornewton, Late Glacial Aveline's Hole, Merlin's, Dowel, Robin Hood's, Pinhole; then Mesolithic Demen's Dale, Neolithic Links of Noltland, Isbister and Dowel Cave, numerous later records.

Red-footed Falcon *Falco vespertinus*

Merlin *Falco columbarius*
Recorded 10 times, from Late Glacial (Soldier's Hole, Robin Hood's, Pinhole, Cat Hole), then Iron Age Bu, Howe, Norse Buckquoy, Mediaeval Lincoln, Copt Hay; also Darfur Crag of uncertain date.

Hobby *Falco subbuteo*
Only three records, from Ipswichian Bacon Hole, Late Glacial Gough's Old Cave, and Mediaeval Stoney Middleton. Presumably always rather scarce, and not a major species in falconry, though a few falconers specialised in using it to hunt larks.

Eleanora's Falcon *Falco eleanorae*

Peregrine *Falco peregrinus*
Around 26 records, from Late Glacial Aveline's and Soldier's Hole, through Mesolithic Gough's and Port Eynon Caves, to a scatter of later sites (Iron Age Barrington, Danebury, Meare, Howe; Roman Heybridge; Saxon Ramsbury, Ipswich; Viking Jarlshof); then 13 Mediaeval or later sites, documenting the increase in falconry (see Tables 6.3 and 6.4).

Gyr Falcon *Falco rusticola*
Possible record (Gyr/Peregrine) from Late Glacial Potter's Cave; two from the royal mews at Mediaeval Winchester, presumably imported.

Baillon's Crake/Little Crake *Porzana pusilla/parva*
A single record of a small crake from Neolithic Tideslow, Derbyshire.

Spotted Crake *Porzana porzana*
Recorded four times, from Ightham Fissures (date uncertain), Neolithic Papa Westray and Iron Age Bu, Howe. As a widely scattered but scarce resident of wetlands, likely to have been more common formerly than now.

Water Rail *Rallus aquaticus*
The 25 records range from Late Glacial Merlin's Cave and Ightham Fissure (uncertain date), through Mesolithic Risga, Dowel Cave, Neolithic Oronsay, Iron Age Skaill, Howe, Old Scatness, Meare to Roman Thenford, Wroxeter, Filey, Caerleon. Only three later records, Buckquoy (Norse), Castle Acre (Norman) and Camber (post-Medieval) Castles. Known in Ireland from four caves, dates uncertain (Newhall, Alice, Catacomb, Barntick) and Iron Age Newgrange.

Corncrake *Crex crex*
Recorded from 24 sites, ranging from Mesolithic Port Eynon, Demen's Dale to late Mediaeval Stafford Castle; many sites northern/islands, but also, for example, Roman Colchester, Rudston, Dorchester, Saxon Wraysbury, Late Christian Lagore. At Raystown, Co. Meath, more common in seventh to eighth century than *Gallus*.

Moorhen *Gallinula chloropus*
Known from 33 sites, back to Cromerian West Runton, post-Cromerian Boxgrove, Ipswichian London. From Late Glacial caves (e.g. Pinhole, Robin Hood's), Mesolithic Soldier's Hole, Neolithic Dowel Cave, to a scatter of later sites (e.g. Iron Age Dinorben, Meare, Howe, Bronze Age Burwell Fen, Roman Lincoln, Cirencester, Saxon West Stow, Mediaeval Norwich, Oxford, Stafford and Barnard Castle). In Ireland, from Catacomb, Newhall and Castlepook Caves and Lagore.

Coot *Fulica atra*
A strong archaeological record, 42 sites, from Ipswichian Crayford, Late Glacial Merlin's Cave, Mesolithic Mt Sandel, frequent Iron Age, Roman and Mediaeval records. Distinctive size, among rails, and presumably eaten widely.

Common Crane *Grus grus*
Well recorded, from 131 sites. Large, osteologically distinctive, and a popular dietary item. Rare/absent in Late Glacial cave sites (Newhall Cave, Catacomb Cave, undated). Present in Mesolithic Star Carr, Thatcham; only Neolithic records, Mount Pleasant and footprints at Formby Point, but more widespread from Bronze Age onwards (see Table 6.5). A supposed Sarus Crane (Longthorpe, Iron Age) is surely a large male Common Crane.

Demoiselle Crane *Anthropoides virgo*
A single record, an unmistakable bill from the mid-Devensian of Pinhole Cave.

Little Bustard *Tetrax tetrax*
A single record, from Devensian Tornewton Cave. In recent history, used to occur as a vagrant regularly, to around 1950, now scarce and irregular.

Great Bustard *Otis tarda*
Despite a strong historical documentary record, only five archaeological sites, three of them Late Glacial; Roman Fishbourne needs confirmation, but the femur from Baynard's Castle, Late Mediaeval, seems secure (see Chapters 3 and 7). Still occurs occasionally as a vagrant, reintroduction being attempted.

Oystercatcher *Haematopus ostralegus*
With 29 reports, from Neolithic Point of Cott, Quanterness, Papa Westray and Isbister, through Bronze Age Midhowe, Iron Age Bu, Old Scatness Broch, Skaill, to 4 Roman, 4 Viking, 1 Saxon and 12 Mediaeval or later sites, well recorded in later times, but strangely absent earlier.

Avocet *Recurvirostra avosetta*
Only two archaeological records, Roman Caerleon and post-Mediaeval Camber Castle.

Black-winged Stilt *Himantopus himantopus*

Stone Curlew *Burhinus oedicnemus*
Only recorded from Bronze Age Nornour, an unlikely provenance.

Little Ringed Plover *Charadrius dubius*

Ringed Plover *Charadrius hiaticula*
Only eight records, from Late Glacial Chudleigh Fissure, Pinhole, Walton, Robin Hood caves, Neolithic Oronsay, Mediaeval Writtle, London (Greyfriars), but too similar in size to several other small waders, probably under-recorded.

Kentish Plover *Charadrius alexandrinus*

Dotterel *Charadrius morinellus*

Golden Plover *Pluvialis apricaria*
With 100 records, and a further 16 "Golden/Grey Plover", well recorded, a testimony to its abundance, wide range and popularity as a food item. A few early records (e.g. Ipswichian Bacon Hole, Late Glacial Robin Hood's, Pinhole and Inchnadamph), but mostly Roman (30) and Mediaeval/post-Mediaeval (38).

Grey Plover *Pluvialis squatarola*
Has 29 records, from Late Glacial Pinhole, Robin Hood's Cave, Aveline's Hole, then Mesolithic Demen's Dale, Inchnadamph, Port Eynon Cave, but mostly Roman to Mediaeval sites. Some bones appreciably longer, more slender than Golden Plover, so id probably secure. By no means confined to coastal sites, so presumably traded, for example, to Mediaeval Lincoln, London Baynard's Castle and Greyfriars.

Lapwing *Vanellus vanellus*
Recorded 78 times, including Late Glacial (Chudleigh Fissure, Gough's Old, Pinhole, Robin Hood's Caves), Mesolithic Star Carr, Neolithic Embo, Point of Cott, Quanterness, 11 Roman sites, 7 Saxon, 35 Mediaeval and later.

Knot *Calidris canuta*
Recorded from 18 sites, from Devensian Pinhole, Late Glacial Chudleigh Fissure, Robin Hood's Cave, then Mesolithic Demen's Dale, Neolithic Westray (Point of Cott), Bronze Age Nornour, through to Roman Camulodunum, several Mediaeval and later sites.

Sanderling *Calidris alba*
Claimed only from Mediaeval Castle Rising Castle, but no doubt difficult to distinguish from other small waders.

Dunlin *Calidris alpina*
With 18 records (plus two uncertain Dunlin/Ringed Plover, Dunlin/Sandpiper), ranging from Ipswichian Bacon Hole, Minchin Hole, Devensian Torbryan to Mesolithic, Iron Age, Roman and Mediaeval, moderately well represented. Probably always the commonest small wader in winter, but confident identification always difficult, in death as in life.

Curlew Sandpiper *Calidris ferruginea*

Purple Sandpiper *Calidris maritima*

Ruff *Philomachus pugnax*
Only reported four or five times, from Mesolithic Demen's Dale, Iron Age Howe, Bronze Age Nornour, Mediaeval Oxford (The Hamel) and possibly from Mediaeval Portchester.

Jack Snipe *Lymnocryptes minimus*
Recorded six times, from Devensian Torbryan Cave, Late Glacial Chudleigh Fissure, Cat Hole and Merlin's Cave, then from Norse Buckquoy and Mediaeval Iona.

Snipe *Gallinago gallinago*
Well represented, 69 records, but few early dates (Devensian Pinhole, Late Glacial Kirkdale, Chudleigh Fissure, Mesolithic Demen's Dale). More common from Neolithic (Quanterness, Links of Noltland, Point of Cott) onwards; 11 Roman, 6 Saxon, 22 Mediaeval and 8 later sites. Does this indicate a N species, spreading S as wet farmland was created, or a record of increasing exploitation?

Great Snipe *Gallinago media*
Reported from post-Cromerian Westbury-sub-Mendip.

Woodcock *Scolopax rusticola*
With 230 records, the most frequently recorded wader; distinctive and convenient size, specialised hunting techniques and culinary value all contribute, but must have been at least as frequent as now. Scant early record (Late Glacial Cat Hole, Neolithic Isbister, Durrington Walls) but regular from Iron Age onwards. Mostly Roman (68) and Mediaeval or later (119).

Black-tailed Godwit *Limosa lapponica*
Only six records of this former, now returned, breeder: Late Glacial Soldier's Hole, Neolithic Quanterness, Roman Colchester, Barnsley Park, and Mediaeval Colchester, Stafford Castle. An additional eight "godwit sp." could belong to this or the next sp.; they fall largely in the same time and space range, though one from Bronze Age Nornour is intriguing.

Bar-tailed Godwit *Limosa limosa*
19 records: from Late Glacial Chudleigh, and then nothing until 1 Iron Age (Scalloway), 2 Roman (Ilchester, Colchester), 1 Saxon (Portchester) and 13 Mediaeval or later records. Mostly in SE England, except Scalloway and Woods Quay, Dublin.

Whimbrel *Numenius phaeopus*
Fourteen records, well scattered in time and space: Devensian Pinhole, Walton, Late Glacial Chudleigh Fissure, then Iron Age Crosskirk, Roman Colchester, Norse Buckquoy, Mediaeval Castle Rising Castle, Castletown (Man), Lincoln (Flaxengate, and three levels (Saxon, Mediaeval, post-Mediaeval) at Portchester.

Curlew *Numenius arquata*
Predictably, this largest of waders has a good representation, 89 records; a few early records (Late Glacial Aveline's Hole, Cat Hole), then a gap till Neolithic Rousay, Isbister, Papa Westray, more numerous and widespread from Iron Age through Roman, Saxon and especially Mediaeval times. Strongly suggests that not so common or widespread in pre-Roman times, but might reflect changes in harvesting techniques.

Redshank *Tringa totanus*
Only 23 records, and none earlier than Neolithic Links of Noltland, Bronze Age Nornour; 4 Iron Age, 1 Roman, 2 Saxon sites. Many (10) records from high-status Mediaeval sites (e.g. Launceston, Castle Rising, Barnard and Baynard's Castles, Portchester, Perth). In Ireland, from Alice and Keshcorran Caves.

Spotted Redshank *Tringa erythropus*
Only reported twice, from Neolithic Papa Westray and Mediaeval Baynard's Castle, London.

Greenshank *Tringa nebularia*
Recorded 11 times, from Devensian Pinhole Cave, then Neolithic Dowel Cave, Links of Noltland and Quanterness, Iron Age Bu and Howe, Roman Ower, Over Purbeck, Saxon West Stow, Norse Buckquoy to Mediaeval Baynard's Castle, London. Also two uncertain Greenshank/Redshank from Saxon and Mediaeval Jarrow.

Green Sandpiper *Tringa ochropus*
Has 10 records, from Cromerian West Runton, Late Glacial Merlin's Cave, Mesolithic Demen's Dale through Iron Age Howe and Roman Thenford to Mediaeval and later Exeter, London – Baynard's Castle and Greyfriars; also one uncertain Green Sandpiper/Turnstone from Roman Waddon Hill.

Wood Sandpiper *Tringa glareolus*

Common Sandpiper *Actitis hypoleucos*

Turnstone *Arenaria interpres*
With 14 records, from Ipswichian Bacon Hole, Late Glacial Walton, Robin Hood, Pinhole Caves, Mesolithic Demen's Dale, Port Eynon Cave, Neolithic Papa Westray, Bronze Age Jarlshof, Iron Age Howe, Bu, Mediaeval Barnard's Castle, Castletown, well scattered in time, mostly in N and coastal sites.

Red-necked Phalarope *Phalaropus lobatus*
Recorded from Iron Age Bu, Orkney, within its recent range.

Grey Phalarope *Phalaropus fulicarius*
Reported from Norse Buckquoy.

Pomarine Skua *Stercorarius pomarinus*
Reported from Norse Buckquoy.

Arctic Skua *Stercorarius parasiticus*
Only recorded from Rousay, of uncertain post-glacial date.

Long-tailed Skua *Stercorarius longicaudus*
A single record, from Soldier's Hole, mid-Devensian.

Great Skua *Catharacta skua*
Only three records, predictably all in N: Neolithic Papa Westray, Iron Age Bu and Viking Old Scatness Broch.

Little Gull *Larus minutus*
Only one record, Mediaeval Baynard's Castle, London. A regular migrant past the coasts, and has attempted to breed four times, not yet successfully.

Black-headed Gull *Larus ridibundus*
Recorded from Cromerian Boxgrove, but then not till Neolithic Orkney (Quanterness, Links of Noltland); also Iron Age Howe, Norse Buckquoy, Jarshof, but mostly Mediaeval sites; only 18 records in all, much less numerous than its present-day status might suggest.

Common Gull *Larus canus*
Slightly better represented than *ridibundus*, 26 records. Late Glacial Walton, Pinhole, Soldier's Hole, but otherwise from Neolithic (Isbister, Links of Noltland), Iron Age (Howe, Pennyland, Poundbury) and later. Geographically, reported widely, from Portchester to Orkney.

Lesser Black-backed Gull *Larus fuscus*
Reported 10 times, from a variety of sites: Late Glacial Castlepook Cave, Neolithic Links of Noltland and Papa Westray, Iron Age Meare, Howe, Saxon Flixborough, Viking Brough of Birsay, Mediaeval Dyserth Castle, Battle Abbey and later Exeter; but unlikely to be distinguishable from Herring Gull skeletally.

Herring Gull *Larus argentatus*
Claimed from 25 sites, but osteologically indistinguishable from Lesser Black-backed Gull, as indicated by the further 36 more cautious "Herring/LBB" records. Only one Late Glacial (Aveline's Hole) and no early post-glacial records – a S species, late returning? Widespread in time and geographically from Mesolithic/Neolithic Ferriter's Cove onwards.

Glaucous Gull *Larus hyperboreus*
Unlikely to be distinguishable osteologically from GBB; two records claimed, Mediaeval Iona, Bronze Age Maitresse Ile, Jersey; and six more cautious Glaucous/GBB all from Orkney.

Great Black-backed Gull *Larus marinus*
Recorded from 33 mostly coastal sites (but note preceding caution), back to Mesolithic Morton, but also at, for example, Iron Age Meare, Roman Exeter.

Kittiwake *Rissa tridactyla*
Thinly recorded, from post-Cromerian Boxgrove, then not till Mesolithic Morton, Neolithic and Bronze Age Westray (Point of Cott, Bu), Iron Age (Scatness, Scalloway, Danebury, Gussage All Saints). No Roman or Saxon records; several Mediaeval and later records include Exeter, Brentford, York – traded?

Little Tern *Sterna albifrons*

Sandwich Tern *Sterna sandvicensis*
Recorded only three times, from Neolithic Papa Westray, Neolithic Bu and Mediaeval London (Baynard's Castle).

Roseate Tern *Sterna dougallii*

Common Tern *Sterna hirundo*
A limited record, five sites, but a good time spread from Late Glacial Merlin's Cave and Mesolithic Risga to Neolithic Oronsay, Roman and Saxon Portchester.

Arctic Tern *Sterna paradisaea*

Guillemot *Uria aalge*
Recorded from 66 sites, plus eight uncertain Guillemot/Razorbill records. Earliest Pastonian Chillesford, then Late Glacial Paviland Cave and Chudleigh Fissure. More evident in Mesolithic (Morton, Risga, Ferriter's Cave), Neolithic (Embo, Point of Cott, Rousay, Links of Noltland, Quanterness, Oronsay) and later. Mostly coastal sites, especially in the N. Three from Anglo-Scandinavian and Mediaeval York suggest trade inland.

Razorbill *Alca torda*
Recorded from 40 sites, from Pastonian Bacton, Ipswichian Minchin Hole, Bacon Hole; surprisingly absent in Late Glacial record, but present again in Mesolithic (Risga, Morton, Port Eynon Cave), Neolithic (Oronsay, Point of Cott, Links of Noltland) and later sites. Present as far S as Jersey (Bronze Age Maitresse Ile) and Guernsey (Le Dehus, undated). Mostly coastal, especially N and W isles, but, for example, three records from York suggest trading, Bronze Age Burwell Fen suggests "wreck".

Great Auk *Pinguinus impennis*
Moderately well recorded, back to Cromerian Boxgrove and Late Glacial St Brelade's Bay, Jersey. Of 40 records, most Neolithic – Iron Age (21), thinning into later times. Regular on N sites, especially on islands, but as far S as Roman Halangy Down, Scilly (see Table 7.3).

Black Guillemot *Cepphus grylle*
Recorded from 11 sites, from Late Glacial caves (Pinhole, Robin Hood's) to several Orkney sites of Neolithic to Viking age (Papa Westray, Bu, Howe, Buckquoy), and as far S as Roman Filey.

Little Auk *Alle alle*
For a relatively small and unfamiliar species, well recorded, 24 times: from Late Glacial Chudleigh Fissure, Cat Hole and Merlin's Caves, through Mesolithic Inchnadamph, Port Eynon Cave to 3 Neolithic, 7 Iron Age sites, all in N or W islands; few later records (Pictish Buckquoy, Viking Old Scatness Broch, Skaill and post-Mediaeval Lindisfarne). Presumably storm-driven, or does the frequency imply that it once bred here? Nearest modern breeding sites are N Iceland, but thought to have retreated from S breeding range.

Puffin *Fratercula arctica*
Recorded from 35 sites, mostly coastal and northern: five Late Glacial caves (Chudleigh Fissure, Potter's, Pinhole, Robin Hood's, Creag nan Uamph); Mesolithic Port Eynon, Morton, Inchnadamph; later sites mostly Orkney, Shetland, but also Roman Filey, sixteenth century Castle Rising Castle.

Pallas' Sandgrouse *Syrrhaptes paradoxus*

Rock Dove *Columba livia*
Impossible to distinguish wild Rock Doves from Domestic Pigeons osteologically, and distinction from Stock Dove difficult. Collectively, some 70 records. Historical evidence indicates that the Romans had domestic doves; the numerical record (see Table 5.1) implies that they might have introduced them to GB, but the historical record implies that dovecotes were a Norman import.

Stock Dove *Columba oenas*
Claimed from 48 sites, from Late Glacial Chudleigh Fissure, Pinhole, Merlin's, Robin Hood's Caves onwards. Common in Roman (18) and Mediaeval (13) sites; but cautions above apply.

Wood Pigeon *Columba palumbus*
With 86 records, the best recorded pigeon. Earliest Hoxnian Barnham, and a few Late Glacial (Pinhole, Potter's Cave) and Mesolithic caves (Port Eynon). Mostly Roman (17), Saxon (10) and Mediaeval (37) sites (see Table 5.1 for combined record of pigeons).

Turtle Dove *Streptopelia turtur*
Only two records, Late Glacial Pinhole and Roman Staines.

Collared Dove *Streptopelia decaocto*

Ringed Dove *Streptopelia risoria*

Rose-ringed Parakeet *Psittacula krameri*

Cuckoo *Cuculus canorus*
Only archaeological record from Roman Exeter; a distinctive skeleton, but not a species likely to be encountered (not food, not a scavenger). The AS name geac, which became the old name gowk, supplanted by cuckoo by time of Chaucer.

Barn Owl *Tyto alba*
Well represented archaeologically with 43 records; from Late Glacial caves (Robin Hood's, Cat Hole, Pinhole, Inchnadamph), and numerous Neolithic, Roman and later sites; no Mesolithic records (too wooded, or too few sites?). Recorded in Ireland from Catacomb Cave and Lagore.

Snowy Owl *Bubo scandiaca*
Surprisingly, only one certain record, Devensian Kent's Cavern, despite abundant records of lemmings, and abundance of this species, for example, in French cave sites.

Eagle Owl *Bubo bubo*
No more than 10 records, but a long history, ranging from Pastonian East Runton, Hoxnian Swanscombe, Wolstonian Tornewton, then Late Glacial Langwith, Ossom's and Merlin's Cave. Status as recent native species best attested from Mesolithic Demen's Dale; possibly Iron Age Meare (see Table 3.1). Present breeding pair or two likely to be descended from escaped aviary birds.

Hawk Owl *Surnia ulula*
Two Late Glacial records, from Pinhole and Robin Hood's Caves, Creswell Crags.

Little Owl *Athene noctua*
One early record of this largely S European species, from post-Cromerian Westbury, is believable but two later ones, from Late Glacial Chudleigh Fissure, Aveline's Hole, need checking. Hoodwinks? or misidentified? Present population dates from introductions in 1870–1880s.

Tawny Owl *Strix aluco*
The 28 records have an odd time spread: post-Cromerian Boxgrove, Devensian/Late Glacial Langwith, Pinhole, Robin Hood's and Ossom's Caves, Mesolithic Demen's Dale, Wetton Mill, Iron Age Howe, Dowel Cave, Slaughterford; then a gap through Roman and early Saxon times, but 3 late Saxon (Faccombe Netherton, Flixborough), 12 Mediaeval or later records. Used as a decoy in hunting or hawking?

Long-eared Owl *Asio otus*
Only four records, from Late Glacial Soldier's Hole, Iron Age Danebury, Roman Wroxeter, and undated Teesdale Cave. Difficult to distinguish osteologically from the next species. Possible Irish record (*Asio* sp.) from Mediaeval Baltrasna.

Short-eared Owl *Asio flammeus*
Recorded 17 times, including post-Cromerian Westbury and 8 Glacial/Late Glacial caves (e.g. Pinhole, Robin Hood, Merlin's, Dowel, Aveline's and Soldier's Holes) when appropriate open conditions, with lemmings, would have suited it. Later, mostly from N sites: Neolithic Isbister, Links of Noltland, Bronze Age Hindlow Cairn, Iron Age Skaill, post-Roman Ossom's Eyrie, but also Norse York (Coppergate). No sure Irish records.

Tengmalm's Owl *Aegolius funereus*
Two Late Glacial records, from Pinhole and Robin Hood Caves, Creswell, of this N species.

Nightjar *Caprimulgus europaeus*
One record from Late Glacial Merlin's Cave, one from Early Christian Raystown, Co. Meath.

Swift *Apus apus*
Only five records, from post-Cromerian Boxgrove, Late Glacial Walton, and Roman Winterbourne, Mediaeval Middleton Stoney, Canterbury Cathedral. When nested principally in old hollow trees (as in Białowieża), unlikely to be fossilised.

Alpine Swift *Apus melba*
A single mid-Devensian record, from Pinhole Cave; unmistakeably large carpometacarpus and tarsometatarsus leave no doubt on id.

Kingfisher *Alcedo atthis*
Only two records, from Devensian Pinhole Cave and Late Glacial Merlin's Cave.

Bee-eater *Merops apiaster*

Hoopoe *Upupa epops*

Wryneck *Jynx torquilla*
Only two records, from Neolithic Quanterness and sixth to seventh century Raystown, Co. Meath.

Green Woodpecker *Picus viridis*
Only two records, from Late Glacial caves: Chelm's Combe and Merlin's.

Great Spotted Woodpecker *Dendrocopos major*
Only nine records, all from cave sites; six of these probably Late Glacial (Chudleigh Fissure, Gough's Old, Dowel, Langwith, Pinhole, Robin Hood's); then Neolithic Fox Hole, but two important Irish records, Alice and Newhall , one with a C^{14} date in the Bronze Age.

Lesser Spotted Woodpecker *Dendrocopos minor*
Only two records, from Late Glacial Pinhole and Robin Hood's Caves, Creswell, Derbyshire.

White-backed Woodpecker *Dendrocopos leucotos*

Black Woodpecker *Dryocopus martius*

Calandra Lark *Melanocorypha calandra*
Mentioned by Chaucer, but presumably from French sources or experience.

Crested Lark *Galerida cristata*
Recorded from five cave sites, probably all Late Glacial (Ightham Fissures, Chudleigh Fissure, Torbryan, Happaway, Merlin's), and Mediaeval Portchester. Given abundance just S of the Channel, surprising that appears never to have colonised in modern times, and only 16 vagrants have been reported. Essentially sedentary, but predicted to be a likely coloniser if climate changes.

Skylark *Alauda arvensis*
Well recorded, 54 sites from Ipswichian Bacon and Minchin Holes, Devensian Bridged Pot, Torbryan and Tornewton caves, to Late Glacial, for example, Pinhole, Robin Hood and Merlin's Caves. Possibly missing in wooded Mesolithic, but then continuously from Neolithic (e.g. Quanterness, Papa Westray, Fox Hole) through to Mediaeval (e.g. Loughor, Barnard's and Rumney Castles) and later (e.g. Exeter, York (Aldwark, Coffee Yard)) (see Table 3.6).

Wood Lark *Lullula arborea*
Reported four times: Devensian Pinhole Cave, Roman Hambledon, post-Roman Ossom's Eyrie and Mediaeval Portchester.

Shore Lark *Eremophila alpestris*
Two records, from Late Glacial Chudleigh Fissure and Ightham Fissures (date uncertain, probably similar), when it might well have been a breeding bird.

Sand Martin *Riparia riparia*
Only claimed from Bronze/Iron Age Wilsford Shaft on Salisbury Plain.

Crag Martin *Ptyonoprogne rupestris*
Recorded from three Late Glacial caves at Creswell Crags: Robin Hood's, Pinhole and Ossiferous Fissure C8. A S species with only four recent records; was it open but warmer at times in Late Glacial?

Swallow *Hirundo rustica*
Recorded 22 times, from post-Cromerian Westbury, Ipswichian Bacon Hole, Devensian Pinhole, Torbryan, Ossiferous Fissure C8, Late Glacial Robin Hood's, Merlin's Caves. In Postglacial, from Neolithic Carding Mill, Links of Noltland, Iron Age Wilsford Shaft, Howe, Roman Cirencester, Tiddington to Saxon York (Fishergate), Ossom's Eyrie. More common in cave sites (mostly early) than buildings, perhaps skewing record.

House Martin *Delichon urbica*
Only five records, including undated Newhall Cave, Neolithic Dowel Cave, post-Roman Ossom's Eyrie, Saxon York – Fishergate, and sixteenth to seventeenth century Naval Victualling Yard. Naturally cliff-dwelling, might have been expected more often in wide-mouthed caves.

Richard's Pipit *Anthus novaeseelandiae*
Reported from Late Glacial Aveline's Hole. A plausible species to have occurred in open tundra-like conditions.

Tree Pipit *Anthus trivialis*
Only four records, from Ipswichian Tornewton Cave, Late Glacial Chudleigh Fissure, Iron Age Bu, post-Roman Ossom's Eyrie.

Meadow Pipit *Anthus pratensis*
Only 15 records, mostly from Devensian/Late Glacial caves (Chelm's Combe, Pinhole, Robin Hood's, Langwith, Neale's, Ossiferous Fissure C8), also Saxon Lewes, Ipswich, Mediaeval Thrislington.

Rock Pipit *Anthus petrosus*
Only claimed from four caves, Late Glacial Chudleigh Fissure, Aveline's Hole and Langwith, and Mesolithic Port Eynon, and (Water Pipit) Mediaeval Portchester.

Yellow Wagtail *Motacilla flava*
Reported from Ightham Fissures, of uncertain date.

Grey Wagtail *Motacilla cinerea*
Reported from Late Glacial Chudleigh Fissure and post-Roman Ossom's Eyrie.

Pied Wagtail *Motacilla alba*
Reported from seven sites, mostly caves: Late Glacial Merlin's, Chudleigh Fissure, Aveline's Hole, undated Ightham, Mesolithic Port Eynon, also Iron Age Bu, Newgrange.

Waxwing *Bombycilla garrulus*
Only two records, from Devensian Pinhole and Iron Age Howe.

Dipper *Cinclus cinclus*
Only seven subfossil records, mostly Late Glacial cave sites (Robin Hood's, Pinhole, Torbryan, Merlin's, Chudleigh Fissure, plus St Brelade's Bay, Jersey) and Neolithic Dowel Cave.

Wren *Troglodytes troglodytes*
A scatter of 16 records, from Late Glacial caves (Pinhole, Robin Hood, Chudleigh Fissure), Mesolithic Hazleton Long Cairn, Neolithic Quanterness, Dowel Cave, Bronze Age Borwick, Iron Age Howe to Roman Filey, Caerwent, Winterbourne, Saxon Wraysbury, York (Fishergate) and Mediaeval Thrislington.

Hedge Sparrow (Dunnock) *Prunella modularis*
A single record of an Accentor sp. claimed from Roman Birdoswald, and unlikely to be any other species. Reported from 18 sites: Cromerian Boxgrove, then several records from Late Glacial cave sites (including Aveline's Hole, Soldier's Hole, Merlins, Pinhole, Robin Hood's) and a scatter of later records. The alternative Saxon name dunnock implies that it was recognised early; Chaucer refers to it as the heysoge, while Shakespeare calls it the hedge-sparrow, and refers to it as a cuckoo's host.

Robin *Erithacus rubecula*
Of 28 records of this distinctively long-legged chat, most are Glacial/Late Glacial cave sites: Torbryan, Hoe Grange, Langwith, Pinhole, Robin Hood, Chudleigh Fissure, Aveline's Hole or later cave sites: Mesolithic Wetton Mill, Neolithic Fox Hole, Dowel Cave, post-Roman Ossom's Eyrie. Recorded at Roman Frocester, Dorchester, Caerwent, Mediaeval London (Greyfriars), later York (Lawrence St).

Nightingale *Luscinia megarhynchos*
Reported twice, from Late Glacial Langwith Cave and Chudleigh Fissure.

Redstart *Phoenicurus phoenicurus*
Reported nine times, mostly from cave sites (better preservation? hardly a habitat choice); mostly Late Glacial (Robin Hood's, Aveline's Hole, Pinhole, Chudleigh Fissure), but also Mesolithic Wetton Mill, Neolithic Dowel; two non-cave sites, Iron Age Dun Mor Vaul and Mediaeval Barnard's Castle.

Whinchat *Saxicola rubetra*
Reported from six sites, mostly Late Glacial caves (Pinhole, Robin Hood's, Merlin's, Aveline's Hole) also Keshcorran (date?) and post-Roman Ossom's Eyrie. A further eight records of Chat sp., which might be this or the next species, also mostly Late Glacial caves, plus Mesolithic and Neolithic Dowel, Neolithic Quanterness.

Stonechat *Saxicola torquata*

Wheatear *Oenanthe oenanthe*
Of 14 records, mostly cave sites: Ipswichian Bacon Hole, Glacial/Late Glacial Walton, Robin Hood, Chudleigh Fissure, Merlin's, Langwith, Ightham Fissure, Mesolithic Port Eynon. Later from Neolithic Dowel Cave, Quanterness, Links of Noltland, Iron Age Glastonbury, post-Roman Ossom's Eyrie.

Ring Ouzel *Turdus torquatus*
Only 11 records, mostly caves: Late Glacial Robin Hood, Pinhole, Chelm's Combe, Merlin's, Soldier's Hole, also Neolithic Dowel, post-Roman Ossom's Eyrie, undated Hathaway. Elsewhere, at Neolithic Quanterness, Iron Age Howe.

Blackbird *Turdus merula*
One of the commonest passerines in archaeological sites, reflecting both its ubiquity and its culinary value. At least 85 records, back to Cromerian West Runton; present from Late Glacial caves (Dowel, Cat Hole, Merlin's, Robin Hood's) and Mesolithic (Demen's Dale, Wetton Mill, Doghole Fissure). Numerous in Roman and Mediaeval sites. Identification difficulties with similar species indicated by an additional Blackbird/Fieldfare (Stafford Castle) and 13 Blackbird/Ring Ouzel records, mostly from Late Glacial cave sites. Present in several Irish cave sites, of uncertain date, and from Iron Age sites at Newgrange and Dalkey Island.

Fieldfare *Turdus pilaris*
Claimed from 34 sites, though identification rarely explained. At least nine Late Glacial records (Aveline's Hole, Pinhole, Soldier's Hole, etc.), through Mesolithic (Port Eynon, Gough's Old Caves), Neolithic (e.g. Dowel Cave), Iron Age (Dorchester, Howe) to Mediaeval Abingdon, Hertford Castle.

Song Thrush *Turdus philomelos*
With 66 records, numerous throughout, from Late Glacial (e.g. Gough's Old, Robin Hood, Merlin's caves, Soldier's and Aveline's Holes), through Mesolithic Wetton Mill to Neolithic Dowel Cave, Grime's Graves, Quanterness, Mount Pleasant; present at 7 Iron Age, 9 Roman, 3 Saxon, 13 Mediaeval or later sites. In Ireland, present at several cave sites, date uncertain (Keshcorran, Newhall, Barntick, etc.), Mesolithic Mount Sandel and Iron Age Newgrange. A further 91 records of "thrush sp." indicate the difficulties of identification in this genus.

Redwing *Turdus iliacus*
Reported 50 times, plus another 12 uncertain Redwing/Song Thrush. From 12 Devensian/Late Glacial caves (e.g. Walton, Pinhole, Robin Hood, Cat Hole, Chudleigh Fissure), Neolithic Quanterness, Dowel Cave, Iron Age Danebury, Slaughterford. At several later high status sites (e.g. Roman Uley Shrines, Cirencester, Shakenoak, Colchester; Saxon Southampton; Mediaeval Dryslwyn and Castle Rising Castles), presumably eaten.

Mistle Thrush *Turdus viscivorus*
With 42 sites, well recorded, due to its distinctive size, probably also its value as food. In numerous Late Glacial caves (e.g. Merlin's, Aveline's Hole, Robin Hood's, Pinhole, Neale's), Mesolithic Demen's Dale, Port Eynon Cave), Neolithic Grime's Graves, Tideslow, Mount Pleasant, Iron Age Bu, Howe, Roman Colchester, Frocester, Chew Valley to Mediaeval London (Greyfriars); also in Ireland, from undated Castlepook and Alice Caves, Iron Age Keshcorran, Neolithic Newgrange.

Cetti's Warbler *Cettia cetti*

Fan-tailed Warbler *Cisticola juncidis*

Savi's Warbler *Locustella luscinoides*

Sedge Warbler *Acrocephalus schoenobaenus*
A single record, from Late Glacial Aveline's Hole; unlikely on climate and habitat.

Dartford Warbler *Sylvia undata*

Lesser Whitethroat *Sylvia curruca*
Reported from Iron Age Bu.

Whitethroat *Sylvia communis*
Reported from Late Glacial Chudleigh Fissure and post-Roman Ossom's Eyrie Cave.

Garden Warbler *Sylvia borin*
Reported from Hoxnian Swanscombe and Mediaeval London, Greyfriars.

Blackcap *Sylvia atricapilla*
Uncertain that the various similar-sized *Sylvia* could be reliably distinguished skeletally, but there are six records from Late Glacial cave sites (Pinhole, Robin Hood's, Aveline's Hole, Neale, Doghole, Chelms Combe).

Wood Warbler *Phylloscopus sibilatrix*

Chiffchaff *Phylloscopus collybita*

Willow Warbler *Phylloscopus trochilus*

Goldcrest *Regulus regulus*
Improbably, there are three records, from Late Glacial Cat Hole and Neolithic Quanterness, Dowel Cave. Given its size, identifications probably correct (would not be possible to rule out Firecrest), but these records say more about the abilities of excavators than about historical ecology of the species.

Firecrest *Regulus ignicapillus*

Spotted Flycatcher *Muscicapa striata*
Reported from Devensian Torbryan Cave, Mesolithic Wetton Mill Rockshelter and post-Roman Ossom's Eyrie.

Pied Flycatcher *Ficedula hypoleuca*

Collared Flycatcher *Ficedula albicollis*

Bearded Tit *Panurus biarmicus*

Long-tailed Tit *Aegithalos caudatus*
Reported four times, from three Late Glacial caves on the Derby/Nottingham border (Pinhole, Robin Hood's, Ossiferous Fissure C8) and Mesolithic Dog Hole Fissure.

Marsh Tit *Poecile palustris*

Willow Tit *Poecile montanus*

Coal Tit *Periparus ater*
One record claimed, from Late Glacial Chudleigh Fissure.

Blue Tit *Cyanistes caeruleus*
Rarely recorded, only three records: Neolithic Dowel Cave, post-Roman Ossom's Eyre and Mediaeval Barnard Castle. Too small to be found routinely, and likely to be confused with other Paridae.

Great Tit *Parus major*
Recorded 18 times, mostly from cave sites: 11 Late Glacial (e.g. Chelm's Combe, Chudleigh Fissure, Robin Hood's), but then from Mesolithic Dowel Cave, Wetton Mill, Iron Age Howe, Roman Chew Valley Lake.

Crested Tit *Parus cristatus*

Nuthatch *Sitta europaea*
Recorded from nine sites: Cromerian West Runton, Devensian Langwith and Ossiferous Fissure C8, Late Glacial Pinhole, Robin Hood's Cave, Merlin's Cave, Aveline's Hole and Chudleigh Fissure, and Neolithic Fox Hole Cave.

Treecreeper *Certhia familiaris*
Only three records, from Late Glacial Chudleigh Fissure, Mesolithic Wetton Mill and post-Roman Ossom's Eyrie.

Golden Oriole *Oriolus oriolus*

Red-backed Shrike *Lanius collurio*
Present at Ightham Fissure (date uncertain) and Neolithic Dowel Cave. Uncertain "shrike sp." at Iron Age Danebury, Mediaeval Launceston Castle may be this species.

Great Grey Shrike *Lanius excubitor*
Recorded from Late Glacial Chudleigh Fissure; also claimed from Neolithic Dowel Cave, but is in fact Red-backed Shrike. Possibly present Neolithic Howe (as an uncertain shrike/thrush), and Roman Colchester (as "grey shrike"). Portrayal in the Selborne Missal and probable mention by Chaucer suggest that it has been a conspicuous if uncommon winter visitor since at least Mediaeval times.

Jay *Garrulus glandarius*
Forty-three records (plus an uncertain Jay/Magpie, Flixborough), from Cromerian West Runton, Late Glacial caves (Pinhole, Robin Hood's, Soldier's Hole, Chudleigh Fissure), Mesolithic (Wetton Mill, Demen's Dale) and later sites. S distribution, no further N than York; no Scottish records. In Ireland, at Catacomb, Keshcorran and Newhall Caves, of uncertain date, and at Iron Age Balinderry crannog, Mediaeval Dublin.

Siberian Jay *Perisoreus infaustus*

Magpie *Pica pica*
Forty-six records, ranging from Late Glacial (e.g. Pinhole, Robin Hood's, Soldier's Hole, Cat Hole, Merlin's Caves) through Neolithic Fox Hole, Roman (e.g. Uley, Wroxeter, Bancroft Villa) to Mediaeval (e.g. Nantwich, Lewes, Middleton Stoney). In Ireland, at undated sites (Castlepook, Catacomb, Newhall, Alice Caves) and late Mediaeval Johnstown.

Azure-winged magpie *Cyanopicus cyanus*

Nutcracker *Nucifraga caryocatactes*
One record, from Late Glacial Robin Hood's Cave.

Chough *Pyrrhocorax pyrrhocorax*
A thin scatter of 15 records, from Ipswichian Kirkdale Cavern, Late Glacial Paviland Cave, Goughs' Old Cave, Cat Hole and Chudleigh Fissure, through Mesolithic Port Eynon, Iron Ae Bu to Mediaeval Exeter. In Ireland, from Lagore, and on Man from Perwick Bay ("Roman") and Castletown (seventeenth century).

Alpine Chough *Pyrrhocorax graculus*

Jackdaw *Corvus monedula*
Well recorded, 177 times, plus 14 less certain Jackdaw/Jay (1), Jackdaw/Magpie (12) and Jackdaw/Chough (1). Numerous in Late Glacial caves (Aveline's Hole, Soldier's Hole, Ossom's, Merlin's); scarce in Mesolithic (Dog Hole Fissure) and Neolithic (Fox Hole); common in Roman and Mediaeval sites. Absent from the early Scottish island and coastal sites.

Rook *Corvus frugilegus*
Most of the 47 records are late; only 2 Late Glacial (Pinhole, Aveline's Hole), no Mesolithic or Neolithic sites. More common as farming increased: Iron Age Meare, Danebury, Budbury, Thornton-le-Dale, 17 Roman, 20 Saxon-Mediaeval and later records.

Carrion Crow *Corvus corone*
Reported from 22 records, including, Devensian Tornewton, then 3 Iron Age (Glastonbury, Dun Mor Vaul, Scatness), and then 9 Roman records, 4 Mediaeval. Symbolic significance, or eaten? Also 3 Carrion/Hooded Crow, which would certainly not be distinguishable osteologically, and another 78 cautious "Crow sp." which presumably reflects this uncertainty. The 159 records of "Crow/Rook" indicate a further uncertainty, but adding all these together confirms the relative abundance and widespread presence of large corvids. Abundance in Roman and Mediaeval sites presumably indicates that they were eaten regularly, though might simply reflect their scavenger status round habitations.

Hooded Crow *Corvus cornix*
Two records of Hooded Crow were presumably identified from their distribution as much as from their bones – Late Iron Age Scalloway and ninth century Jarlshof.

Raven *Corvus corax*
Among the best recorded of wild birds: 267 records. Earliest Wolstonian and Ipswichian Tornewton; only 5 Late Glacial, no Mesolithic records, but 23 Iron Age, 93 Roman (symbolic significance?), 106 later records (scavenger?).

Starling *Sturnus vulgaris*
With 109 records, well documented. Earliest Cromerian West Runton, post-Cromerian Boxgrove, Ipswichian Bacon and Minchin Holes. Numerous Late Glacial records (e.g. Chudleigh Fissure, Merlin's, Robin Hood's, Pinhole Caves). In post-Glacial, continuous record, from Mesolithic Port Eynon Cave, Hazleton Long Cairn, Wetton Mill; Neolithic Embo, Point of Cott, Papa Westray, Links of Noltland; 10 Iron Age, 22 Roman, 9 Saxon/Viking, 35 Mediaeval or later records. Many Neolithic-Iron Age and later records from N isles, perhaps supporting the notion that colonised mainland Scotland in nineteenth century from them.

House Sparrow *Passer domesticus*
With 42 records, remarkably well represented. Four Late Glacial records (Aveline's Hole, Merlin's Cave, Pinhole, Robin Hood's; also perhaps Langwith) imply an early presence, but might be misidentified. Apparently absent in Mesolithic, Neolithic. Numerous from Iron Age (Abingdon, Harston Mill, Danebury, Old Scatness Broch), Roman and later times. A further 10 Bronze Age-Mediaeval "sparrow sp." are as likely to be this sp. as not.

Tree Sparrow *Passer montanus*

Snow Finch *Montifringilla nivalis*
Claimed only from Devensian Pinhole Cave.

Chaffinch *Fringilla coelebs*
Known from 24 records, including several Late Glacial cave sites (Robin Hood's, Pinhole, Inchnadamph, Aveline's Hole, Merlin's, Langwith), Mesolithic Wetton Mill, Neolithic Dowel Cave, Roman and later York.

Brambling *Fringilla montifringilla*
Claimed from four sites: Late Glacial Pinhole, Robin Hood's Cave, Neolithic Quanterness, and Mediaeval Baynard's Castle; possibly also (Brambling/Chaffinch) from Port Eynon Cave. Bones perceptibly larger than Chaffinch, nearer Greenfinch, so identifications plausible.

Serin *Serinus serinus*
Reported from Hoxnian Swanscombe.

Greenfinch *Chloris chloris*
One of the better recorded small passerines, 16 sites from Late Glacial Walton, Pinhole, Chelm's Combe, Neolithic Dowel Cave to Roman York (colonia, General Accident) and Mediaeval Battle Abbey; in Ireland, from Neolithic Newgrange, later Keshcorran Caves.

Goldfinch *Carduelis carduelis*
Only recorded six times (Late Glacial Chudleigh Fissure, Neolithic Dowel Cave, Iron Age Gussage All Saints, post-Roman Ossom's Eyre, post-Mediaeval Peel, Beverley), but probably under-recorded, due to small size and potential confusion with other similar-sized finches, especially Linnet.

Siskin *Carduelis spinus*

Linnet *Carduelis cannabina*
Recorded only seven or eight times, from three Late Glacial caves (Pinhole, Aveline's Hole, Chudleigh Fissure), Neolithic Quanterness, Dowel Cave and possibly Links of Noltland, and Mediaeval Stoney Middleton; also Keshcorran Caves (date?) in Co Sligo.

Twite *Carduelis flavirostris*
A single record, from Neolithic Quanterness.

Redpoll *Carduelis flammea*
Reported from three caves, Late Glacial Merlin's, Mesolithic Dog Hole and post-Roman Ossom's Eyrie.

White-winged Crossbill *Loxia leucoptera*

Common Crossbill *Loxia curvirostra*
Only four records, assumed to be this species, from Wolstonian Tornewton Cave, Devensian Creswell Crags, Merlin's Cave and post-Roman Ossom's Eyrie (see discussion in Chapter 3).

Scottish Crossbill *Loxia scotica*

Parrot Crossbill *Loxia pytyopsittacus*

Pine Grosbeak *Pinicola enucleator*
Only three records, from Devensian Torbryan and Late Glacial Merlin's and Robin Hood's Caves. A Boreal species whose presence in colder times is perhaps as expected.

Bullfinch *Pyrrhula pyrrhula*
Only nine records, mostly Late Glacial cave sites (Aveline's Hole, Pinhole, Robin Hood's) or uncertainly dated (Catacomb, Newhall, Keshcorran, Ireland), but also Neolithic Dowel Cave, post-Roman Ossom's Eyrie, Mediaeval Caerleon.

Hawfinch *Coccothraustes coccothraustes*
Recorded from 16 sites, mostly caves; mostly Late Glacial (9 records, including Robin Hood's, Merlin's, Pinhole, Chudleigh), but also Mesolithic Demen's Dale, Neolithic Dowel Cave, Iron Age Abigndon, Roman Oxford. Also Newhall Cave, Ireland, of uncertain date, but another indication of loss of woodlands there and their birds.

Lapland Bunting *Calcarius lapponicus*
Recorded from Late Glacial Chudleigh Fissure.

Snow Bunting *Plectrophenax nivalis*
Ten records, all from Devensian/Late Glacial caves (Aveline's and Soldier's Holes, Pinhole, Robin Hood's, etc), except Iron Age Howe.

Yellowhammer *Emberiza citrinella*
Only nine records, from Late Glacial Chudleigh Fissure and Aveline's Hole, Roman Uley Shrines, post-Roman Ossom's Eyrie, Norse York (Coppergate) and Mediaeval Portchester, Middleton Stoney and Lincoln. Another seven records of "bunting sp. (4), bunting/finch (2), bunting/lark" could belong anywhere among the short-winged passerines.

Cirl Bunting *Emberiza cirlus*

Reed Bunting *Emberiza schoeniclus*
Only four records, from three Late Glacial caves (Pinhole, Robin Hood, Chudleigh Fissure) and Iron Age Howe.

Corn Bunting *Miliaris calandra*
Only five records claimed, four Late Glacial (Chudleigh Fissure, Aveline's Hole, Pinhole, Robin Hood's Cave) and Iron Age Howe. Distinctively bigger than other buntings.

Rose-breasted Grosbeak *Pheuticus ludovicianus*
Reported from two Late Glacial cave sites at Creswell Crags, Ossiferous Fissure C8 and Robin Hood's, but an American vagrant seems an unlikely species, and a misidentification must be suspected.

References

Acobas, P. (1993). Shakespeare's Ornithology. http://perso.wanadoo.fr/acobas.net/english/shakespeare/masters/.

Adams, J.M. & Faure, H. (1997). Preliminary vegetation maps of the World since the Last Glacial Maximum: an aid to archaeological understanding. *Journal of Archaeological Science* **24**: 623–647.

Albarella, U. (2005). Alternate fortunes? The role of domestic ducks and geese from Roman to Medieval times in Britain. In *Feathers, Grit and Symbolism. Birds and Humans in the Ancient Old and New Worlds* (eds G. Grupe & J. Peters), pp. 249–258. Verlag Marie Leidorf, Rahden, Westphalia.

Albarella, U. & Davies, S.J.M. (1996). Mammals and birds from Launceston Castle, Cornwall: decline in status and the rise of agriculture. *Circaea* **12**: 1–156.

Albarella, U. & Thomas, R. (2002). They dined on crane: bird consumption, wild fowling and status in medieval England. *Acta Zoologica Cracoviensa* **45** (Special issue): 23–38.

Albarella, U., Beech, M. & Mulville, J. (1997). *The Saxon, Medieval and Post-medieval Mammal and bird bones excavated 1989–91 from Castle Mall, Norwich, Norfolk.* London: AML Report New Series **72/97**.

Albarella, U., Marrazzo, D., Spinetti, A. & Viner, S. In preparation. The animal bones from Welland Bank Quarry (Lincolnshire).

Allen, D.E. (1978). *The Naturalist in Britain. A social history.* Paperback edn. Pelican, London.

Allen, D. & Green, C.S. (1998). The Fir Tree Field Shaft: the date and archaeological and paleoenvironmental potential of a chalk swallowhole feature. *Proceedings of the Dorset Natural History and Archaeology Society* **120**: 25–37.

Allison, E. (1986). An archaeozoological study of the bird bones from seven sites in York. Ph.D., York University, York.

Allison, E.P. (1987). *Bird Bones from the Quayside, Queen St, Newcastle-upon-Tyne.* Ancient Monuments Lab Report **96/87**.

Allison, E.P. (1988). The bird bones. In *The Origin of the Newcastle Quayside* (eds C. O'Brien, L. Bown, S. Dixon & R. Nicholson). *Society of Antiquaries, Newcastle, Monograph* **3**: 133–137.

Allison, E.P. (1988b) *The Bird Bones from Hardendale Quarry, Shap, Cumbria.* Ancient Monuments Laboratory Report, 51/88.

Allison, E.P. (1989). The bird bones. In *The Birsay Bay Project Volume 1: Coastal Sites beside the Brough Road, Birsay, Orkney. Excavations 1976–1982* (ed. C.D. Morris), pp. 235–239, 247–248. University of Durham Department of Archaeology Monograph Series.

Allison, E.P. (1990). The Bird Bones, in R. Daniels, R., The development of Medieval Hartlepool: excavations at Church Close, 1984–85. *Archaeology Journal* **147** 337–410.

Allison, E. (1991). *Bird Bones from Annetwell Street, Carlisle, Cumbria, 1980–84.* Ancient Monuments Lab Report, 36/91, 1–10.

Allison, E.P. (1997a). Birds. In *The Romano-British Villa at Castle Copse, Great Bedwyn* (eds E. Houteller & T.N. Howe). Indiana University Press, Bloomington, IN.

Allison, E.P. (1997b). Bird bones. In *Settlements at Skail, Deerness, Orkney. Excavations by Peter Gelling of the Prehistoric, Pictish, Viking and later periods, 1963–1981* (ed. S. Buteux). BAR British Series 260. Archaeopress, Oxford.

References

Allison, E.P. (2000). The bird bone. In *Roman and Medieval Carlisle: the Southern Lanes. Excavations 1981–2* (ed. M. McCarthy). Department of Archaeological Science, University of Bradford Research Report 1, pp. 89–90.

Allison, E.P. & Rackham, D.J. (1996). The bird bones. In *The Birsay Bay Project. Volume 2: Sites in Birsay Village and on the Brough of Birsay* (ed. C.D. Morris). University of Durham Department of Archaeology Monograph Series 2.

Allison, E.P., Locker, A. & Rackham, D.J. (1985). The animal remains. In *An excavation in Holy Island Village 1977* (ed. D. O'Sullivan). *Archaeologia Aeliana* (5th series) **15**: 83–96.

Andrews, P. (unpublished). The microfauna from Longstone Edge.

Andrews, C.W. (1917). Report on the remains of birds. In *The Glastonbury Lake Village: A full description of the excavations and the relics discovered 1892–1907* (eds A. Bulleid & H.S.G. Gray), **2**: 631–637. Glastonbury Antiquarian Society.

Andrews, P. (1990). *Owls, Caves and Fossils*. British Museum (Natural History), London.

Armitage, P. & West, B. (1984). The Faunal Remains, in Thompson, A., Grew, F. & Schofield, J., Excavations at Aldgate, 1974. *Post-Medieval Archaeology* **18**: 1–148.

Armour-Chelu, M. (1988). *Taphonomic and cultural information from an assemblage of Neolithic bird bones from Orkney*. BAR British Series 186, pp. 69–76. Archaeopress, Oxford.

Armour-Chelu, M. (1991). The faunal remains. In *Maiden Castle, Excavations and field survey 1985–6* (ed. N.M. Sharples), English Heritage Archaeology Report **19**: 139–148.

Armstrong, A.L. (1928). Excavations in Pin Hole Cave, Creswell Crags, Derbyshire. *Proceedings of the Prehistoric Society* **6**: 330–334.

Ashdown, R.R. (1979). The avian bones from Station Road, Puckering. In *Excavations at Puckeridge and Braughing 1975–9* (ed. C. Partridge). *Hertfordshire Archaeology* **7**: 92–96.

Ashdown, R.R. (1993). The avian bones. In *Pennyland and Hartigans. Two Iron Age and Saxon sites in Milton Keynes* (ed. R.J. Williams). *Buckinghamshire Archaeology Society Monograph Series* **4**: 154–158.

Backhouse, J. (2001). *Medieval birds in the Sherborne Missal*. British Library, London.

Baker, P. (1998). *The Vertebrate Remains from Scole-Dickleburgh, Excavated in 1993 (Norfolk and Suffolk), A140 and A143 Road Improvement Project*. Ancient Monuments Laboratory Report, 29/98.

Baker, H., Stroud, D.A., Aebischer, N.J., Cranswick, P.A., Gregory, R.D., McSorley, C.A., Noble, D.G. & Rehfisch, M.M. (2006). Population estimates of birds in Great Britain and the United Kingdom. *British Birds* **99**: 25–44.

Balch, H.E. & Troup, R. (1910). A late Celtic & Romano-British Cave Dwelling at Wookey Hole. *Archaeologia* **62**: 565–592.

Barclay, A. & Halpin, C. (1998). *Excavations at Barrow Hills, Radley, Oxfordshire*. Vol I: *The Neolithic and Bronze Age Monument Complex*. Oxford Archaeology Unit, Thames Valley Landscapes.

Barker, G. (1983). The animal bones. In *Isbister: A Chambered Tomb in Orkney* (ed. J.W. Hedges). BAR British Series 115. Archaeopress, Oxford.

Barker, F.K., Barrowclough, G.F. & Groth, J.G. (2002). A phylogenetic hypothesis for passerine birds: taxonomic and biogeographic implications of an analysis of nuclear DNA sequence data. *Proceedings of the Royal Society, London, B* **269**: 295–308.

Bate, D.A. (1934). The Domestic fowl in pre-Roman Britain. *Ibis* **13**: 390–395.

Bate, D.A. (1966). Bird bones. In *The Meare Lake Village* (ed. H.S.G. Gray), **3**, pp. 408–410. Taunton Castle, Taunton.

Baxter, I.L. (1993). An eagle skull from an excavation in High Street, Leicester. *The Leicestershire Archaeological and Historical Society Transactions* **67**: 101–105.

Bedwin, O. (1975). Animal bones. In *Further excavations in Lewes* (ed. D.J. Freke). *Sussex Archaeological Collections* **114**: 189–190.

Bell, A. (1915). Pleistocene and later bird faunas of Great Britain and Ireland. *Zoologist* (srs. 4) **19**: 401–412.

Bell, A. (1922). Pleistocene and later birds of Great Britain and Ireland. *Naturalist* **1922**: 251–253.

Beneke, N. (1999). The evolution of the vertebrate fauna in the Crimean mountains from the Late Pleistocene to the mid-Holocene. In *The Holocene History of the European Vertebrate Fauna* (ed. N. Benecke), pp. 43–57. Marie Leidorf, Rahden, Westphalia.

Bennett, K.D. (1988). A provisional map of forest types for the British Isles 5000 years ago. *Journal of Quaternary Science* **4**: 141–144.

Bent, D.C. (1978). The Animal Remains, in Liddle, P., A late medieval enclosure in Donington Park. *Transactions of the Leicestershire Archaeological and Historical Society* **53**: 14–15.

Benton, M.J. (1999). Early origins of modern birds and mammals: molecules vs. morphology. *BioEssays* **21**: 1043–1051.

Benton, M.J. & Cook, E. (2005). British Tertiary fossil bird GCR sites. In *Mesozoic and Tertiary Fossil and Birds of Great Britain*, pp. 125–159. Geological Conservation Review Series No. 32. Joint Nature Conservation Committee, Peterborough.

Bezzel, E. & Wildner, H. (1970). Zur Ernährung bayerischer Uhus (*Bubo bubo*). *Vogelwelt* **91**: 191–198.

Biddick, K. (1984). Bones from the Iron Age Cat's Water subsite. In *Excavations at Fengate, Peterborough: Fourth Report* (ed. F. Pryor), pp. 217–225. Royal Ontario Museum Archaeological Monograph 7.

Bidwell, P.T. (1980). *Roman Exeter: Fortress and Town*. Exeter City Council, Exeter.

Bland, R.L., Tully, J. & Greenwood, J.J.D. (2004). Birds breeding in British gardens: an underestimated population? *Bird Study* **51**: 97–106.

Blondel, J. & Mourer-Chauvire, C. (1998). Evolution and history of the western Palearctic avifauna. *Trends in Ecology and Evolution* **13**: 488–492.

Boisseau, S. (1995). Former distribution of some extinct and declining British birds using place-name evidence. Unpublished B.Sc. (Zoology) project, University of Manchester.

Boisseau, S. & Yalden, D.W. (1999). The former status of the Crane *Grus grus* in Britain. *Ibis* **140**: 482–500.

Boles, D. (1995). The world's oldest songbird. *Nature* **374**: 21–22.

Bond, S. (1988). A bibliographic investigation into the rates and times of extinction of the Red Kite (*Milvus milvus*), Goshawk (*Accipiter gentilis*) and Raven (*Corvus corax*) from the counties of England, Scotland and Wales. B.Sc. (Hons.), University of Manchester, Manchester.

BOU (British Ornithologists' Union) (1998). *The British List*, 1st edn. BOU, Tring.

Bourdillon, J. & Coy, J. (1980). The animal bones. In *Excavations at Melbourne Street, Southampton, 1971–1976* (ed. P. Holdsworth), Southampton Archaeology Research Committee Report/CBA Research Report 1, pp. 79–118.

Bourne, W.R.P. (2003). Fred Stubbs, Egrets, Brewes and climatic change. *British Birds* **96**: 332–339.

Bramwell, D. (1954). Report on work at Ossom's Cave for 1954. *Peakland Archaeological Society Newsletter* **11** (unpaginated).

Bramwell, D. (1955). Second report on the excavation of Ossum's Cave. *Peakland Archaeological Society Newsletter* **12**: 13–16.

Bramwell, D. (1956). Third report on excavations at Ossum's Cave. *Peakland Archaeological Society Newsletter* **13**: 7–9.

Bramwell, D. (1960a). Some research into bird distribution in Britain during the Late Glacial and Post-Glacial periods. *Bird Report, Merseyside Naturalist's Association*, 51–58.

Bramwell, D. (1960b). The vertebrate fauna of Dowel Cave. *Peakland Archaeological Society Newsletter* **17**: 9–12.

Bramwell, D. (1967). *Report on the Bird Bones from the Roman Villa, Great Staughton*. Ancient Monuments Laboratory Report 1547.

Bramwell, D. (1969). Birds, in Rahtz, P.A., Excavations at King John's Hunting Lodge, Writtle, Essex, 1955–57. *Society of Medieval Archaeology Monograph Series* **3**: 114–115.

Bramwell, D. (1970). Bird remains. In *An Iron Age Promontory Fort at Budbury, Bradford-upon-Avon, Wiltshire* (ed. G.J. Wainwright). *Wiltshire Archaeology and Natural History Magazine* **65**: 154.

Bramwell, D. (1971). Capercaillie (*Tetrao urogallus*) remains from Peak District Caves. *Pengelly Newsletter* **17**: 10.

Bramwell, D. (1974). Bird bones. In *Dun Mor Vaul: an Iron Age brock on Tiree* (ed. E.W. Mackie), pp. 199–200. University of Glasgow Press, Glasgow.

Bramwell, D. (1975a). Bird remains from Medieval London. *London Naturalist*, **54**, 15–20.

Bramwell, D. (1975c). The bird bones. In *Excavations in Medieval Southampton* (eds C. Platt & G. Coleman-Smith), Vol. 1, pp. 340–341. Leicester University Press, Leicester.

Bramwell, D. (1975f). Bird remains. In *Excavations on the site of St. Mildred's Church, Bread Street, London, 1973–74* (eds P. Marsden, T. Dyson & M. Rhodes). *Transactions of the London and Middlesex Archaeological Society* **26**: 207–208.

Bramwell, D. (1976a). The vertebrate fauna at Wetton Mill Rock Shelter. In *The excavation of Wetton Mill Rock Shelter, Manifold Valley, Staffs* (ed. J.H. Kelly), pp. 40–51. City Museum & Art Gallery, Stoke on Trent.

Bramwell, D. (1976b). Report on the bird bones from Walton, Aylesbury. In *Saxon and Medieval Walton, Aylesbury. Excavations 1973–74* (ed. M. Farley). *Records of Buckinghamshire* **20**: 287–289.

Bramwell, D. (1977a). Bird bones. In *Excavations in King's Lynn 1963–1970* (eds H. Clarke & A. Carter). *Society of Medieval Archaeology Monograph Series* **7**: 399–402.

Bramwell, D. (1977b). Bird and vole bones from Buckquoy, Orkney. In *A Pictish and Viking Age Farmstead at Buckquoy, Orkney* (ed. A. Ritchie). *Proceedings of the Society of Antiquaries of Scotland* **108**: 209–211.

Bramwell, D. (1978a). The bird bones. In *The excavation of an Iron Age Settlement, Bronze Age Ring-Ditches, and Roman features at Ashville Trading Estate, Abingdon (Oxfordshire). 1974–76* (ed. M. Parrington), Vol. 1/28, pp. 133. Oxford Archaeology Unit Report/CBA Research Report.

Bramwell, D. (1978c). The fossil birds of Derbyshire. In *Birds of Derbyshire* (ed. R.A. Frost), pp. 160–163. Moorland Publishing Company.

Bramwell, D. (1979a). Bird remains. In *Investigations in Orkney* (ed. C. Renfrew). *Report of the Research Committee of the Society of Antiquity* **38**: 138–143.

Bramwell, D. (1979b). Bird bones. In *Frocester Roman Court Villa* (eds B.S. Smith & N.M. Herbert). *Transactions of the Bristol and Gloucestershire Archaeological Society* **97**: 61–62.

Bramwell, D. (1979e). The bird bones. In *St Peters Street, Northampton: Excavations 1973–76* (ed. J.H. Williams). *Northampton Development Corporation Archaeological Monograph* **2**: 399.

Bramwell, D. (1980a). Identification and interpretation of bird bones. In *Excavations in North Elham Park 1967–1972* (ed. P. Wade-Martins). *East Anglian Archaeological Report* **9**: 377–409.

Bramwell, D. (1981a). Report on Bones of Birds. In *Excavations in Iona 1964 to 1974* (ed P. Reece), Vol. 5, pp. 45–46. University of London Institute of Archaeology Occasional Publication.

Bramwell, D. (1981b). The bird remains from the Hindlow Cairn. In *A Cairn on Hindlow, Derbyshire: Excavations 1953* (eds P. Ashbee & R. Ashbee). *Derby Archaeological Journal* **101**: 39.

Bramwell, D. (1983a). Bird remains. In *Isbister: A chambered tomb in Orkney* (ed. J.W. Hedges), BAR British Series 115, pp. 159–170. Archaeopress, Oxford.

Bramwell, D. (1983c). Bird bones from Knap of Howar. In *Excavations of a Neolithic Farmstead at Knap of Howar, Papa Westray, Orkney* (ed. A. Ritchie), *Proceedings of the Society of Antiquaries of Scotland* **113**: 40–121.

Bramwell, D. (1983d). Bird and amphibian remains. In *Caerwent (Venta Silurum): The Excavations of the North West Corner Tower and Analysis of the Structural Sequence of the defences* (ed. P. Casey). *Archaeologia Cambrensis* **132**: 49–77.

Bramwell, D. (1984). The birds of Britain – when did they arrive? In *In the Shadow of Extinction* (eds R.D. Jenkinson & D.D. Gilbertson). J.R.Collis, Sheffield.

Bramwell, D. (1985). Identification of bird bones. In *The Excavation of a Romano-British Rural Establishment at Barnsley Park, 1961–1979: Part III* (eds G. Webster, P. Fowler, B. Noddle & L. Smith). *Transactions of the Bristol and Gloucestershire Archaeological Society* **103**: 96–97.

Bramwell, D. (1986a). Report on the Bird Bone, in Bateman, J. & Redknap, M., Coventry: Excavations on the Town Wall 1976–78. *Coventry Museum Monograph Series* **2**.

Bramwell, D. (1987). The bird remains. In *Bu, Gurness and the brochs of Orkney. Part 1: Bu* (ed. J.W. Hedges). BAR British Series 163. Archaeopress, Oxford.

Bramwell, D. (1994). Bird bones. In *Howe: four millennia of Orkney prehistory. Excavations 1978–82* (ed. B.B. Smith). Society of Antiquaries of Scotland Monograph Series 9. Edinburgh.

Bramwell, D. & Wilson, R. (1979). The bird bones. In *Excavations at Broad Street, Abingdon* (ed. M. Parrington). *Oxoniensia* **44**: 20–21.

Bramwell, D. & Yalden, D.W. (1988). Birds from the Mesolithic of Demen's Dale, Derbyshire. *Naturalist* **113**: 141–147.

Bramwell, D., Yalden, D.W. & Yalden, P.E. (1990). Ossom's Eyrie Cave: an archaeological contribution to the recent history of vertebrates of Britain. *Zoological Journal of the Linnean Society* **98**: 1–25.

Brothwell, D. (1993). Animal bones. In *Excavations at Loughor Castle, West Glamorgan* (ed. J.M. Lewis). *Archaeologia Cambrensis* **142**: 170–171.

Brown, A. & Grice, P. (2005). *Birds in England*. T.&A.D. Poyser, London.

Brown, L. (1976). *British Birds of Prey*. Collins, London.

Browne, S. (1988). Animal bone evidence. In *Hen Domen, Montgomery. A Timber castle on the English-Welsh border. Excavations 1960–1988. A summary report* (eds P. Barker & R.A. Higham), p.14. The Hen Domen Archaeological Project.

Browne, S. (2000). The animal bones. In *Hen Domen, Montgomery. A Timber castle on the English-Welsh Border. A final report* (eds R.A. Higham & P. Barker), pp. 126–134. University of Exeter Press.

Bryant, D.M. & Westerterp, K.R. (1980). The energy budget of the House Martin (*Delichon urbica*). *Ardea* **68**: 91–102.

Buckland-Wright, J.C. (1987) Animal Bones in Green, C.S Excavations at Poundbury I, The Settlements. *Dorset Natural History and Archaeological Society Monograph Series* **7** (supplementary microfiche).

Buckland-Wright, J.C. (1993). The animal bones. In *Excavations at Poundbury 1966–80. Vol II: The cemeteries* (eds D.E. Farwell & T.L. Molleson). *Dorset Natural History and Archaeology Society Monograph Series* **11**: 110–111.

Bulleid, A. & Gray, H. St. G. (1948) *The Meare Lake Village*. Vol 1. Taunton Castle, Taunton, Somerset.

Burenhult, G. (1980). *The archaeological excavation at Carrowmore, County Sligo. Excavation seasons 1977–1979*. Theses and papers in North European Archaeology, Institute of Archaeology, University of Stockholm.

Campbell, J.B. (1977). *The Upper Palaeolithic of Britain*. Clarendon Press, Oxford.

Carrott, J., Dobney, K., Hall, A.R., Irving, B., Issitt, M., Jaques, D., Kenward, H.K., Large, F., McKenna, B., Milles, A., Shaw, T. & Usai, R. (1995). *Assessment of Biological Remains and Sediments From Excavations at the Magistrates' Court Site, Hull (site code HMC94)*. Report from the EAU, York, 95/17.

Carrott, J., Dobney, K., Hall, A.R., Issitt, M., Jaques, D., Johnstone, C., Kenward, H.K., Large, F. & Skidmore, P. (1997). *Environment, land use and activity at a medieval and post-medieval site at North Bridge, Doncaster, South Yorkshire*. Report from the EAU, York, 97/16.

Carss, D.N. & Ekins, G.R. (2002). Further European integration: mixed subspecies colonies of Great Cormorants *Phalacocorax carbo* in Britain – colony establishment, diet, and implications for fisheries management. *Ardea* **90**: 23–41.

Charles, R. & Jacobi, R.M. (1994). The Lateglacial fauna from Robin Hood Cave, Creswell: a re-assessment. *Oxford Journal of Archaeology* **13**: 1–32.

Cherryson, A.K. (2002). The identification of archaeological evidence for hawking in medieval England. *Acta Zoologica Cracoviensia* **45**: 307–314.

Clark, J.G.D. 1954. *Excavations at Star Carr.* Cambridge University Press, Cambridge.

Clark, J.D.G. & Fell, C.I. (1953). The Early Iron Age site at Micklemoor Hill, West Harling, Norfolk, and its pottery. *Proceedings of the Prehistoric Society* **19**: 1–36.

Clarke, A.S. (1965). The animal bones. In *The excavation of a chambered cairn at Embo, Sutherland* (eds A.S. Henshall & J.C. Wallace). *Proceedings of the Society of Antiquaries of Scotland* **96**: 35–36.

Clark, J.A., Robinson, R.A., Balmer, D.E., Blackburn, J.R., Grantham, M.J., Griffin, B.M., Marchant, J.H., Risley, K. & Adams, S.Y. (2005) Bird ringing in Britain and Ireland in 2004. *Ringing and Migration* **22**: 85–127.

Clot, A. & Mourer-Chauviré, C. (1986). Inventaire systematique des oiseaux quaternaires des Pyrenées Francaises. *Munibe (Antropologia y Arqueologia)* **38**: 171–184.

Clutton-Brock, J. (1979). Report of the Mammalian remains other than Rodents from Quanterness. In *Investigations in Orkney* (ed. C. Renfrew). *Reports of the Research Committee of the Society of Antiquaries, London* **38**: 112–133.

Cohen, A.S., D. (1986). *A Manual for the Identification of Bird Bones from Archaeological Sites.* Alan Cohen, London.

Coles, B. (2006). *Beavers in Britain's Past.* Oxbow Press, Oxford.

Coles, J.M. (1971). Birds in the Early Settlement of Scotland: Excavations at Morton, Fife. *Proceedings of the Prehistoric Society* **37**: 350–351.

Connell, B. & Davis, S.J.M. (Unpublished) *Animal bones from Roman Carlisle, Cumbria; The Lanes (2) excavations, 1978–1982.* Ancient Monuments Laboratory Report.

Cooper, J.H. (2005). Pigeons and pelagics: interpreting the Late Pleistocene avifaunas of the continental 'island of Gibraltar'. *Proceedings of the International Symposium Insular Vertebrate Evolution: the Palaeontological Approach* **12**: 101–112.

Cooper, J.H. (2000). First fossil record of Azure-winged Magpie *Cyanopica cyanus* in Europe. *Ibis* **142**: 150–151.

Cooper, A. & Penny, D. (1997). Mass survival of birds across the Cretaceous-Tertiary boundary: molecular evidence. *Science* **275**: 1109–1113.

Cowles, G.S. (1973). Bird bones excavated from East Gate, Lincoln. In *The Gates of Roman Lincoln* (eds F.H. Thompson & J.B. Whitwell). Society of Antiquaries of London, Oxford.

Cowles, G.S. (1978). Bird Bones, p. 146 in Canham, R., *2000 Years of Brentford.* London.

Cowles, G.S. (1980a). Bird bones. In *Excavations at Billingsgate Buildings 'Triangle', Lower Thames Street, 1974* (ed. D.M. Jones), London and Middlesex Archaeological Society Special Paper **4**, p. 163. London.

Cowles, G.S. (1980b). Bird bones. In *Southwark Excavations 1972–1974*, **1**, pp. 231–232. London.

Cowles, G.S. (1981). The first evidence of Demoiselle Crane *Anthropoides virgo* and Pygmy Cormorant *Phalacrocorax pygmaeus* in Britain. *Bulletin of the British Ornithological Club*, **10**: 383.

Cowles, G.S. (1993). Vertebrate remains. In *The Uley Shrines: Excavation of a ritual complex on West Hill, Uley, Gloucestershire, 1977–9* (eds A. Woodward & P. Leach), English Heritage Archaeological Report 17, pp. 257–303.

Coy, J. (1980). *Bird Bones from Westgate, Southampton.* Ancient Monuments Laboratory, English Heritage, London.

Coy, J. (1981a). Animal husbandry and faunal exploitation in Hampshire. In *The Archaeology of Hampshire – from the Palaeolithic to the Industrial Revolution* (eds S.J. Shennan & R.T. Schadla-Hall). *Hampshire Field Club and Archaeological Society Monograph* **1**: 95–103.

Coy, J. (1981b). *Bird Bones from Chalk Lane*. AML Report OS No. 3450. London.

Coy, J. (1982). The animal bones. In *Excavations of an Iron Age Enclosure at Groundwell Barn, Blunsdon St Andrews 1976–7* (ed. C. Gingell). *Wiltshire Archaeology & Natural History Magazine* **76**: 68–73.

Coy, J. (1983a). The animal bone, in Jarvis, K.S., Excavations in Christchurch 1969–1980. *Dorset Natural History and Archaeology Society Monograph Series* **5**: 91–97.

Coy, J. (1984a). The bird bones. In *Danebury: an Iron Age hill fort in Hampshire vol 2, The excavations 1969–1978: The finds* (ed. B. Cunliffe). *CBA Research Report* **52**: 527–531.

Coy, J. (1984b). *Animal Bones from Saxon, Medieval and post-Medieval phases (10–18 C.) of Winchester Western Suburbs*. Ancient Monuments Laboratory, English Heritage, London.

Coy, J. (1987a). Animal bones. In *A Banjo Enclosure in Micheldever Wood, Hampshire* (ed. P.J. Fasham). *Trust for Wessex Archaeology/Hampshire Field Club Monograph* **5**: 45–47.

Coy, J. (1987b). The animal bones. In *Romano-British Industries in Purbeck. Excavations at Ower and Rope Lake Hole* (ed. P.J. Woodward). *Dorset Natural History and Archaeology Society Monograph Series* **6**: 114–118, 177–179.

Coy, J. (1991). Bird Bones. In *Report on the Faunal remains from Wirral Park Farm (The Mound), Glastonbury* (eds T. Darvill & J. Coy), Unpublished Report to the Ancient Monuments Laboratory **245**.

Coy, J. (1992) Faunal remains, in Butterworth, C.A. & Lobb, S.J. (eds) Excavations in the Burghfield Area, Berkshire. Developments in the Bronze Age and Saxon Landscapes. *Wessex Archaeology Report* **1**: 128–130.

Coy, J. (1995). Animal Bones, in Fasham, P.J. & Reevill, G., Brighton Hill South (Hatch Warren): An Iron Age Farmstead and Deserted Medieval Village in Hampshire. *Wessex Archaeology Report* **7**: 132–135.

Coy, J. (1997). Comparing bird bones from Saxon sites: problems of interpretation. *International Journal of Osteoarchaeology* **7**: 415–421.

Coy, J. & Hamilton-Dyer, S. (1993). The bird and fish bone. In *Excavations at Iona 1988* (ed. F. McCormick). *Ulster Journal of Archaeology* **56**: 100–101.

Crabtree, P.A. (1983) *Report on the animal bones from the Chapter House at St.Alban's Abbey*. Unpublished Report.

Crabtree, P.A. (1985). The faunal remains. In *West Stow. Anglo-Saxon Village* Vol. I: *Text* (ed. S. West). *East Anglian Archaeology Report* **24**: 85–95.

Crabtree, P.A. (1989a). West Stow, Suffolk: Early Anglo Saxon animal husbandry. *East Anglian Archaeology Report* **47**: 27.

Crabtree, P.A. (1989b). Faunal remains from Iron Age and Romano-British features. In *West Stow, Suffolk: The Prehistoric and Romano-British Occupations* (ed. S. West). *East Anglian Archaeology Report* **28**: 101–105.

Crabtree, P.A. (1994). The animal bones present from Ipswich, Suffolk, recovered from 16 sites excavated between 1974–1988. Unpublished Report.

Cramp, S., Simmons, K.E.L. & others, (1977–1994) (HBWP) *Handbook of the Birds of Europe, the Middle East and North Africa*. (9 vols). Oxford University Press, Oxford.

Currant, A.P. (1989). The Quaternary origins of the British mammal fauna. *Biological Journal of the Linnean Society* **38**: 23–30.

Currant, A.P. & Jacobi, R. (2001). A formal mammalian biostratigraphy for the Late Pleistocene of Britain. *Quaternary Science Reviews* **20**: 1707–1716.

D'Arcy, G. (1999). *Ireland's Lost Birds*. Betaprint Ltd, Dublin.

D'Arcy, G. (2006). Little bird bone: long story. *Irish Wildlife* **2**: 10–12.

Darby, H.C. (1976). *A New Historical Geography of England after 1600*. Cambridge University Press, Cambridge.

Darby, H.C. (1983). *The Changing Fenland*. Cambridge University Press, Cambridge.

Darby, H.C. & Versey, G.R. (1975). *Domesday Gazetteer*. Cambridge University Press, Cambridge.

Darvill, T. & Coy, J. (1985). Report on the faunal remains from the Mound, Glastonbury. In *Excavations on the Mound, Glastonbury, Somerset, 1971* (ed. J. Carr), *Proceedings of the Somerset Archaeological and Natural History Society* **129**: 56–60.

David, A. (1991). Late Glacial archaeological residues from Wales: a selection, pp. 141–159 in *The Late Glacial in north-west Europe: Human adaptation and environmental change at the end of the Pleistocene*. (eds N. Barton, A.J. Roberts & D.A. Roe). CBA Research Report **77**.

Davidson, J.L. & Henshall, A.S. (1989). *The Chambered Cairns of Orkney. An Inventory of the Structures and their Contents*. Edinburgh University Press.

Davis, S.J.M. (1981). The effects of temperature change and domestication on the body size of Late Pleistocene to Holocene mammals of Israel. *Paleobiology* **7**: 101–114.

Davis, S.J.M. (1997). *Animal bones from the Roman site Redlands Farm, Stanwick, Northamptonshire, 1990 excavation*. Ancient Monuments Laboratory Report, 106/97.

Deane, C.D. (1979). The Capercaillie as an Irish species. *Irish Birds* **1**: 364–369.

Dissaranayake, R. (1992). An analysis of passerine humeri. Unpublished B.Sc. (Environmental Biology) thesis, University of Manchester.

Dobney, K. & Jaques, D. (2002). Avian signatures for identity and status in Anglo-Saxon England. *Acta Zoologica Cracoviensia* **45**: 7–21.

Dobney, K., Milles, A., Jaques, D. & Irving, B. (1994). *Material Assessment of the Animal Bone Assemblage from Flixborough*. Unpublished Reports from the Environmental Archaeology Unit, York, 94/6, 1–7.

Dobney, K., Jaques, D. & Irving, B. (1996). Of butchers and breed: report on vertebrate remains from various sites in the City of Lincoln. *Lincoln Archaeological Studies* **5**: 1–215.

Dobney, K., Jaques, D., Carrott, J., Hall, A.R., Issitt, M. & Large, F. (2000). Biological remains. In *Excavations on the site of the Roman signal station at Carr Naze, Filey, 1993–94* (ed. P. Ottaway). *Archaeology Journal* **157**: 148–179.

Dobney, K., Jaques, D., Barrett, J. & Johnstone, C. (2007). *Farmers, Monks and Aristocrats. The Environmental Archaeology of Anglo-Saxon Flixborough. Excavations at Flixborough*, Vol. 3. Oxbow Books, Oxford.

Driesch, A.v.d. (1999). The crane, *Grus grus*, in prehistoric Europe and its relation to the Pleistocene crane, *Grus primigenia*. In *The Holocene History of the European Vertebrate Fauna* (ed. N. Benecke), pp. 201–209. Marie Leidorf, Rahden, Westphalia.

Drovetski, S.V. (2003). Plio-Pleistocene climatic oscillations, Holarctic biogeography and speciation in an avian subfamily. *Journal of Biogeography* **30**: 1173–1181.

Dyke, G.J. (2001). The evolutionary radiation of modern birds: systematics and patterns of diversification. *Geological Science* **36**: 305–315.

Dyke, G.J., Dortangs, R.W., Jagt, J.W.M., Schulp, A.S., Mulder, E.W.A., & Chiappe, L.M. (2002) Europe's last Mesozoic bird. *Naturwissenschaften* **89**: 408–411.

Dyke, G.J. & Gulas, B.E. (2002). The fossil Galliform bird *Paraortygoides* from the Lower Eocene of the United Kingdom. *American Museum Novitates* **3360**: 1–14.

Dyke, G.J., Nudds, R.L. & Benton, M. J. (2007a). Modern avian radiation across the Cretaceous-Paleogene boundary. *Auk* **124**: 339–341.

Dyke, G.J., Nudds, R.L. & Walker, C.A. (2007b). The Pliocene *Phoebastria* ('*Diomedea*') *anglica*: Lydekker's English fossil albatross. *Ibis* **149**: 626–631.

Eastham, A. (1971). The bird bones. In *Excavations at Fishbourne, 1961–1969* (ed. B. Cunliffe), **2**, pp. 388–393.

Eastham, A. (1975). The bird bones. In *Excavations at Portchester.* Vol. I: *Roman* (ed. B. Cunliffe), pp. 409–415. Report of the Research Committee of the Society of Antiquaries of London.

Eastham, A. (1976). The bird bones. In *Excavations at Portchester Castle.* Vol. II: *Saxon* (ed. B. Cunliffe), pp. 287–296. Report of the Research Committee of the Society of Antiquaries of London 33.

Eaton M.A., Ausden M., Burton N., Grice P.V., Hearn R.D., Hewson C.M., Hilton G.M., Noble D.G., Ratcliffe N. and Rehfisch M.M. (2006). The value of volunteers in monitoring birds in the UK, p. 31 in *The state of the UK's birds 2005.* RSPB, BTO, WWT, CCW, EN, EHS and SNH, Sandy, Bedfordshire.

Edwards, A. & Horne, M. (1997). Animal bones. In *Sacred Mound, Holy Rings* (ed. A. Whittle), Oxbow Monograph 74, pp. 117–129. Oxford.

Ekwall, E. (1936). *Studies on English Place-names.* Kunglinga Vitterhets Historie och Antikvitets Akademiens handlingar, Stockholm.

Ekwall, E. (1960). *The Concise Oxford Dictionary of English Place-names*, 4th edn. Oxford University Press, Oxford.

Elzanowski, A. (2002a). Archaeopterygidae (Upper Jurassic, Germany). In *Mesozoic Birds: Above the Heads of Dinosaurs* (eds L.M. Chiappe & L.M. Witmer), pp. 129–159. University of California Press, Los Angeles, CA.

Elzanowski, A. (2002b). Biology of basal birds and the origin of bird flight. In *Proceedings of the 5th Symposium of the Society of Avian Paleontology and Evolution* (eds Zhonge Zhou & Fucheng Zhang), pp 211–226. Science Press, Beijing.

Erbersdobler, K. (1968). *Vergleichend morphologische Untersuchungen an Einzelknochen des postcranialen Skeletts in Mitteleuropea vorkommender mittelgrosser Huhnervogel.* Universität Munchen, Munich.

Ericson, P.G.P. & Tyrberg, T. (2004). *The Early History of the Swedish Avifauna.* Kunglinga Vitterhets Historie och Antikvits Akademiens Handlingar, Stockholm.

Ericson, P.G.P., Tyrberg, T., Kjellberg, A.S., Jonsson, L. & Ullén, I. (1997). The earliest record of House Sparrows (*Passer domesticus*) in Northern Europe. *Journal of Archaeological Science* **24**: 183–190.

Ericson, P.P., Christids, L., Cooper, A., Irestedt, M., Jackson, J., Johansson, U.S. & Norman, J.A. (2002). A Gondwanan origin of passerine birds supported by DNA sequences of the endemic New Zealand wrens. *Proceedings of the Royal Society, London, B*, **269**: 235–241.

Evans, C. & Serjeantson, D. (1988). The back water economy of a fen-edge community in the Iron Age: the Upper Delphs, Haddenham. *Antiquity* **62**: 360–370.

Ewart, J.C. (1911). Animal remains. In *A Roman frontier post and its people. The fort of Newstead – in the parish of Melrose* (ed. J. Curle), pp. 362–377. Maclehose & Sons, Glasgow.

Faegri, K. & Iversen, J. (1975). *Textbook of Pollen Analysis.* Blackwell, Oxford.

Feduccia, A. (1995). Explosive evolution in Tertiary birds and mammals. *Science* **267**: 637–638.

Feduccia, A. (1996). *The Origin and Evolution of Birds.* Yale University Press, New Haven.

Fick, O.K.W. (1974). *Vergleichend morphologische Untersuchungen an Einzelknochen europaischer Taubenarten.* Universitat Munchen, Munchen.

Field, D. (1999). The animal bones. In *Iron Age and Roman Quinton. The evidence for the ritual use of the site (Site 'E' 1978–1981)* (ed. R.M. Friendship-Taylor), Vol. 5, pp. 57–60. Upper Nene Archaeology Society Fascicule.

Finlay, J. (1991). Animal bone. In *Excavations of the Wheelhouse and other Iron Age structures at Sollas, North Uist, by R.J.C. Atkinson in 1957* (ed. E. Cambell). *Proceedings of the Society of Antiquaries of Scotland* **121**: 147–148.

Finlay, J. (1996). Human and animal bone. In *The excavation of a succession of prehistoric roundhouses at Cnoc Stanger, Reay, Caithness, Highland, 1981–2* (ed. R.J. Mercer). *Proceedings of the Society of Antiquaries of Scotland* **126**: 157–189.

Finney, S.K., Pearce-Higgins, J.W. & Yalden, D.W. (2005). The effect of recreational disturbance on an upland breeding bird, the Golden Plover *Pluvialis apricaria. Biological Conservation* **121**: 53–63.

Fisher, C.T. (1986). Bird bones from the excavations at Crown Car Park, Nantwich, Cheshire. *Circaea* **4**: 55–64.

Fisher, C.T. (1996). Bird bones. In *Excavations in Castletown, Isle of Man 1989–1992* (eds P.J. Davey, D.J. Freke & D.A. Higgins), pp. 144–151. Liverpool University Press.

Fisher, C.T. (2002). The bird bones. In *Excavations on St Patrick's Isle, Peel, Isle of Man, 1982–99. Prehistoric, Viking, Medieval and Later* (ed. D.J. Freke). Liverpool University Press.

Fisher, J. (1966). *The Shell Bird Book*. Ebury Press & Michael Joseph, London.

Fisher, J.H. (1977). *The Complete Poetry and Prose of Geoffrey Chaucer*. Holt, Reinhart and Winston, New York.

Fisher, J. & Lockley, R.M. (1954). *Sea-Birds*. Collins, London.

Fitter, R.S.R. (1959). *The Ark in our Midst*. Collins, London.

Fok, K.W., Wade, C.M. & Parkin, D. (2002). Inferring the phylogeny of disjunct populations of the azure-winged magpie *Cyanopica cyanus* from mitochondrial control sequences. *Proceedings of the Royal Society of London* B, **269**: 1671–1679.

Forbes, C.L., Joysey, K.A. & West. R.G. (1958) On Post-Glacial Pelicans in Britain. *Geological Magazine* **95**: 153–160.

Fountaine, T.M.R., Benton, M.J., Dyke, G.J. & Nudds, R.L. (2005). The quality of the fossil record of Mesozoic birds. *Proceedings of the Royal Society, London, B*, **272**: 289–294.

Fraser, F.C. & King, J.E. (1954). Birds. In *Excavations at Star Carr* (ed. J.G.D. Clark). Cambridge University Press.

Fuller, E. (1999). *The Great Auk*. Errol Fuller, Southborough, Kent.

Fuller, R.J. (2000). Influence of treefall gaps on distributions of breeding birds within interior old-growth stands in Białowieża Forest, Poland. *Condor* **102**: 267–274.

Fuller, R.J. (2002). Spatial differences in habitat selection and occupancy by woodland bird species in Europe: a neglected aspect of bird-habitat relationships. *Avian Landscape Ecology, IALE (UK)* 101–110.

Fuller, R.J. & Gough, S.J. (1999). Changes in sheep numbers in Britain: implications for bird populations. *Biological Conservation* **91**: 7–89.

Fumihito, A., Miyake, T., Sumi, S.-I., Takada, M., Ohno, S. & Kondo, N. (1994). One subspecies of the red junglefowl (*Gallus gallus gallus*) as the matriarchic ancestor of all domestic breeds. *Proceedings of the National Academy of Sciences* **91**: 12505–12509.

Fumihito, A., Miyake, T., Takada, M., Shingu, R., Endo, T., Gojobori, T., Kondo, N. & Ohno, S. (1996). Monophyletic origin and unique dispersal patterns of domestic fowls. *Proceedings of the National Academy of Sciences* **93**: 6792–6795.

Galbraith, C.A., Groombridge, R. & Tucker, C. (2000). *Report of the UK Raptor Working Group*. DETR and JNCC, Bristol and Peterborough.

Galton, P. & Martin, L.D. (2002). *Enaliornis*, an early Cretaceous Hesperornithiform Bird from England, with comments on other Hesperornithiformes. In *Mesozoic Birds: Above the Heads of Dinosaurs* (eds L.M. Chiappe & L.M. Witmer), pp. 317–338. California University Press, Berkeley, CA.

Gardner, N. (1997). Vertebrates and small vertebrates. In *Sacred mound, Holy Rings. Silbury Holl and the West Kennet Palisade enclosures: a later Neolithic complex in North Wiltshire* (ed. A. Whittle), Oxbow Monograph 74, pp. 47–49.

Garmonsway, G.N. (1947). *Aelfric's Colloquy*, 2nd. edn. Methuen, London.

Garrad, L.S. (1972). Bird remains, including those of a Great Auk *Alca impennis,* from a midden deposit in a cave at Perwick Bay, Isle of Man. *Ibis* **114**: 258–259.

Garrad, L.S. (1978). Evidence for the history of the vertebrate fauna of the Isle of Man, pp. 61–76 in *Man and Environment in the Isle of Man* (ed P.J. Davey). BAR British Series 54(1).

Geikie, J. (1881). *Prehistoric Europe: A Geological Sketch.* Edward Stanford, London.

Gelling, M. (1987). Anglo-Saxon Eagles. *Leeds Studies in English* (n.s.) **XVIII**: 173–181.

Gelling, M. & Cole, A. (2000). *The Landscape of Place-Names.* Shaun Tyas, Stamford, CA.

Genoud, M. (1985). Ecological energetics of two European shrews: *Crocidura russula* and *Sorex coronatus* (Soricidae: Mammalia). *Journal of Zoology, London (A)* **207**: 63–85.

Gentry, A., Clutton-Brock, J. & Groves, C.P. (2003). The naming of wild animal species and their domestic derivatives. *Journal of Archaeological Science* **31**: 645–651.

Gibbons, D.W., Reid, J.B. & Chapman, R.A. (1993). *The New Atlas of Breeding Birds in Britain and Ireland: 1988–1991.* T.&A.D. Poyser, London.

Gidney, L. (1991). *Leicester, The Shires 1988 excavations: the animal bones from the Medieval deposits at St Peter's Lane.* Ancient Monuments Laboratory Report **116/91**.

Gidney, L. (1992). The animal bone. In *Excavations at Brougham Castle, 1987* (ed. J.H. Williams). *Transactions of the Cumberland and Westmorland Antiquarian and Archaeological Society* **92**: 120–121.

Gidney, L. (1993). *Leicester, The Shires 1988 Excavations: Further Identifications of Small Mammal and Bird Bones.* Centre For Archaeology Report 92/93, pp. 1–16.

Gidney, L. (1995) *The Cathedral, Durham City. An assessment of the animal bones.* Durham Environmental Archaeology Report, 4/95.

Gidney, L. (1996). *Housesteads 1974–1981: Animal Bone Assessment.* Department of Archaeology, Durham University.

Gilbert, B.M., Martin, L.D. & Savage, H.G. (1996). *Avian Osteology.* Missouri Archaeological Society, Columbia, MO.

Gilmore, F. (1969). The animal and human skeletal remains. In *Excavations at Hardingstone, Northants 1967–8* (ed. P.J. Woods). Northamptonshire County Council.

Godwin, H. (1975). *The History of the British Flora.* Cambridge University Press.

Gosler, A.G., Greenwood, J.J.D. & Perrins, C. (1995). Predation risk and the cost of being fat. *Nature* **377**: 621–623.

Gray, H.S.G. (1966). *The Meare Lake Village: A Full Description of the Excavations and Relics from the Eastern Half of the West Village, 1910–1933.* Taunton Castle, Taunton.

Greene, J.P. (1989). *Norton Priory. The Archaeology of a Medieval Religious House.* Cambridge University Press.

Greenoak, F. (1979). *All the Birds of the Air.* Andre Deutsch, London.

Greenwood, J.J.D. & Carter, N. (2003). Organisation eines nationalen Vogelmonitorings durch den British Trust for Ornithology – Erfahrungsbericht aus Grossbritannien (Organising national bird monitoring by the BTO – experiences from Britain). Berichte des Landesamtes fur umweltschutz Sachsen-Anhalt 1/2003, pp. 14–26.

Greenwood, J.J.D., Gregory, R.D., Harris, S., Morris, P.A. & Yalden, D.W. (1996). Relationships between abundance, body size and species number in British birds and mammals. *Philosophical Transactions of the Royal Society of London,* B **351**: 265–278.

Grieve, S. (1882). Notice on the discovery of remains of the great auk or garefowl (*Alca impennis*) on the Island of Oronsay, Argyllshire. *Journal of the Linnean Society (Zoology)* **16**: 479–487.

Grigson C. 1999. The mammalian remains. In *The Harmony of Symbols. The Windmill Hill Causewayed Enclosure* (eds. A.Whittle, J. Pollard & C. Grigson), pp.164–252. Oxbow Books, Oxford.

Grigson, C. & Mellars, P.A. (1987). The mammalian remains from the middens. In *Excavations on Oronsay* (ed. P.A. Mellars), pp. 243–289. Edinburgh University Press, Edinburgh.

Groom, D.W. (1993). Magpie *Pica pica* predation on Blackbird *Turdus merula* nests in an urban area. *Bird Study* **40**: 55–62.

Gruar, D., Peach, W. & Taylor, R. (2003). Summer diet and body condition of Song Thrushes *Turdus philomelos* in stable and declining farmland populations. *Ibis* **145**: 637–649.

Gurney, J.H. (1921). *Early Annals of Ornithology*. H.F. & G. Witherby, London.

Hall, J.J. (1982). The cock of the wood. *Irish Birds* **2**: 38–47.

Hallen, Y. (1994). The use of bone and antler at Foshigarry and Bac Mhic Connain, two Iron Age sites on North Uist, Western Isles. *Proceedings of the Society of Antiquaries of Scotland* **124**: 189–231.

Hamilton, R. (1971). Animal remains. In *Latimer: Belgic, Roman, Dark Age, and Early Modern Farm* (ed. K. Branigan), pp. 163–166. Chess Valley Archaeology and History Society, Bristol.

Hamilton-Dyer, S. (1993). The animal bones. In *Excavations in the Scamnum Tribunorum at Caerleon. The Legionnary Museum Lite 1983–5* (ed. V.D. Zienkiewicz). *Britannia* **24**: 132–136.

Hamilton-Dyer, S. (1999). Animal bones. In *A35 Tolpuddle to Puddletown Bypass DBFO, Dorset, 1996–8. Incorporating excavations at Tolpuddle Ball 1993* (eds C.M. Hearne & V. Birbeck). *Wessex Archaeological Report* **15**: 188–202.

Hamilton-Dyer, S. & McCormick, F. (1993). The animal bones. In *Excavation of a shell midden site at Carding Mill Bay, near Oban, Scotland* (eds K.D. Connock, B. Finlayson & C.M. Mills). *Glasgow Archaeological Journal* **17**: 34.

Hammon A. (2005). Late Romano-British – early medieval socio-economic and cultural change: Analysis of the mammal and bird bone assemblages from the Roman city of Viroconium Cornoviorum, Shropshire. Unpublished PhD thesis, University of Sheffield.

Harcourt, R.A. (1969a) *Animal bones from South Witham*. Ancient Monuments Laboratory Report, 1556.

Harcourt, R.A. (1971a). The animal bones from Durrington Walls. In *Durrington Walls: Excavations 1966–68* (eds G.J. Wainwright & I.H. Longworth). *Report of the Research Committee of the Society of Antiquaries of London* **29**: 188–191.

Harcourt, R.A. (1971b). The animal bones. In *Mount Pleasant, Dorset, Excavations 1970–71. Incorporating an account of excavations undertaken at Woodhenge in 1970* (ed. G.J. Wainwright). *Report of the Research Committee of the Society of Antiquaries of London* **37**: 214–215.

Harcourt, R.A. (1979a). The animal bones. In *Gussage All Saints; an Iron Age Settlement in Dorset* (ed. G.J. Wainwright), pp. 150–160. HMSO, London.

Harcourt, R. (1979b). The animal bones. In *Mount Pleasant, Dorset, Excavations 1970–71* (ed. G.J. Wainwright), pp. 214–215. Society of Antiquaries, London.

Hare, J.N. (1985). *Battle Abbey. The Eastern Range and the Excavations of 1978–80*. Historic Buildings and Monuments Commission for England Archaeological Report.

Harkness (1871). The discovery of a Kitchen-Midden at Ballycotton in County Cork. *Report of the British Association*, 150–151.

Harman, M. (1983). Animal remains from Ardnave, Islay. In *Excavations at Ardnave, Islay* (eds G. Ritchie & H. Welfare). *Proceedings of the Society of Antiquaries of Scotland* **113**: 343–347.

Harman, M. (1993a). Mammalian and bird bones. In The Fenland Project Number 7: Excavations in Peterborough and the Lower Welland Valley 1960-1969 (eds W.G. Simpson, D. Gurney, J. Neve & F.M.M. Pryor). East Anglian Archaeological Report. **61**: 98–123.

Harman, M. (1993b). The animal bones. In *Caister-on-Sea Excavations by Charles Green 1951–55* (eds M.J. Darling & D. Gurney). *East Anglian Archaeology Report* **66**: 223–236.

Harman, M. (1994). Bird bones. In *Excavations at the Romano-British settlement at Pasture Lodge Farm, Long Bennington, Lincolnshire 1975–77 by H.M. Wheeler* (ed. R.S. Leary). *Occasional Papers in Lincolnshire History and Archaeology* **10**: 52.

Harman, M. (1996a). Birds. In *Dragonby*, Vols I and II (ed. J. May). Oxbow Monograph 61.

Harman, M. (1996b). Mammal and Bird Bones. In *The archaeology and ethnology of St Kilda. Number 1. Archaeological excavations on Hirta 1986–1990* (ed N. Emery). H.M.S.O., Edinburgh.

Harman, M. (1997). Bird bone. In *The excavations of a stalled cairn at the Point of Cott, Westray, Orkney* (ed. J. Barber). Scottish Trust for Archaeological Research.

Harris, S., Morris, P., Wray, S. & Yalden, D.W. (1995). *A Review of British Mammals: population estimates and conservation status of British mammals other than cetaceans.* JNCC, Peterborough.

Harrison, C.J.O. (1978). A New Jungle-fowl from the Pleistocene of Europe. *Journal of Archaeological Science* **5**: 373–376.

Harrison, C.J.O. (1979). Pleistocene birds from Swanscombe, Kent. *London Naturalist* **58**: 6–9.

Harrison, C.J.O. (1980a). A re-examination of British Devensian and earlier Holocene bird bones in the British Museum (Natural History). *Journal of Archaeological Science* **7**: 53–68.

Harrison, C.J.O. (1980b). Pleistocene bird remains from Tornewton Cave and the Brixham Windmill Hill Cave in south Devon. *Bulletin of the British Museum (Natural History), Geology* **33**: 91–100.

Harrison, C.J.O. (1980c). Additional birds from the Lower Pleistocene of Olduvai, Tanzania: and potential evidence of Pleistocene bird migration. *Ibis* **122**: 530–532.

Harrison, C.J.O. (1985). The Pleistocene birds of South East England. *Bulletin of the Geological Society of Norfolk* **35**: 53–69.

Harrison, C.J.O. (1986). Bird remains from Gough's Cave, Somerset. *Proceedings of the University of Bristol Spelaeological Society* **17**: 305–310.

Harrison, C.J.O. (1987a). A re-examination of the Star Carr birds. *Naturalist* **112**: 141.

Harrison, C.J.O. (1987b). Pleistocene and Prehistoric birds of South-West Britain. *Proceedings of the University of Bristol Spelaeological Society* **18**: 81–104.

Harrison, C.J.O. (1988). Bird Bones from Soldier's Hole, Cheddar, Somerset. *Proceedings of the University of Bristol Spelaeological Society* **18**: 258–264.

Harrison, C.J.O. (1989a). Bird bones from Chelm's Combe Shelter, Cheddar, Somerset. *Proceedings of the University of Bristol Spelaeological Society* **18**: 412–414.

Harrison, C.J.O. (1989b). Bird remains from Gough's Old Cave, Cheddar, Somerset. *Proceedings of the University of Bristol Spelaeological Society* **18**: 409–411.

Harrison, C.J.O. & Cowles, G.J. (1977). The extinct large cranes of the North-West Palaearctic. *Journal of Archaeological Society* **4**: 25–27.

Harrison, C.J.O. & Stewart, J.R. (1999). Avifauna. In *Boxgrove. A middle Pleistocene Hominid site at Eartham Quarry, Boxgrove, West Sussex* (eds M.B. Roberts & S.A. Parfitt). English Heritage Archaeological Report 17.

Harrison, C.J.O. & Walker, C.A. (1977). A re-examination of the fossil birds from the Upper Pleistocene in the London Basin. *London Naturalist* **56**: 6–9.

Harrison, C.J.O. & Walker, C.A. (1978). The North Atlantic Albatross, *Diomedea anglica*, a Pliocene-Lower Pleistocene species. *Tertiary Research* **2**: 45–46.

Harrison, T.P. (1956). *They Tell of Birds: Chaucer, Spenser, Milton, Drayton.* Greenwood Press, Westport, CT.

Harting, J.E. (1864). *The Ornithology of Shakespeare.* 1978 Reprint edn. Gresham Books, Old Woking, Surrey.

Hatting, T. (1968). Animal bones from the basal middens. In *Excavations at Dalkey Island, Co Dublin 1956–59* (ed. G.D. Liversage). *Proceedings of the Royal Irish Academy*, **66**, 172–174.

HBWP (*Handbook of the Birds of the Western Palearctic*) see Cramp *et al.* 1977–1994.

Hedges, J.W. (1984). *Tomb of the Eagles.* John Murray, London.

Hencken, H. (1950). Lagore crannog: an Irish royal residence of the seventh to tenth century AD. *Proceedings of the Royal Irish Academy* **53**: 1–247.

Henderson-Bishop, A. (1913). An Oronsay shell-mound – A Scottish pre-Neolithic site. *Proceedings of the Society of Antiquaries of Scotland* **48**: 52–108.

Henshall, A.S. & Ritchie, J.N.G. (1995). *The Chambered Cairns of Sutherland – an inventory of the structures and their contents*. Edinburgh University Press, Edinburgh.

Hewlett, J. (2002). *The Breeding Birds of the London Area*. London Natural History Society.

Hickling, R. (1983). *Enjoying Ornithology*. Calton, Staffordshire: T. & A.D. Poyser.

Hitosugi, S., Tsuda, K., Okabayashi, H. & Tanabe, Y. (2007). Phylogenetic relationships of mitochondrial DNA cytochrome b gene in east Asian ducks. *Journal of Poultry Science* **44**: 141–145.

Holloway, S. (1996). *The Historical Atlas of breeding Birds in Britain and Ireland 1875–1900*. T.&A.D. Poyser, London.

Hou, L., Martin, L.D., Zhou, Z. & Feduccia, A. (1996). Early adaptive radiation of birds: evidence from fossils from Northeastern China. *Science* **274**: 1164–1167.

Houlihan, P.F. (1996). *The Animal World of the Pharaohs*. Thames and Hudson, London.

Hudson, P.J. (1992). *Grouse in Space and Time*. Game Conservancy Trust, Fordingbridge.

ICZN (2003). Opinion 2027. Usage of 17 specific names based on wild species which are predated by or contemporary with those based on domestic animals. *Bulletin of Zoological Nomenclature* **60**: 81–84.

Izard, K. (1997). The animal bones. In *Birdoswald. Excavations of a Roman fort on Hadrian's Wall, and its successive settlements 1987–92* (ed. T. Wilmott), English Heritage Archaeology Report 14, pp. 363–370. London.

Jackson, J.W. (1953). Archaeology and palaeontology. **In** *British Caving. An introduction to speleology* (ed. C.H.D. Cullingford). Routledge & Kegan Paul Ltd.

Jacobi, R. (2004). The Late Upper Palaeolithic lithic collection from Gough's Cave, Cheddar, Somerset and Human use of the cave. *Proceedings of the Prehistoric Society* **70**: 1–92.

Janossy, D. (1986). *Pleistocene Vertebrate Faunas of Hungary*. Elsevier, Amsterdam.

Jaques, D. & Dobney, K. (1996). Animal Bone, in Zeepvat, R. & Copper-Reade, H., Excavations within the outer bailey of Hertford Castle. *Hertfordshire Archaeology* **12**: 33–37.

Jędrzejewska, B. & Jędrzejewski, W. (1998). *Predation in Vertebrate Communities*. Springer Verlag, Berlin.

Jefferies, D.J. (2003). *The Water Vole and Mink Survey of Britain 1996–1998 with a History of the Long-term Changes in the Status of both Species and their Causes*. Vincent Wildlife Trust, London.

Jenkinson, R.D. (1984). *Creswell Crags. Late Pleistocene Sites in the East Midlands*. BAR British Series 122. Archaeopress, Oxford.

Jenkinson, R.D. & Bramwell, D. (1984). The birds of Britain: When did they arrive? In *In the Shadow of Extinction: A Late Quaternary Archaeology and Palaeoecology of the Lake, Fissures and Smaller Caves at Cresswell Crags SSSI* (eds D.D. Gilbertson & R.D. Jenkinson), pp. 89–99. Department of Prehistory and Archaeology, University of Sheffield.

Johnstone, C. & Albarella, U. (2002). *The Late Iron Age and Romano-British Mammal and Bird Bone Assemblage from Elms Farm, Heybridge, Essex (Site Code: HYEF93–95)*. Centre For Archaeology Report, 45/2002.

Jones, E.L. (1966). Lambourn Downs. In *The Birds of Berkshire and Oxfordshire* (ed. M.C. Radford), pp. 16–24. Longmans, London.

Jones, E.V. & Horne, B. (1981). Analysis of skeletal material. In *A Romano-British inhumation cemetery at Dunstable* (ed. C.L. Matthews). *Bedfordshire Archaeological Journal* **15**: 69–72.

Jones, G. (1984). Animal bones. In *Excavations in Thetford 1948–59 and 1973–80* (eds A. Rogerson & C. Dallas). *East Anglian Archaeology Report* **22**: 187–191.

Jones, G. (1993). Animal and bird bone. In *Excavations in Thetford by B.K. Davidson between 1964 and 1970* (ed. C. Dallas). *East Anglian Archaeology Report* **62**: 176–189.

Jones, R.T. (1978). The animal bones. In *Excavations at Wakerley, Northants, 1972–75* (eds D.A. Jackson & T.M. Ambrose). *Britannia* **9**: 115–242.

Jones, R.T. & Serjeantson, D. (1983) *The animal bones from five sites at Ipswich*. Ancient Monuments Laboratory Report, 13/83.

Jones, R.T., Reilly, K. & Pipe, A. (1997). The animal bones. In *Castle Rising Castle, Norfolk* (eds B. Morley & D. Gurney). *East Anglian Archaeology Report* **81**: 123–131.

Jones, R.T., Sly, J., Beech, M., & Parfitt, S. (1988). Animal Bones: Summary, in Martin, E., Burgh: The Iron Age and Roman Enclosure. *East Anglian Archaeology Report* **40**: 66–67.

Joysey, K.A. (1963) A scrap of bone, pp. 197-203 in Brothwell, D. & Higgs, E. (eds) *Science in Archaeology* (1st ed.). Thames & Hudson, London.

Kear, J. (1990). *Man and Wildfowl*. T.&A.D. Poyser, London.

Kear, J. (2003). Cavity nesting ducks: why woodpeckers matter. *British Birds* **96**: 217–233.

Kendrick, T.D. (1928) *The Archaeology of the Channel Islands Vol I: The Bailiwick of Guernsey*. Methuen & Co Ltd., London.

King, A. & Westley, B. (1989). The animal bones. In *Pentre Farm, Flint 1976–81. An official building in the Roman lead mining district* (ed. T.J. O'Leary), BAR British Series 207. Archaeopress, Oxford.

King, J.E. (1962). Report on Animal bones. In *Excavations at the Maglemosian site at Thatcham, Berkshire, England* (ed. J.J. Wymer). *Proceedings of the Prehistoric Society* **28**: 329–361.

King, J. (1965). Bones: bird. In *Romano-British settlement at Studland, Dorset* (ed. B.A. Field). *Proceedings of the Dorset Natural History and Archaeology Society* **87**: 49–50.

King, J.M. (1987). The bird bones. In *Longthorpe II: The Military Works-Depot: an episode in landscape history* (eds G.B. Dannell & J.P. Wild). *Britannia Monograph Series* **8**.

Kitson, P.R. (1997). Old English bird-names (I). *English Studies* **78**: 481–505.

Kitson, P.R. (1998). Old English bird-names (II). *English Studies* **79**: 2–22.

Kraaijeveld, K. & Nieboer, E.N. (2000). Late Quaternary palaeogeography and evolution of Arctic breeding waders. *Ardea* **88**: 193–205.

Kraft, E. (1972). *Vergleichend Morphologische Untersuchungen an Einzelknochen Nord- und Mitteleuropaischer Kleinerer Huhnervogel*. Institut fur Palaeoanatomie, Domesticationsforschung und Gesichte der Tiermedizen, University of Munich.

Kurochkin, E.V. (1985). A true carinate bird from the Lower Cretaceous deposits in Mongolia and other evidence of Early Cretaceous birds in Asia. *Cretaceous Research* **6**: 271–278.

Lacaille, A.D. (1954). *The Stone Age in Scotland*. Oxford University Press, London.

Lack, P.C. (1986). *The Atlas of Wintering Birds in Britain and Ireland*. T. & A.D. Poyser, Calton.

Lambert, R.A. (2001). The osprey on Speyside: an environmental history. In *Contested Mountains*. White Horse Press, Cambridge.

Lawrance, P. (1982). Animal Bones, in Coad, J.G. & Streeten, A.D.F., Excavations at Castle Acre Castle, Norfolk, 1972–77. Country House and Castle of the Norman Earls of Surrey. *Archaeology Journal* **139**: 138–301.

Legge, A.J. (1991). The animal bones. In *Papers on the Prehistoric Archaeology of Cranborne Chase* (eds J. Barrett, R. Bradley & M. Hall), Oxbow Monograph 11, p. 77.

Legge, A.J. & Rowley-Conwy, P.A. (1988). *Star Carr Revisited: a re-analysis of the large mammals*. Centre for Extra-Mural Studies, Birkbeck College, University of London.

Lever, C. (1977). *The Naturalized Animals of the British Isles*. Hutchinson, London.

Levine, M.A. (1986). The vertebrate fauna from Meare East 1982. *Somerset Levels Papers* **12**.

Levitan, B. (1984b). Faunal remains from Priory Barn and Benham's Garage, in Leach, P., The Archaeology of Taunton. Excavations and fieldwork to 1980. *Western Archaeological Trust Executive Monograph* **8**: 167–192.

Levitan, B. (1990). The vertebrate remains. In *Brean Down Excavations 1983–1987* (ed. M. Bell). *English Heritage Archaeology Report* **15**: 220–239.

Levitan, B. (1994a). Birds. In *Ilchester* Vol. 2: *Archaeology, excavation and fieldwork to 1984* (ed. P. Leach), pp. 179–190. Sheffield Excavation Reports.

Levitan, B. (1994b). Vertebrate Remains from the Villa, in Williams, R.J. & Zeepvat, R., Bancroft. A Late Bronze Age/Iron Age Settlement, Roman Villa and Temple-Mausoleum. Volume II - Finds and Environmental Evidence. *Buckinghamshire Archaeology Society Monograph Series* **7**: 536–538.

Liley, D. & Clarke, R.T. (2003). The impact of urban development and human disturbance on the numbers of nightjar *Caprimulgus europaeus* on heathlands in Dorset, England. *Biological Conservation* **114**: 219–230.

Lindow, B.E.K. & Dyke, G. (2006). Bird evolution in the Eocene: climate change in Europe and a Danish fossil fauna. *Biological Reviews* **81**: 483–499.

Locker, A. (1977). Animal Bones and Shellfish, in Neal, D.E., Excavations at the Palace of Kings Langley, Hertfordshire, 1974–76. *Medieval Archaeology* **21**: 160–162.

Locker, A. (1988). Animal Bones, in Hinton, P., Excavations in Southwark 1973–76, Lambeth 1973–79. *London and Middlesex Archaeological Society Publication* **3**: 427–442.

Locker, A. (1990). The bird remains. In *Excavation of the Iron Age, Roman and Medieval settlement at Gorhambury, St Albans* (eds D.S. Neal, A. Wardle & J. Hunn). *English Heritage Archaeological Report* **14**: 210–211.

Locker, A. (1991). The Group 'B' Bone. In *The Roman Villa site at Keston. First Report (Excavations 1968–1978)* (eds B. Philip, K. Parfitt, J. Willson, M. Dutto & W. Williams). *Kent Research Report Monograph Series* **6**: 286–288.

Locker, A. (1999). The bird bones. In *Halangy Down, Isles of Scilly* (ed. P. Ashbee), pp 113–115. *Cornish Archaeology* **35**: 5–201.

Locker, A. (2000). Animal bone. In *Potterne 1982–5: Animal Husbandry in Later Prehistoric Wiltshire* (ed. A.J. Lawson). *Wessex Archaeology Report* **17**: 107–109.

Lockwood, W.B. (1993). *The Oxford Dictionary of British Bird Names*. Ppb edn. Oxford University Press, Oxford.

Love, J.A. (1983). *The Return of the Sea Eagle* Cambridge University Press, Cambridge.

Lovegrove, R. (1990). *The Red Kite's Tale: the story of the Red Kite in Wales*. RSPB, Sandy.

Lovegrove, R. (2007). *Silent Fields: the long decline of a nation's wildlife*. Oxford University Press, Oxford.

Lowe, P.R. (1933). The differential characters in the Tarso-Metatarsi of *Gallus* and *Phasianus* as they bear on the Problem of the Introduction of the Pheasant into Europe and the British Isles. *Ibis* **1933**: 333–343.

Lucchini, V., Hoglund, J., Klaus, S., Swenson, J & Randi, E. (2001). Historical Biogeography and a Mitochondrial DNA Phylogeny of Grouse and Ptarmigan. *Molecular Phylogenetics and Evolution* **20**: 149–162.

Luff, R.M. (1982). *A zooarchaeological Study of the Roman N.W. Provinces*. BAR International Series. Archaeopress, Oxford.

Luff, R.M. (1985). The fauna. In *Sheepen: an early Roman industrial site at Camulodunum* (ed. R. Niblett), CBA Research Report 57, pp. 143–149.

Luff, R.M. (1993). Poultry and game. In *Animal Bones from Excavations in Colchester 1971–85* (ed. R.M. Luff), Colchester Archaeological Report 12, pp. 83–98. Colchester Archaeology Trust, Colchester.

MacCartney, E. (1984). Analysis of faunal remains. In *Excavations at Crosskirk Broch, Caithness* (ed. H. Fairhurst), Vol. 3. Society of Antiquaries of Scotland Monograph Series.

MacDonald, K.C. (1992). The Domestic Chicken (*Gallus gallus*) in Sub-Saharan Africa: A Background to its Introduction and its Osteological Differentiation from Indigenous Fowls (Numidinae and *Francolinus* sp.). *Journal of Archaeological Science* **19**: 303–318.

MacGregor, A. (1996). Swan rolls and beak markings. Husbandry, exploitation and regulation of *Cygnus olor* in England c. 1100–1900. *Anthropozoologica* **22**: 39–68.

Mainland, I. & Stallibrass, S. (1990). *The Animal Bone from the 1984 Excavations of the Romano-British Settlement at Papcastle, Cumbria*. Ancient Monuments Laboratory Report, 4/90.

Maltby, M. (1979). Faunal studies on urban sites: the animal bones from Exeter 1971–75. *Exeter Archaeological Reports* **2**: 1–210.

Maltby, M. (1982). Animal and bird bones. In *Excavations at Okehampton Castle, Devon. Part 2: The Bailey* (eds R.A. Higham, J.P. Allen & S.R. Blaylock). *Devon Archaeological Society Proceedings* **40**: 114–135.

Maltby, M. (1983). The animal bones. In *Wigber Lowe, Derbyshire: a Bronze Age and Anglian burial site in the White Peak* (ed. J. Collis), pp. 47–51. Department of Prehistory & Archaeology, University of Sheffield.

Maltby, M. (1984). The animal bones. In *Silchester: Excavations on the Defence 1974–1980* (ed. M.G. Fulford), pp. 199–212. London.

Maltby, M. (1985). The animal bones. In *The prehistoric settlement at Winnall Down, Winchester* (ed. P.J. Fasham). *Trust for Wessex Archaeology/Hampshire Field Club Monograph* **2**: 97–109.

Maltby, M. (1990). Animal bones from Coneybury Henge. In *The Stonehenge Environs Project* (ed. J. Richards), English Heritage Archaeological Report 16, pp. 150–154. London.

Maltby, M. (1992). The animal bone. In *The Marlborough Downs: A later Bronze Age Landscape and its origins* (ed. C. Gingell). *Wiltshire Archaeological and Natural History Society Monograph* **1**: 137–139.

Maltby, M. (1993). Animal bones. In *Excavations at the Old Methodist Chapel and Greyhound Yard, Dorchester* (eds P.J. Woodward, S.M. Davies & A.H. Graham). *Dorset Natural History and Archaeology Society Monograph Series* **12**: 315–340.

Maltby, M. & Coy, J. (1982). Bones, in Potter, T.W. & Potter, C.F., A Romano-British Village at Grandford, March, Cambridgeshire. *British Museum Occasional Paper* **35**: 98–122. London.

Manegold, A., Mayr, G. & Mourer-Chauviré, C. (2004). Miocene songbirds and the composition of the European passeriform avifauna. *Auk* **121**: 1155–1160.

Maroo, S. & Yalden, D.W. (2000). The Mesolithic mammal fauna of Great Britain. *Mammal Review* **30**: 243–248.

Marren, P. (2002). *Nature Conservation*. Harper Collins, London.

Marquiss, M. & Rae, R. (2002) Ecological differentiation in relation to bill size amongst sympatric, genetically undifferentiated crossbills *Loxia* spp. *Ibis* **144**: 494–508.

Mayr, G. (2000). Die Vogel der Grube Messel – ein Einblick in die Vogelwelt Mitteleuropas vor 49 Millionen Jahren. *Natur und Museum* **130**: 365–378.

Mayr, G. (2005). The Paleogene fossil record of birds in Europe. *Biological Reviews* **80**: 515–542.

McCarthy, M. (1995). Faunal Hunting, fishing and fowling in the late prehistoric Ireland: the scarcity of the bone record. pp. 107–119 in A. Desmond, G. Johnson, M. McCarthy, J. Sheehan and E. Shee Twohig (eds.), New Agendas in Irish Prehistory. Wordwell, Bray.

McCarthy, M. (1999). Faunal remains. In *Excavations at Ferriter's Cove, 1983–95: Last foragers, first farmers in the Dingle peninsula* (eds P.C. Woodman, E. Anderson & N. Finlay), p. 203. Wordwell Ltd.

McCormick, F. (1984). Small animal bones. In Excavations at Pierowall Quarry, Westray, Orkney (ed. N.M. Sharples). *Proceedings of the Society of Antiquaries of Scotland* **114**: 111.

McCormick, F., Hamilton-Dyer, S. & Murphy, E. (1997). The animal bones. In *Excavations at Caldicot, Gwent: Bronze Age Palaeochannels in the Lower Nedern Valley* (eds N. Nayling & A. Caseldine), pp. 218–235. CBA Research Report 108.

McEvedy, C. & Jones, R. (1978). *Atlas of World Population History*. Penguin Books, London.

McGhie, H. (1999). Persecution of birds of prey in north Scotland as evidenced by taxidermists' stuffing books. *Scottish Birds* **20**: 98–110.

Mead, C. (2000). *The State of the Nation's Birds*. Whittet Books, Stowmarket.

Meddens, B. (1987). *Assessment of the Animal Bone Work from Wroxeter Roman City, Shropshire: from sites Wroxeter Barker (AML Site 49) and Wroxeter Webster (AML 340)*. English Heritage Ancient Monuments Laboratory Report 171. London.

Mellars, P.A. (1987). *Excavations on Oronsay.* Edinburgh University Press, Edinburgh.
Mills, A.D. (2003). *Oxford Dictionary of British Place Names.* Oxford University Press, Oxford.
Moore, P.G. (2002). Ravens (*Corvus corax corax* L.) in the British Landscape: a thousand years of ecological biogeography in place-names. *Journal of Biogeography* **29**: 1039–1054.
Moore, N.W. (1957). The past and present status of the Buzzard in the British Isles. *British Birds* **50**: 173–197.
Moreau, R.E. (1972). *The Palaearctic-African Bird Migration Systems.* Academic Press, London.
Morris, G. (1990). Animal bone and shell. In *Excavations at Chester. The lesser Medieval religious houses. Sites investigated 1964–1983* (ed. S.W. Ward). *Grosvenor Museum Archaeological Excavations and Survey Report* **6**: 178–189.
Morris, P.A. (1981). The Antiquity of the Duchess of Richmond's Parrot. *Museums Journal* **81** 153–154.
Morris, P.A. (1990). Examination of a preserved Hawfinch (*Coccothraustes coccothraustes*) attributed to Gilbert White. *Archives of Natural History* **17**: 361–366.
Morris, P.A. (1993). An historical review of bird taxidermy in Britain. *Archives of Natural History* **20**: 241–255.
Mourer-Chauviré, C. (1993). The Pleistocene avifaunas of Europe. *Archaeofauna* **2**: 53–66.
Mulkeen, S. & O'Connor, T.P. (1997). Raptors in towns: towards an ecological model. *International Journal of Osteoarchaeology* **7**: 440–449.
Mullins, E.H. (1913). The Ossiferous Cave at Langwith. *Journal of the Derbyshire Archaeological and Natural History Society* **35**: 137–158.
Mulville, J.A. (1995) *Faunal remains from the moat at Wood Hall, Womersley, Yorkshire.* Unpublished report by Sheffield Environmental Facility.
Mulville, J. & Grigson, C. (2007). The animal bones. In *Building Memories. The Neolithic Cotswold Long Barrow at Ascott-Under-Wychwood, Oxfordshire* (ed. D. Benson & A. Whittle). Oxbow Books, Oxford.
Murison, G., Bullock, J.M., Underhill-Day, J. & Langston, R. (2007). Habitat type determines the effects of disturbance on the breeding productivity of the Dartford Warbler *Sylvia undata. Ibis* **149**: 16–26.
Murray, E. & Albarella, U. (2005). Mammal and avian bone. In *Dragon Hall, King Street, Norwich: excavation and survey of a late medieval merchant's trading complex* (ed. A. Shelley). *East Anglian Archaeology* **112**: 158–167.
Murray, E. & Hamilton-Dyer, S. (2007). *The Animal Bones from Raystown, Co. Meath.* Unpublished report.
Murray, E., McCormick, F. & Plunkett, G. (2004). The food economies of Atlantic island monasteries: the documentary and archaeo-environmental evidence. *Environmental Archaeology* **9**: 179–188.
Murray, H.K. & Murray, J.C. (1993) Excavations at Rattray, Aberdeenshire. *Medieval Archaeology* **37**: 101–218.
Nelson, T.H. (1907). *The Birds of Yorkshire.* Brown & Sons, London.
Nethersole-Thompson, D. (1973). *The Dotterel.* Collins, Glasgow.
Newton, E.T. (1905). (Animal bones from Silchester). *Archaeologia* **59**: 369.
Newton, E.T. (1906a). Birds. In *The Exploration of the Caves of County Clare* (ed. R.F. Scharff). *Transactions of the Royal Irish Academy* **33**: 53–57.
Newton, E.T. (1906b). (mammalian and other bones from Silchester). *Archaeologia* **60**: 164–167.
Newton, E.T. (1908). (bones from Silchester). *Archaeologia* **61**: 213–214.
Newton, E.T. (1917). Notes on Bones found in the Creag nan Uamh Cave, Inchnadamff, Assynt, Sutherland, in Peach, B.N. & Horne, J. The bone cave in the valley of Allt nan Uamh (Burn of the Caves), near Inchnadamff, Assynt, Sutherlandshire. *Proceedings of the Royal Society of Edinburgh* **1916–1917**: 344–349.

Newton, E.T. (1921b) Note on the remains of birds obtained from Aveline's Hole, Burrington Combe, Somerset. *Proceedings of the University of Bristol Spelaeological Society* **1**: 73.

Newton, E.T. (1922) List of Avian Species Identified from Aveline's Hole, Burrington. *Proceedings of the University of Bristol Spelaeological Society* **1**: 119–121.

Newton, E.T. (1923). The common crane fossil in Britain. *Naturalist* **1923**: 284–285.

Newton, E.T. (1924a). Note on bird's bones from Merlin's Cave. *Proceedings of the University of Bristol Spelaeological Society* **2**: 159–161.

Newton, E.T. (1924b) Note on additional Species of Birds from Aveline's Hole. *Proceedings of the University of Bristol Spelaeological Society* **2**: 121.

Newton, E.T. (1926). Review of the species. In *Excavations at Chelm's Combe, Cheddar* (ed. H.E. Balch). *Proceedings of the Somerset Archaeological and Natural History Society* **72**: 115–123.

Newton, E.T. (1928) Pelican in Yorkshire peat. *Naturalist* **1928**: 167.

Newton, I. (1986). *The Sparrowhawk*. T. & A.D. Poyser, Calton.

Nicholson, R.A. (2003). Bird bones. In *Old Scatness and Jarlshof Environs Project: field season 2002* (eds S.J. Dockrill, J.M. Bond & V.E. Turner). Shetland Amenity Trust/University of Bradford.

Noddle, B. (1983). The animal bones from Knap of Howar. In *Excavation of a Neolithic farmstead at Knap of Howar, Papa Westray, Orkney* (ed. A. Ritchie), *Proceedings of the Society of Antiquaries of Scotland* **113**: 92–100.

Norberg, R.A. (1985). Function of vane asymmetry and shaft curvature in bird flight feathers: inferences on flight ability of *Archaeopteryx*. In *The Beginnings of Birds* (eds M.K. Hecht, J.H. Ostrom, G. Viohl & P. Wellnhofer), pp. 303–318. Freunde des Jura-Museums Eichstatt, Eichstatt.

Northcote, E.M. (1980). Some Cambridgeshire Neolithic to Bronze Age birds and their presence or absence in England in the Late-Glacial and Early Flandrian. *Journal of Archaeological Science* **7**: 379–383.

Northcote, E.M. & Mourer-Chauviré, C. (1988). The extinct crane *Grus primigenia* Milne-Edwards in Majorca, Spain. *Geobios* **21**: 201–208.

O'Connor, T.P. (1982). Animal bones from Flaxengate, Lincoln c 870–1500. *Archaeology of Lincoln* **18**: 1–52.

O'Connor, T.P. (1984a) *Bones from Aldwark, York*. Ancient Monuments Laboratory Report, 3491.

O'Connor, T.P. (1984b). *Selected Groups of Bones from Skeldergate and Walmgate*. York Archaeological Trust for the CBA.

O'Connor, T.P. (1985). Hand-collected bones from Roman to Medieval Deposits at the General Accident Site, York. Report to the Ancient Monuments Laboratory, November 1985.

O'Connor, T.P. (1986). The animal bones. In *The Legionary Fortress Baths at Caerleon. Part II: The Finds* (ed. J.D. Zienkiewicz), pp. 224–248. National Museum of Wales, Cardiff.

O'Connor, T.P. (1987b) *Bones from Roman to Medieval deposits at the City Garage, 9 Blake Street, York (1975–6)*. Ancient Monuments Laboratory Report, 196/87.

O'Connor, T.P. (1988). *Bones from the General Accident Site, Tanner Row*. The Archaeology of York, 15/2. CBA, London.

O'Connor, T.P. (1989). *Bones from Anglo-Scandinavian Levels at 16–22 Coppergate*. The Archaeology of York, 15/3. CBA, London.

O'Connor, T.P. (1991). *Bones from 46–54 Fishergate, York*. The Archaeology of York, 15/4. CBA, London.

O'Connor, T.P. (1993). Bird bones. In *Excavations at Segontium (Caernarfon) Roman Fort 1975–1979* (eds P.J. Casey, J.L. Davies & J. Evans). *CBA Research Report* **90**: 119.

O'Connor, T.P. & Bond, J.M. (1999). *Bones from Medieval Deposits at 16–22 Coppergate and Other Sites in York*. Council for British Archaeology.

O'Sullivan, D. & Young, R. (1995). *Book of Lindisfarne, Holy Island*. English Heritage, London.

O'Sullivan, T. (1996). Faunal remains. In *The excavation of two Bronze Age burial cairns at Bu farm, Rapness, Westray, Orkney* (eds J. Barber, A. Duffy & J. O'Sullivan). *Proceedings of the Society of Antiquaries of Scotland* **126**: 103–120.

O'Sullivan, T. (1998). Birds. In *Scalloway: a broch, Late Iron Age settlement and Medieval cemetry in Shetland* (ed. N.M. Sharples), pp. 116–117. Oxbow Monographs **82**. Oxford.

Osborne, P.E. (2005). Key issues in assessing the feasibility of reintroducing the great bustard *Otis tarda* to Britain. *Oryx* **39**: 22–29.

Otto, C. (1981). *Vergleichend morphologische Untersuchungen an Einzelknochen in Zentraleuropa vorkommender mittelgrosser Accipitridae 1. Schadel, Brustbein, Schultergurtel und Vorderextremitat*. Ludwigs-Maximilians-Universität, Munich.

Palmer, L.S. & Hinton, M.A.C. (1928). Some gravel deposits at Walton, near Clevedon. *Proceedings of the University of Bristol Spelaeological Society* **3**: 154–161.

Parker, A.J. (1988). The birds of Roman Britain. *Oxford Journal of Archaeology* **7**: 197–226.

Parry, S. (1996). The avifaunal remains. In *Excavations at Barnfield Pit, Swanscombe, 1968–72.* (eds B. Conway, J. McNabb & N. Ashton), Vol. 94, pp. 137–143. British Museum, Department of Prehistoric and Romano-British Antiquities, London.

Parsons, D. & Styles, T. (1996). Birds in *Amber*: the nature of English place-name elements. *English Place-name Society Journal* **28**: 5–31.

Pearsall, W.H. (1950). *Mountains and Moorlands*. Collins, London.

Perrins, C. & Greer, T. (1980). The effect of Sparrowhawks of Tit populations. *Ardea* **68**: 133–142.

Phillips, G. (1980). The Fauna, in Armstrong, P., Excavations in Scale Lane/Lowgate 1974. Hull Old Town Report Series 4. *East Riding Archaeologist* **6**: 77–82.

Pickles, I. (2002). *Place names and domestic birds*. Unpub. B.Sc. project report, School of Biological Sciences, University of Manchester.

Pienkowski, M.W. (1984). Breeding biology and population dynamics of Ringed plovers *Charadrius hiaticula* in Britain and Greenland: nest predation asa possible factor limiting distribution and timing of breeding. *Journal of Zoology, London* **202**: 83–114.

Piertney, S.B., Summers, R. & Marquiss, M. (2001) Microsatellite and mitochondrial DNA homogeneity among phenotypically diverse crossbill taxa in the UK. *Proceedings of the Royal Society of London* B **268**: 1511–1517.

Platt, M. (1933a). Report on the animal bones from Jarlshof, Sumburgh, Shetland. In *Further excavations at Jarlshof, Shetland* (ed. A.O. Curle). *Proceedings of the Society of Antiquaries of Scotland* **67**: 127–136.

Platt, M. (1933b). Report on Animal and other bones. In *The Broch of Midhowe, Rousay, Orkney* (eds J.G. Callander & W.G. Grant), *Proceedings of the Society of Antiquaries of Scotland* **68**: 514–515.

Platt, M. (1956). The animal bones. In *Excavations at Jarlshof, Shetland* (ed. J.R.C. Hamilton), 1, pp. 212–215. HMSO, London.

Poole, A.F. (1989). *Ospreys. A natural and unnatural history*. Cambridge University Press, Cambridge.

Price, C.R. (2003). *Late Pleistocene and early Holocene small mammals in South West Britain*. BAR British Series 347. Archaeopress, Oxford.

Price, D.T. (1983). The European Mesolithic. *American Antiquity* **48**: 761–78.

Prummel, W. (1997). Evidence of hawking (falconry) from bird and mammal bones. *International Journal of Osteoarchaeology* **7**: 333–8.

Rackham, J. (1979). Animal resources. In *Three Saxo-Norman tenements in Durham City* (ed. M.O.H. Carver). *Medieval Archaeology* **23**: 47–54.

Rackham, J. (1980) *Carlisle, Fisher Street, 1977–Central Unit Excavation 11. Animal bone report.* Ancient Monuments Laboratory Report, 3221.

Rackham, J. (1987). The animal bones. In *The Excavation of an Iron Age settlement at Thorpe Thewles, Cleveland, 1980–1982* (ed. D.H. Heslop), pp. 99–10. CBA Research Report 65.

Rackham, J. (1988). The mammal bones from medieval and post-medieval at Queen Street. In The origins of the Newcastle Quayside. Excavations at Queen Street and Dog Bank. (eds C. O'Brien, L. Bown, S. Dixon & R. Nicholson). *The Society of Antiquaries of Newcastle upon Tyne, Monograph Series*, **3**: 120–132.

Rackham, J. (1995). Animal bone from post-Roman contexts. In *Excavations at York Minster* Volume I. *From Roman fortress to Norman Cathedral. Part 2: The Finds* (eds D. Phillips & B. Heywood). HMSO, London.

Rackham, O. (1986). *The History of the Countryside*. Dent, London.

Rackham, O. (2003). *Ancient Woodland: its history, vegetation and uses in England.*, 2nd edn. Castlepoint Press, Colvend, Dalbeattie.

Ratcliffe, D.A. (1980). *The Peregrine Falcon*. T.&A.D. Poyser, London.

Rebecca, G.W. & Bainbridge, I.P. (1998). The breeding status of the Merlin *Falco columbarius* in Britain in 1993–94. *Bird Study* **45**: 172–187.

Redpath, S.M. & Thirgood, S.J. (1997). *Birds of Prey and Red Grouse*. Stationery Office, London.

Reichstein, H. & Pieper, H. (1986). Untersuchungen an Skelettresten von Vogeln aus Haithabu (Ausgrabung 1966–1969). Karl Wachhholtz, Neumunster.

Reynolds, S.H. (1907). A bone cave at Walton, near Clevedon. *Bristol Naturalists' Society Proceedings* (4th srs) **1**: 183–187.

Richards, M.P., Schulting, R.J. & Hedges, R.E.M. (2003). Sharp shift in diet at the onset of the Neolithic. *Nature* **425**: 366.

Ritchie, J. (1920). *The Influence of Man on Animal Life in Scotland*. Cambridge University Press, London.

Ritchie, J.N.G. (1974). Iron Age finds from Dun an Fheurain, Gallanach, Argyll. *Proceedings of the Society of Antiquaries of Scotland* **103**: 110.

Roberts, G., Gonzales S. & Huddart, D. (1996). Intertidal Holocene footprints and their archaeological significance. *Antiquity* **70**: 647–651.

Roberts, M.B. & Parfitt, S. (1999). *The Middle Pleistocene Hominid Site at ARC Eartham Quarry, Boxgrove, West Sussex, UK*. English Heritage, London.

Sadler, P. (1990). Bird bones. In *Faccombe Netherton. Excavations of a Saxon and Medieval Manorial Complex*, Vol II (ed. J.R. Fairbrother). *British Museum Occasional Paper* **74**: 500–506.

Sadler, P. (2007). The bird bone, pp. 172–179 in Stafford Castle: Survey, Excavations and Research 1978–98, Vol. 2–The Excavations, (ed I. Soden),. Stafford Borough Council.

Saetre, G.-P., Borge, T., Lindell, J., Moum, T., Primeer, C.G., Sheldon, B.C., Haavie, J., A., J. & Ellegren, H. (2001). Speciation, introgressive hybridization and non-linear rate of molecular evolution in flycatchers. *Molecular Ecology* **10**: 737–749.

Salvin, F.H. & Brodrick, W. (1855). *Falconry in the British Isles*, 1980 reprint, Windward. edn. John van Voorst, London.

Schmidt-Burger, P. (1982). Vergleichend morphologische Untersuchungen an Einzelknochen in Zentraleuropa vorkommender mittelgrosser Accipitridae 2. Becken und Hinderextremitat. Ludwigs-Maximilians-Universität, Munich.

Scott, S. (1984). *The Animal Bones from Eastgate, Beverley (1984)*. Environmental Archaeology Unit, University of York.

Scott, K. (1986). Man in Britain in the Late Pleistocene: evidence from Ossom's Cave. In *Studies in the Upper Palaeolithic of Britain and Northwest Europe* (ed. D.A. Roe), BAR International Series 296, pp. 63–87. Archaeopress, Oxford.

Scott, S. (1991). The animal bone. In *Excavations at Lurk Lane, Beverley 1979–82* (eds P. Armstrong, D. Tomkinson & E. D.H.). *Sheffield Excavation Reports* **1**: 216–227.

Scott, S. (1992a). The animal bones. In *Excavations at 33–35 Eastgate, Beverley, 1983–86* (eds D.H. Evans & D.G. Tomlinson). *Sheffield Excavation Reports* **3**: 236–249.

Scott, S. (1992b). The Animal Bone. In Roman Sidbury, Worcester: Excavations 1959–1989. (eds J. Darlington & J. Evans). *Transactions of the Worcestershire Archaeological Society* **13**: 88–92.

Scott, S. (1993). The animal bone from Queen Street Gaol, in Evans, D.H., Excavations in Hull 1975–76. *East Riding Archaeologist* **4**: 192–194.

Seagrief, S.C. (1960). Pollen diagrams from southern England: Crane's Moor, Hampshire. *New Phytologist* **59**: 73–83.

Sereno, P.C. & Chenggang, R. (1992). Early evolution of avian flight and perching: new evidence from the Lower Cretaceous of China. *Science* **255**: 845–848.

Serjeantson, D. (1991). The bird bones. In *Danebury: An Iron Age Hillfort in Hampshire*. Vol 5 *The Excavations 1979–1988: the finds* (eds B. Cunliffe & C. Poole), pp. 479–480. CBA Research Report 73. CBA, London.

Serjeantson, D. (1995). Animal bone. In *Stonehenge in its Landscape. 20th Century Excavations* (eds R.M.J. Cleal, K.E. Walker & R. Montague), pp. 437–451. English Heritage Archaeology Report 10.

Serjeantson, D. (1996). The animal bones. In *Refuse and Disposal at Area 16 East Runnymede, Runnymede Bridge Research Excavations* Vol II (eds S. Needham & T. Spence), pp. 194–219. British Museum, London.

Serjeantson, D. (2001). The Great Auk and the Gannet: a prehistoric perspective on the extinction of the Great Auk. *International Journal of Osteoarchaeology* **11**: 43–55.

Serjeantson, D. (2005). Archaeological records of a gadfly petrel *Pterodroma* sp. from Scotland in the first millennium AD. *Documenta Archaeobiologiae* **3**: 233–244.

Serjeantson, D. (2006). Birds: food and a mark of status. In *Food in Medieval England* (eds C.M. Woolgar, D. Serjeantson & T. Waldron), pp. 131–147. Oxford University Press, Oxford.

Shackleton, N.J. (1977). The oxygen isotope record of the Late Pleistocene. *Philosophical Transactions of the Royal Society of London B* **280**: 169–182.

Shackleton, N.J., Berger, A. & Peltier, W.R. (1991). An alternative astronomical calibration of the lower Pleistocene timescale based on ODP site 677. *Transactions of the Royal Society of Edinburgh* **81**: 252–261.

Sharrock, J.T.R. (1976). *The Atlas of Breeding Birds in Britain and Ireland*. British Trust for Ornithology, Tring, Hertfordshire.

Sheppard, J. (1922). Vertebrate remains from the peat of Yorkshire: new records. *Naturalist* **1922**: 187–188.

Shrubb, M. (2003). *Birds, Scythes and Combines: a history of birds and agricultural change*. Cambridge University Press, Cambridge.

Simms, C. (1974). Cave research at Teesdale Cave 1878–1971. Post-glacial fauna from Upper Teesdale. *Yorkshire Philosophical Society Annual Report* **1974**: 34–50.

Simms, E. (1971). *Woodland Birds*. Collins, London.

Slack, K.E., Delsuc, F., Mclenachan, P.A., Arnason, U. & Penny, D. (2007). Resolving the root of the avian mitogenic tree by breaking up long branches. *Molecular Phylogenetics and Evolution* **42**: 1–13.

Smith, C. (2000). Animal bone. In *Castle Park, Dunbar. Two thousand years on a fortified headland*. (ed D.R. Perry). Society of Antiquaries of Scotland Monograph Srs. **16**: 194–297. Edinburgh.

Smith, I. & Lyle, A. (1979). *Distribution of Freshwaters in Great Britain*. I.T.E., Edinburgh.

Smout, T. C. (2003) Highland land-use before 1800: misconceptions, evidence and realities, pp. 5–23 in Smout, T. C. (ed.) *Scottish woodland history*. Scottish Cultural Press, Edinburgh.

Speakman, J.R. & Racey, P.A. (1991). No cost of echolocation for bats in flight. *Nature* **350**: 421–423.

Speakman, J.R. & Thomas, D.W. (2003). Physiological ecology and energetics of bats. In *Bat Ecology* (eds TH Kunz & M.B. Fenton). University of Chicago Press.

Stallibrass, S. (1982). Faunal Remains. In *A Romano-British village at Grandford, March, Cambridgeshire.* (eds T.W. Potter & C.F. Potter), Vol. 35, pp. 98–122. British Museum Occasional Paper.

Stallibrass, S. (1993). *Animal Bones from the Excavations in the Southern Area of the Lanes, Carlisle, Cumbria 1981–82.* Ancient Monuments Laboratory Report 96. London.

Stallibrass, S. (1996). Animal bones. In *Excavations at Stonea, Cambridgeshire 1980–85* (eds R.P.J. Jackson & T.W. Potter), pp. 587–612. British Museum, London.

Stallibrass, S. (2002). Animal bones. In *Cataractonium: Roman Catterick and its hinterland. Excavations and research, 1958–1997. Part II* (ed. P.R. Wilson), pp. 392–435. CBA Research Report 129. CBA, London.

Stallibrass, S. & Nicholson, R. (2000). Animal and fish bone. In *Bremetenacum. Excavations at Roman Ribchester 1980, 1989–1990* (eds K. Buxton & C. Howard-Davis), Vol. 9, pp. 375–378. Lancaster Imprints Series.

Steadman, D.W. (1981). (Review of) Birds of the British Lower Eocene. *Auk* **98**: 205–207.

Stelfox, A.W. (1938). The birds of Lagore about one thousand years ago. *Irish Naturalists' Journal* **7**: 37–43.

Stelfox, A.W. (1942). Report on the animal remains from Ballinderry 2 Crannog. In *Ballinderry Crannog No. 2* (ed. H. Hencken). *Proceedings of the Royal Irish Academy* **47C**: 20–21, 60–74.

Stevenson, H. (1870). *Birds of Norfolk.* van Voorst, London.

Stevenson, H. & Southwold, T. (1890). *The Birds of Norfolk.* Gurney & Jackson, London.

Stewart, J.R. (1998). The Avifauna. In *Excavations at the Lower Palaeolithic Site at East Farm, Barnham, Suffolk, 1989–94* (eds N. Ashton, S.G. Lewis & S. Parfitt), British Museum Occasional Paper 125, pp. 107–109.

Stewart, J.R. (2002a). Sea-birds from coastal and non-coastal, archaeological and 'natural' Pleistocene deposits, or not all unexpected deposition is of human origin. *Acta Zoologica Cracoviensia* **45** (Special issue): 167–178.

Stewart, J.R. (2002b). The evidence for the timing of speciation of modern continental birds and the taxonomic ambiguity of the Quaternary fossil record. In *Proceedings of the 5th Symposium of the Society of Avian Paleontology and Evolution.* (eds Z. Zhou & F. Zhang), pp. 259–280. Science Press, Beijing.

Stewart, J.R. (2004). Wetland birds in the recent fossil record of Britain and northwest Europe. *British Birds*, **97,** 33–43.

Stewart, J.R. (2007a). *An evolutionary study of some archaeologically significant avian taxa in the Quaternary of the western Palaearctic.* BAR International Series 1653, pp. 1–272. Archaeopress, Oxford.

Stewart, J.S. (2007b). The fossil and archaeological record of the Eagle Owl in Britain. *British Birds* **100**: 481–486.

Stroud, D., Reed, T.M., Pienkowski, M.W. & Lindsay, R.A. (1987). *Birds, Bogs and Forestry. The peatlands of Caithness and Sutherland.* Nature Conservancy Council, Peterborough.

Stuart, A.J. (1982). *Pleistocene Vertebrates in the British Isles.* Longman, London.

Svenning, J.-C. (2002). A review of natural vegetation openness in north-western Europe. *Biological Conservation* **104**: 133–148.

Swanton, M. (1975). *Anglo-Saxon Prose.* Dent, London.

Sykes, N. (2004). The dynamics of status symbols: wildfowl exploitation in England AD 410–1550. *Archaeological Journal* **161**: 82–105.

Sykes, N.J. (2007). *The Norman conquest: a zooarchaeological perspective.* BAR International Series 1656. Archaeopress, Oxford.

Tagliacozzo, A. & Gala, M. (2002). Exploitation of Anseriformes at two Upper Palaeolithic sites in southern Italy: Grotta Romanelli (Lecce, Apulia) and Grotta del Santuario della Madonna a Praia a Mare (Cosenza, Calabria). *Acta Zoologica Cracoviensia* **45** (Special issue): 117–131.

Tapper, S. (1992). *Game Heritage*. Game Conservancy, Fordingbridge.

Tapper, S.C., Potts, G.R. & Brockless, M.H. (1999). The effect of experimental reductions in predator presence on the breeding success and population density of grey partridges. *Journal of Applied Ecology* **33**: 968–979.

Tatner, P.A. (1983). The diet of urban Magpies *Pica pica*. *Ibis* **125**: 90–107.

Tatner, P.A. & Bryant (1986). Flight cost of a small passerine measured using doubly labelled water: implications for energetics studies. *Auk* **103**: 169–180.

Taylor, I. (1998). *The Barn Owl*. Cambridge University Press, Cambridge.

Thawley, C. (1981). The mammal, bird and fish bones, in Mellor, J.E. & Pearce, T., The Austin Friars, Leicester. *CBA Research Report* **35**: 173–175.

Thomas, C.D. & Lennon, J.J. (1999). Birds extend their ranges northwards. *Nature* **399**: 213.

Thomas, J. (2000). The Great Bustard in Wiltshire: flight into extinction? *Wiltshire Archaeological and Natural History Magazine* **93**: 63–70.

Ticehurst, N.F. (1920). On the former abundance of the Kite, Buzzard and Raven in Kent. *British Birds* **14**: 34–37.

Ticehurst, N.F. (1957). *The Mute Swan in England*. Cleaver-Hume Press, London.

Tinbergen, L. (1946). Sperver als Roofvijand van Zangvogels. *Ardea* **34**: 1–123.

Tomek, T.B. & Bocheński, Z.M. (2000). *The Comparative Osteology of Europaean Corvids (Aves: Corvidae), with a Key to the Identification of their Skeletal Elements*. Institute of Systematics and Evolution of Animals, Polish Academy of Sciences, Krakow.

Tomiałojc, L. (2000). Did White-backed Woodpeckers ever breed in Britain? *British Birds* **93**: 453–456.

Tomiałojc, L., Wesołowski, T. & Walankiewicz, W. (1984). Breeding bird community of a primaeval temperate forest (Białowieża National Park, Poland). *Acta Ornithologica* **20**: 241–310.

Toynbee, J.M.C. (1973). *Animals in Roman Life and Art*. John Hopkins University Press paperback, 1996. edn. Thames & Hudson, London.

Tubbs, C.R. (1974). *The Buzzard*. David & Charles, Newton Abbot.

Turk, F.A. (1971). Notes on Cornish Mammals in prehistoric and historic times No.4: a report on the Animal Remains from Nor-Nour, Isles of Scilly. *Cornish Archaeology* **10**: 79–91.

Turk, F.A. (1978). The animal remains from Nor-Nour: a synoptic view of the finds. In *Excavations at Nornour, Isles of Scilly, 1969–73: the Pre-Roman settlement* (ed. S.A. Butcher). *Cornish Archaeology* **17**: 99–103.

Tyrberg, T. (1991a). Arctic, Montane and Steppe birds as Glacial relicts in the West Palearctic. *Ornithologische Verhandlungen* **25**: 29–49.

Tyrberg, T. (1991b). Crossbill (genus *Loxia*) evolution in the West Palearctic – a look at the fossil evidence. *Ornis Svecica* **1**: 3–10.

Tyrberg, T. (1995). Palaeobiogeography of the genus *Lagopus* in the West Palearctic. *Courier Forschungsinstitut Senckenberg* **181**: 275–291.

Tyrberg, T. (1998). *Pleistocene Birds of the Palearctic: a catalogue*. Nuttall Ornithological Club, Cambridge, MA.

Van Wijngaarden-Bakker, L.H. (1974). The animal remains recovered from the Beaker Settlement at Newgrange, County Meath: first report. *Proceedings of the Royal Irish Academy* **74C**: 313–383.

Van Wijngaarden-Bakker, L.H. (1982). The faunal remains. In *Newgrange. Archaeology, Art and Legend* (ed. M.J. O'Kelly). Thames & Hudson, London.

Van Wijngaarden-Bakker, L.H. (1985). The faunal remains. In *Excavations at Mount Sandel 1973–77* (ed. P.C. Woodman), pp. 71–76. HMSO, Belfast.

Van Wijngaarden-Bakker, L.H. (1986). The animal remains from the Beaker settlement at Newgrange, Co. Meath: Final Report. *Proceedings of the Royal Irish Academy* **86C**: 1–111.

Vera, F.W.M. (2000). *Grazing Ecology and Forest History.* CABI, Oxford.

Village, A. (1990). *The Kestrel.* T.&A.D. Poyser, London.

Waters, E. & Waters, D. (2005). The former status of the Great Bustard in Britain. *British Birds* **98**: 295–305.

Watson, J. (1997). *The Golden Eagle.* T.&A.D. Poyser, London.

Watson, A. & Moss, R. (2004). Impacts of ski-development on ptarmigan (*Lagopus mutus*) at Cairn Gorm, Scotland. *Biological Conservation* **116**: 267–275.

Weinstock, J. (2002). *The Animal Bone Remains from Scarborough Castle, North Yorkshire.* Centre For Archaeology Report, 21/2002, 1–30.

Wenink, P.W., Baker, A.J., Rosner, H.-U. & Tilanus, M.G.J. (1996). Global mitochondrial DNA phylogeny of Holarctic breeding dunlins (*Calidris alpina*). *Evolution* **50**: 318–330.

Wernham, C., Toms, M., Marchant, J., Clark, J., Siriwardena, G. & Baillie, S. (2002). *The Migration Atlas. Movements of the Birds of Britain and Ireland.* T.&A.D. Poyser, London.

West, B. (1994). Birds and mammals. Cl. 6 in *Saxon Environment and Economy in London* (ed. D.J. Rackham & M.D. Culver). C.B.A., London.

West, B. (1995). The case of the missing victuals. *Historical Archaeology* **29**: 20–42.

West, B. & Zhou, B.-X. (1988). Did chickens go north? New evidence for domestication. *Journal of Archaeological Science* **15**: 515–533.

Wheeler, A. (1978). Why were there no fish bones at Star Carr? *Journal of Archaeological Science*, **5**: 85–89.

White, G. (1789). *The Natural History and Antiquities of Selbourne.* B. White, London.

Williams, P.H. (1982). The distribution and decline of British bumble bees (*Bombus* Latr.). *Journal of Apicultural Research* **21**: 236–245.

Wilson, R. (1983). *Piazza Armerina.* Granada, London.

Wilson, R., Allison, E., & Jones, A. (1983). Animal bones & shells. In Late Saxon evidence and excavation of Hinxey Hall, Queen Street, Oxford (ed C. Halpin*). Oxoniensia.* **48**: 68.

Wilson, R., Locker, A. & Marples, B. (1989). Medieval animal bones and marine shells from Church Street and other sites in St Ebbes, Oxford. In *Excavations in St Ebbes, Oxford 1967–76* (eds T.G. Hassal, C.E. Halpin & M. Mellor). *Oxoniensia,* **54**, 258–268.

Woelfle, E. (1967). *Vergleichend morphologische Untersuchungen an Einzelknochen des postcranialen Skelettes in Mitteleuropa vorkommender mittelgrosser Enten, Halbgänse und Säger.* Ludwigs-Maximilians-Universität, Munich.

Wójcik, J.D. (2002). The comparative osteology of the humerus in European thrushes (Aves: *Turdus*) including a comparison with other similarly sized genera of passerine birds – preliminary results. *Acta Zoologica Cracoviensia* **45** (Special issue): 369–381.

Woodman, P.C., McCarthy, M.R. & Monaghan, N. (1997). The Irish Quaternary fauna project. *Quaternary Science Reviews* **7**: 129–159.

Wright, T. (1884). *Anglo-Saxon and Old English Dictionaries,* 2nd edn. Trubner & Co., London.

Yalden, D.W. (1984). What size was *Archaeopteryx*? *Zoological Journal of the Linnean Society* **82**: 177–188.

Yalden, D.W. (1985). Forelimb function in *Archaeopteryx.* In *The Beginnings of Birds* (eds M.K. Hecht, J.H. Ostrom, G. Viohl & P. Wellnhofer), pp. 91–97. Freunde des Jura-Museums Eichstatt, Eichstatt.

Yalden, D.W. (1987). The natural history of Domesday Cheshire. *Naturalist* **112**: 125–131.

Yalden, D.W. (1999). *The History of British Mammals.* T.&A.D. Poyser, London.

Yalden, D.W. (2003). Mammals in Britain – a historical perspective. *British Wildlife* **14**: 243–251.

Yalden, D.W. (2007). The older history of the White-tailed Eagle in Britain. *British Birds* **100**: 471–480.

Yalden, D.W. & Carthy, R.I. (2004). The archaeological record of birds in Britain and Ireland: extinctions, or failures to arrive? *Environmental Archaeology* **9**: 123–126.

Yalden, P.E. & Yalden, D.W. (1989). Small vertebrates. In *Wilsford Shaft. Excavations 1960–62* (eds P. Ashbee, M. Bell & E. Proudfoot), pp. 103–106. English Heritage Archaeology Report 11.

Yapp, W.B. (1962). *Birds and Woods*. Oxford University Press, Oxford.

Yapp, W.B. (1981a). Gamebirds in medieval England. *Ibis* **125**: 218–221.

Yapp, W.B. (1981b). *The Birds of Medieval Manuscripts*. The British Library, London.

Yapp, W.B. (1982a). Birds in captivity in the Middle Ages. *Archives of Natural History* **10**: 479–500.

Yapp, W.B. (1982b). The Birds of the Sherborne Missal. *Proceedings of the Dorset Natural History and Archaeology Society* **104**: 5–15.

Yapp, W.B. (1983). The Illustrations of birds in the Vatican Manuscript of *De arte venandi cum avibus* of Frederick II. *Annals of Science* **40**: 597–634.

Yapp, W.B. (1987). Medieval knowledge of birds as shown in bestiaries. *Archives of Natural History* **14**: 175–210.

Zeuner, F.E. (1963). *A History of Domesticated Animals*. Hutchinson, London.

Zhou, Z. & Zhang, F. (2002) A long-tailed, seed-eating bird from the Early Cretaceous of China. *Nature* **418**: 405–409.

Zhou, Z. & Zhang, F. (2007). Discovery of an ornithurine bird and its implications for Early Cretaceous avian evolution. *Proceedings of the National Academy of Sciences* **102**: 18998–19002.

Index

Acanthisittidae 33
Accipiter gentilis Goshawk 38, 39, 44, 50, 57, 65–6, 68, 71, 74, 78, 87, 96–7, 109, 124, 130, 133, 135–9, 144, 156, 159, 163, 172–3, 187, 192, 194–5, 197–201, 212, 233
Accipiter nisus Sparrowhawk 44, 58, 65–8, 71, 124, 131, 133, 135–8, 156, 179, 190, 198, 200–1, 212, 249
Aelfric 129, 136, 238, 242
Aix galericulata Mandarin 6–7, 9, 34, 66, 187, 191, 194, 205
Alauda arvensis Skylark 16, 21, 36, 38–9, 56, 68–70, 78, 95, 97, 111, 157, 180–3, 189, 222
Albatross 30, 33–4, 188
Alca torda Razorbill 10–1, 34, 36, 57, 58, 75, 77, 79–81, 96, 110, 112, 133, 167, 189, 196, 219
Alcedo atthis Kingfisher 38, 39, 150, 154, 157, 177, 192, 222
Alectoris rufa Red-legged Partridge 13, 36, 111, 187–9, 208
Alle alle Little Auk 10, 11, 40, 44, 55, 58, 75–6, 78–9, 96, 109, 112, 220
Alpine Swift *Apus melba* 38–9, 46, 222
Ambiortus 18, 27
Anas acuta Pintail 10, 56, 57, 77, 79, 87, 101, 109, 131, 132, 192, 206
Anas clypeata Shoveler 10, 35, 56, 87, 90, 101, 109, 133, 134, 191, 206
Anas crecca Teal 10, 34–35, 38–9, 44, 50, 56–7, 74, 78–9, 87, 95–6, 101, 104, 109, 130–4, 191, 205–6
Anas platyrhynchos Mallard 10, 34–6, 38–9, 44, 50, 56–8, 74, 77, 78–1, 87, 89, 95, 101, 104–5, 108–9, 130–4, 156, 180, 182–3, 189, 194, 205
Anas platyrhynchos domesticus Domestic Duck 10, 101, 104, 105, 113, 134, 205, 231
Anas strepera Gadwall 10, 36, 56, 87, 95, 101, 109, 191, 205
Anser anser Greylag Goose 35, 38, 44, 50, 58, 74, 78–80, 131–2, 187, 204
Anser anser domesticus Domestic Goose 101, 102, 104, 113, 130–4, 204
Anser fabalis Bean Goose 36, 76, 103, 203
Anthus pratensis Meadow Pipit 15–6, 38, 69, 156, 158, 175, 180–2, 189, 199, 201, 223
Anthus trivialis Tree Pipit 15, 36, 62, 69, 158, 191, 200, 223
Apus apus Swift 30–2, 35, 44, 46, 179, 190, 221
Apus melba Alpine Swift 38–9, 46, 222

Aquila chrysaetos Golden Eagle 12, 14, 38, 44, 50, 66–7, 76, 78, 88, 96, 108, 111, 124, 137, 147, 173–4, 184, 191, 198, 211, 213, 255
Ardea cinerea Grey Heron 32, 38, 79, 82, 87–8, 90, 95, 111, 119, 121, 130, 132, 134–5, 138–9, 145, 147, 151, 154–5, 157, 160–1, 178, 181–2, 190, 211
Archaeopteryx 18, 25, 26, 27, 46, 249, 255
Asio flammeus Short-eared Owl 6–7, 9, 38, 45, 50, 75, 97, 111, 134, 173, 191, 221
Avocet *Recurvirostra avosetta* 32, 109, 111, 159, 188, 192, 194, 202, 215

Bacon Hole 36, 203, 210, 212–3, 215–7, 219, 223–4
Barbary Dove *Streptopelia risoria* 105, 220
Barn Owl *Tyto alba* 38, 66–7, 78–9, 87, 108, 111, 132–3, 151, 173, 177, 191, 194, 220, 254
Barnacle Goose *Branta leucopsis* 38, 55, 58, 74, 95, 110, 132, 150, 191, 204
Bass Rock 153, 158
Bean Goose *Anser fabalis* 36, 77, 103, 203
Białowieża 61–8, 221, 240, 254
Bittern *Botaurus stellaris* 74, 79, 80–1, 87, 89–91, 111, 129, 131, 134, 138–9, 145, 155, 157, 192, 195, 202, 211
Blackbird *Turdus merula* 15, 16, 34, 36, 38, 44–5, 50, 55–58, 62–4, 71, 77–9, 95–7, 112, 133, 150–1, 157, 180–4, 189, 194, 201, 225, 242
Black Grouse *Tetrao tetrix* 12–3, 38, 40, 44–5, 50, 54, 56, 58, 60, 78–9, 88, 95, 97, 102, 111–2, 132–4, 139, 175, 190, 199, 200, 208
Black Guillemot *Cepphus grylle* 10, 34, 38, 75, 96, 112, 133–4, 190, 196, 219
Black Kite *Milvus nigra* 197, 212
Black Stork *Ciconia nigra* 35, 60, 145, 211
Black Woodpecker *Dryocopus martius* 60, 65, 195, 222
Boke of St Albans 136
Bonasia bonasus Hazel Hen 12, 40, 50, 54–55, 60, 71, 185, 208
Botaurus stellaris Bittern 74, 79, 80–1, 87, 89–91, 111, 129, 131, 134, 138–9, 145, 155, 157, 192, 195, 202, 211
Boxgrove 21, 35, 54, 59, 168–9, 203–8, 214, 218–9, 221, 224, 228, 243, 251
Branta leucopsis Barnacle Goose 38, 55, 58, 74, 95, 110, 132, 150, 191, 204
Branta bernicla Brent Goose 36, 38, 57, 79, 131–2, 204

Branta canadensis Canada Goose 182, 183, 187, 189, 194, 204
Brent Goose *Branta bernicla* 36, 38, 57, 79, 131–2, 204
Brewe 92, 139, 145, 211, 233
Bu, Orkney 79, 96, 217, 218, 219, 228, 235, 250
Bubo bubo Eagle Owl 34–5, 40, 44, 46, 55–6, 58–9, 60, 66, 71, 88, 137–8, 151, 180, 185, 194, 221, 253
Bubo scandiaca Snowy Owl 39, 44, 56, 59, 173, 193, 221
Bucephalus clangula Goldeneye 34–5, 50, 56–7, 65, 87, 132–3, 192, 207
Burhinus oedicnemus Stone Curlew 29, 31, 68, 79, 161, 192, 215
Burwell Fen 80, 84, 91, 140, 148, 207, 214, 219
Buteo buteo Buzzard 34, 39, 44, 57–8, 65–6, 68, 71, 74, 78, 88, 95, 97, 108, 110, 125, 127, 129, 130–4, 137–8, 156, 160, 169, 173–4, 198–9, 213, 248, 254
Buzzard *Buteo buteo* 34, 39, 44, 57–8, 65–6, 68, 71, 74, 78, 88, 95, 97, 108, 110, 125, 127, 129, 130–4, 137–8, 156, 160, 169, 173–4, 198–9, 213, 248, 254

Caldicot, Gwent 79, 140, 204, 206, 209, 211, 247
Calidris alpina Dunlin 36, 46–7, 56, 69, 96, 111, 119, 129, 131, 133, 176, 191, 216, 255
Calonectris diomedea Cory's Shearwater 36, 210
Cambridgeshire Fens 78, 88
Canada Goose *Branta canadensis* 182, 183, 187, 189, 194, 204
Capercaillie *Tetrao urogallus* 12–3, 35, 40, 44, 56–58, 77–8, 102, 112, 121, 124, 146–7, 151, 154, 163, 170, 187, 190, 194–5, 208, 234, 238
Carlisle 110–2, 140–2, 148, 231, 253
Carpometacarpus 4, 5, 7, 14, 27, 37, 222
Cepphus grylle Black Guillemot 10, 34, 38, 75, 96, 112, 133–4, 190, 196, 219
Chaoyungia 27
Chaffinch *Fringilla coelebs* 21, 38, 58, 61–4, 70, 97, 150, 154, 156, 180–4, 189, 194, 200, 229
Chelm's Combe Shelter, Cheddar 55, 59, 222
Chough *Pyrrhocorax pyrrhocorax* 44, 45, 46, 50, 58, 97, 133, 155, 156, 157, 160, 191, 228
Ciconia ciconia White Stork 38–9, 57, 79, 82, 111, 145, 151, 185, 211
Ciconia nigra Black Stork 35, 60, 145, 211
Coccothraustes coccothraustes Hawfinch 38, 45, 56, 58, 62, 71, 77, 162, 192, 230, 248
Collared Dove *Streptopelia decaocto* 7, 105, 159, 189, 202, 220
Columba livia Rock Dove 44, 50, 57, 101, 105–6, 133, 183, 189, 220
Columba livia domestica Domestic Dove 101, 105, 106, 132, 220
Columba palumbus Wood Pigeon 35, 38, 44, 46, 57–8, 64, 68, 95, 101, 106, 131–3, 155, 162, 180–4, 189, 220
Confuciusornis 27, 28

Coracoid 3, 4, 12–4, 25, 27–8, 47, 134
Cormorant *Phalacrocorax carbo* 10, 16, 34–6, 57, 73, 74, 77, 79, 87, 90, 96, 112, 132–4, 150, 154, 156, 190, 193, 210, 236
Corncrake *Crex crex* 44, 47, 58, 69, 70, 87, 95, 109–11, 132–3, 192, 195, 214
Corvus corax Raven 2, 10, 15, 32, 36, 38, 39, 40, 44, 50, 56, 58, 67, 76–9, 96–7, 108, 109, 113, 119–20, 124, 127–8, 131–4, 151, 154–58, 160, 171–3, 190, 228, 233, 248, 254
Corvus corone Crow 5, 15, 36, 38–9, 40, 57–8, 77–9, 87, 95, 108, 117, 124, 131–4, 151, 154–6, 159, 173, 181, 183, 189, 196, 199, 228
Corvus frugilegus Rook 15, 26, 38, 58, 77–9, 87–8, 95–6, 108, 119–20, 124, 133, 155–7, 159, 181–3, 189, 228
Corvus monedula Jackdaw 15–6, 38, 46, 50, 56–7, 78, 95, 117, 120, 132–4, 156–8, 161, 182–3, 184, 190, 195, 229
Cory's Shearwater *Calonectris diomedea* 36, 210
Coturnix coturnix Quail 13, 44, 95, 97–8, 109, 132–3, 150–1, 156–7, 192, 208
Crane *Grus grus* 2, 10, 36–7, 44, 57, 69, 70, 77–9, 80–2, 87–8, 93, 97, 109–11, 116–7, 120–4, 130–3, 138–9, 140–145, 149, 150–1, 153, 155, 159, 163, 184, 192, 195, 214, 231, 233, 238, 243, 249
Crested Lark *Galerida cristata* 45, 70, 222
Creswell Crags 20, 21, 38, 39, 50, 57, 145, 221, 223, 230, 232, 244
Cretaceous 18, 27, 28, 29, 30, 31, 32, 33, 47, 236, 238, 240, 245, 256
Crex crex Corncrake 44, 47, 58, 69, 70, 87, 95, 109–11, 132–3, 192, 195, 214
Cromer Forest Beds 34
Cromerian 19, 34–5, 168, 203, 205–7, 209, 214, 216–9, 221, 224–5, 227–8
Crossbill *Loxia* sp. 36, 38, 41–3, 56, 175, 185, 192, 230, 254
Crosskirk Broch, Caithness 75, 96, 103, 168, 246
Crow *Corvus corone* 5, 15, 36, 38–9, 40, 57–8, 77–9, 87, 95, 108, 117, 124, 131–4, 151, 154–6, 159, 173, 181, 183, 189, 196, 199, 228
Curlew *Numenius arquata* 38, 70, 74, 79, 96, 111, 131–4, 151, 153, 190, 196, 217
Cygnus cygnus Whooper Swan 10–1, 34–6, 44, 57, 74, 80, 82, 84, 86–7, 109, 130, 132–4, 153, 184, 192, 203
Cygnus olor Mute Swan 10, 78, 80–4, 87–8, 109, 129–30, 134, 138, 182–4, 189, 203, 254

Dabchick *Tachybaptus ruficollis* 6, 57, 87, 156, 209
Dalmatian Pelican *Pelecanus crispus* 80, 81, 87–9, 91, 93, 185, 195, 211
Danebury, Hampshire 95–6, 203, 206–7, 212–3, 219, 221, 225, 227–9, 237
Demen's Dale, Derbyshire 56, 58, 65, 205–7, 214–7, 221, 225, 227, 230

Dendrocopus leucotos White-backed Woodpecker 60, 64, 222, 254
Dendrocopus major Great Spotted Woodpecker 44, 50, 78, 191, 222
Diomedea anglica 33, 242
Diver *Gavia* sp. 30, 35, 44, 96, 131, 209
Domestic Dove *Columba livia domestica* 101, 105, 106, 132, 220
Domestic Duck *Anas platyrhynchos domesticus* 10, 101, 104, 105, 113, 134, 205, 231
Domestic Fowl *Gallus domesticus* 12, 13, 57, 79, 85, 96–9, 100–5, 107, 110, 113, 121, 130–4, 153, 182, 204, 208, 232, 240
Domestic Goose *Anser anser domesticus* 101, 102, 104, 113, 130, 131, 132, 133, 204
Dowel Cave, Derbyshire 56, 77, 146, 205, 208, 213–4, 217, 221, 223–7, 229, 230, 233
Dryocopus martius Black Woodpecker 60, 65, 195, 222
Dunlin *Calidris alpina* 36, 46–7, 56, 69, 96, 111, 119, 129, 131, 133, 176, 191, 216, 255
Dunnock *Prunella modularis* 35, 38, 50, 62, 64, 77, 96, 118, 154–5, 181–2, 189, 224

Eagle Owl *Bubo bubo* 34–5, 40, 44, 46, 55–6, 58–9, 60, 66, 71, 88, 137–8, 151, 180, 185, 194, 221, 253
Egretta garzetta Little Egret 90, 93, 110, 145, 154, 159, 180, 193, 202, 211
Enaliornis 18, 27, 240
Eocene 18, 29, 30–3, 47, 238, 246, 253
European Crane *Grus primigenia* 36, 40, 140, 238, 249
Exeter 84, 110, 141, 209, 213, 217–9, 220, 222, 228, 233, 247

Falco columbarius Merlin 14, 38, 44, 50, 66, 96, 133, 136, 137, 154, 155, 173, 192, 198, 213, 251
Falco peregrinus Peregrine 14, 44, 58, 66–7, 88, 95–6, 110, 133–4, 136–9, 144, 154–5, 157, 173, 177–80, 191, 198, 200, 213, 251
Falco rusticola Gyr Falcon 14, 137, 138, 150, 154, 173, 214
Falco subbuteo Hobby 14, 36, 44, 50, 66 , 136, 137, 154, 173, 192, 198, 213
Falco tinnunculus Kestrel 6–7, 9, 14, 16, 35–6, 38, 44–6, 50, 56, 66, 70, 74, 78, 95–6, 120, 130, 134, 136–7, 157, 173, 180, 182, 185, 187, 190, 198, 213, 255
Fea's Petrel *Pterodroma feae* 134, 210
Ferriter's Cove 77, 218
Ficedula hypoleuca Pied Flycatcher 40–1, 62–4, 191, 226
Fieldfare *Turdus pilaris* 15, 16, 38, 39, 44, 50, 58, 78, 96, 112, 150, 193, 225
Fishbourne 111, 145, 163, 206, 208, 215, 239
Fox Hole Cave, Derbyshire 77–8, 146, 227

Fratercula arctica Puffin 6–7 , 9–11, 38, 57, 75, 79, 87, 96, 110, 112, 133–4, 181, 183, 189, 196, 220
Fringilla coelebs Chaffinch 21, 38, 58, 61–4, 70, 97, 150, 154, 156, 180–4, 189, 194, 200, 229
Frocester, Gloucestershire 110, 112, 208, 224–5, 234
Fulmar *Fulmarus glacialis* 57, 73–5, 77, 93, 96, 112, 133–4, 158, 180–1, 183, 189, 196, 209
Fulmarus glacialis Fulmar 57, 73–5, 77, 93, 96, 112, 133–4, 158, 180–1, 183, 189, 196, 209

Gadwall *Anas strepera* 10, 36, 56, 87, 95, 101, 109, 191, 205
Galerida cristata Crested Lark 45, 70, 222
Galliformes 6, 12, 13, 31–2, 97–102
Gallinula chloropus Moorhen 7, 16, 34, 35, 38, 80, 87–8, 130, 132, 150, 183, 189, 214
Gallus domesticus Domestic Fowl 12, 13, 57, 79, 85, 96–9, 100–5, 107, 110, 113, 121, 130–4, 153, 182, 204, 208, 232, 240
Gallus gallus Jungle Fowl 85, 97, 98
Gannet *Morus bassanus* 10, 57–8, 73–4, 77, 79, 88, 96, 112, 133–4, 150, 153, 158, 181, 183, 189, 196, 210, 252
Garrulus glandarius Jay 15, 34, 38, 44–5, 50, 56, 58, 78, 95–6, 129, 151, 155–7, 160, 190, 227
Gavia immer Great Northern Diver 79, 96, 109, 131, 133, 209
Gavia sp. Diver 30, 35, 44, 96, 131, 209
Glastonbury 21, 36, 83–8, 90–2, 95, 104, 140, 148, 206–7, 209, 210, 212, 225, 228, 232, 237
Goldcrest *Regulus regulus* 16, 62, 64, 181–2, 184, 190, 226
Golden Eagle *Aquila chrysaetos* 12, 14, 38, 44, 50, 66–7, 76, 78, 88, 96, 108, 111, 124, 137, 147, 173–4, 184, 191, 198, 211, 213, 255
Goldeneye *Bucephalus clangula* 34–5, 50, 56–7, 65, 87, 132–3, 192, 207
Golden Oriole *Oriolus oriolus* 62, 64, 71, 154–5, 193, 227
Golden Plover *Pluvialis apricaria* 6–7, 9, 14, 16, 35, 37–8, 46, 56, 58, 69, 78–9, 95–6, 108, 110, 131, 133, 176, 190, 215, 240
Goosander *Mergus merganser* 35–6, 38–9, 44, 50, 56, 65, 74, 88, 95–6, 109, 190, 207
Goshawk *Accipiter gentilis* 38, 39, 44, 50, 57, 65–6, 68, 71, 74, 78, 87, 96–7, 109, 124, 130, 133, 135–9, 144, 156, 159, 163, 172–3, 187, 192, 194–5, 197–201, 212, 233
Gough's Cave 55, 103, 203, 206–7, 243
Great Auk *Pinguinus impennis* 75, 77–8, 93, 96, 112, 133, 158, 166–70, 185, 219, 240–1, 252
Great Bustard *Otis tarda* 44, 50, 58, 68, 111, 145, 161, 163–7, 176, 185, 195, 215, 250, 254–5
Great Crested Grebe *Podiceps cristatus* 57, 87, 132, 177, 190, 209
Great Northern Diver *Gavia immer* 79, 96, 109, 131, 133, 209

Great Spotted Woodpecker *Dendrocopus major* 44, 50, 78, 191, 222
Green River 32
Green Woodpecker *Picus viridis* 6–7, 9, 120, 150, 154, 155, 191, 222
Grey Partridge *Perdix perdix* 6–7, 9, 12–3, 35, 38, 44, 50, 56, 58, 68, 70, 78, 88, 96, 111, 132, 138, 147, 189, 199, 208, 254
Greylag Goose *Anser anser* 35, 38, 44, 50, 58, 74, 78–80, 131–2, 187, 204
Grey Plover *Pluvialis squatorola* 14, 38, 50, 56, 58, 74, 108, 130, 215
Grus grus Crane 2, 10, 36–7, 44, 57, 69, 70, 77–9, 80–2, 87–8, 93, 97, 109–11, 116–7, 120–4, 130–3, 138–9, 140–145, 149, 150–1, 153, 155, 159, 163, 184, 192, 195, 214, 231, 233, 238, 243, 249
Grus primigenia 36, 40, 140, 238, 249
Guillemot *Uria aalge* 10, 11, 34, 57–8, 73, 75, 77–9, 96, 110, 112, 133–4, 167, 180–1, 183, 189, 219
Guineafowl *Numida meleagris* 12, 13, 107, 208
Gussage All Saints, Dorset 96, 100, 103, 104, 140, 212, 219, 229, 242
Gyr Falcon *Falco rusticola* 14, 137, 138, 150, 154, 173, 214

Haliaeetus albicilla White-tailed Eagle 12, 14, 36, 44, 50, 57–8, 66–7, 74, 76–8, 80, 87–8, 96, 108, 110, 116, 118, 120, 124–6, 133–4, 137, 148, 151, 159, 163, 172, 173, 174, 184, 187, 192, 194–8, 211, 232, 243, 246, 256
Hamwic 23, 131
Hawfinch *Coccothraustes coccothraustes* 38, 45, 56, 58, 62, 71, 77, 162, 192, 230, 248
Hazel Hen *Bonasia bonasus* 12, 40, 50, 54–55, 60, 71, 185, 208
Hedge Sparrow *Prunella modularis* 35, 38, 50, 62, 64, 77, 96, 118, 154–5, 181–2, 189, 224
Heron (Grey Heron) *Ardea cinerea* 32, 38, 79, 82, 87–8, 90, 95, 111, 119, 121, 130, 132, 134–5, 138–9, 145, 147, 151, 154–5, 157, 160–1, 178, 181–2, 190, 211
Hesperornis 18, 27–29
Hirundo rustica Swallow 15–6, 36, 38, 77, 79, 97, 120, 149, 151, 154, 156–8, 180–2, 190, 223
Hobby *Falco subbuteo* 14, 36, 44, 50, 66, 136, 137, 154, 173, 192, 198, 213
Honey Buzzard *Pernis apivorus* 66–7, 71, 129, 170, 173, 192, 197, 211
Hongshanornis 27–8
Hoodwink 102, 106, 108, 209, 221
House Sparrow *Passer domesticus* 16, 38, 56, 96, 111, 112, 150, 158, 160–1, 180–3, 189, 194, 200, 229, 239
Howe, Orkney 75, 84, 96–7, 100, 103–4, 111, 113, 141, 149, 169, 170, 205, 207–8, 214–5, 217–20, 222, 224, 226, 228
Hoxnian 19, 35, 46, 59, 207–87, 211, 221, 222, 227, 230
Humerus 3–5, 10, 14–6, 25, 29, 33, 35, 86–8, 113, 132–3, 255

Iberomesornis 27–8
Ichthyornis 27, 29
Ipswichian 19, 36, 46, 84, 101, 106, 141, 149, 204–8, 211, 213–8, 220, 223–5, 229
Isbister, Orkney 73–4, 76, 149, 208, 213–4, 216–19, 222, 232, 234

Jackdaw *Corvus monedula* 15–6, 38, 46, 50, 56–7, 78, 95, 117, 120, 132–4, 156–8, 161, 182–3, 184, 190, 195, 229
Jarlshof, Shetland 79, 91, 107, 135, 140, 169, 206, 208, 210–2, 214, 218, 229, 249–50
Jay *Garrulus glandarius* 15, 34, 38, 44–5, 50, 56, 58, 78, 95–6, 129, 151, 155–7, 160, 190, 227
Jeholornis 27–8
Jungle Fowl *Gallus gallus* 85, 97, 98
Jurassic 18, 25, 27, 239

Kestrel *Falco tinnunculus* 6–7, 9, 14, 16, 35–6, 38, 44–6, 50, 56, 66, 70, 74, 78, 95–6, 120, 130, 134, 136–7, 157, 173, 180, 182, 185, 187, 190, 198, 213, 255
Kingfishers Alcediniformes 31–2
Kingfisher *Alcedo atthis* 38, 39, 150, 154, 157, 177, 192, 222
Kittiwake *Rissa tridactyla* 35, 57, 75, 79, 95, 133–4, 153, 181, 183, 189, 196–7, 219
Knap of Howar, Papa Westray 73, 74, 168, 169, 234

Lagopus 50, 52, 54, 207, 254
Lagopus lagopus Willow Grouse 44, 50–2, 54
Lagopus lagopus scotica Red Grouse 12–3, 38–9, 50, 52–8, 69, 74, 79, 95, 97, 112, 134, 162, 175, 196, 199, 207, 251
Lagopus muta Ptarmigan 12–3, 38–40, 44, 50, 52–8, 69, 190, 195–6, 207, 246, 254
Lagore 23, 90, 132, 142, 149, 203–7, 209, 212, 214, 220, 228, 243, 253
Lapwing *Vanellus vanellus* 15–16, 38, 50, 57, 70, 74, 77–80, 95–6, 108, 120, 130, 132, 157, 189, 196, 215
Links of Notland, Westray 73–4
Lithornis 31–2
Little Auk *Alle alle* 10, 11, 40, 44, 55, 58, 75–6, 78–9, 96, 109, 112, 220
Little Bustard *Tetrax tetrax* 39, 44, 215
Little Egret *Egretta garzetta* 90, 93, 110, 145, 154, 159, 180, 193, 202, 211
London Clay 18, 30–3, 47
Loxia sp. Crossbill 36, 38, 41–3, 56, 175, 185, 192, 230, 254

Magpie *Pica pica* 6–7, 9, 15–6, 25, 38, 45, 50, 77–8, 134, 151, 155, 157, 160, 181, 183, 188–9, 200–1, 227, 242, 254
Mallard *Anas platyrhynchos* 10, 34–6, 38–9, 44, 50, 56–8, 74, 77, 78–1, 87, 89, 95, 101, 104–5, 108–9, 130–4, 156, 180, 182–3, 189, 194, 205

Mandarin *Aix galericulata* 6–7, 9, 34, 66, 187, 191, 194, 205
Manx Shearwater *Puffinus puffinus* 58, 79, 87, 96, 133–4, 181, 183, 189, 210
Meadow Pipit *Anthus pratensis* 15–6, 38, 69, 156, 158, 175, 180–2, 189, 199, 201, 223
Meare 59, 83, 84, 86–7, 90, 91–2, 140, 148, 203–10, 212–4, 218, 221, 228, 232, 241, 245
Meleagris gallopavo Turkey 12, 13, 97, 101, 103, 107, 157, 170, 204, 209
Mergus albellus Smew 34, 36, 44, 56, 80, 87, 96, 207
Mergus merganser Goosander 35–6, 38–9, 44, 50, 56, 65, 74, 88, 95–6, 109, 190, 207
Mergus serrator Red-breasted Merganser 34–5, 44, 57, 80, 87, 109, 132, 190, 207
Merlin *Falco columbarius* 14, 38, 44, 50, 66, 96, 133, 136, 137, 154, 155, 173, 192, 198, 213, 251
Merlin's Cave, Wye Valley 56, 106, 205, 207–8, 214, 216–7, 219–3, 225, 227, 229–0
Messel 32, 33, 247
Milvus milvus Red Kite 36, 66–7, 70, 77, 87, 95–6, 108, 131–2, 137–8, 171–4, 192, 197–9, 212, 233, 246
Milvus nigra Black Kite 197, 212
Minchin Hole 36, 216, 219, 222, 228
Mohenjo-daro, Pakistan 99–100
Moorhen *Gallinula chloropus* 7, 16, 34, 35, 38, 80, 87–8, 130, 132, 150, 183, 189, 214
Morton 57, 75, 209–10, 219–20, 236
Morus bassanus Gannet 10, 57–8, 73–4, 77, 79, 88, 96, 112, 133–4, 150, 153, 158, 181, 183, 189, 196, 210, 252
Mount Pleasant, Dorset 76–7, 140, 206, 214, 225, 242,
Mute Swan *Cygnus olor* 10, 78, 80–4, 87–8, 110, 129–30, 134, 138, 182–4, 189, 203, 254

Night Heron *Nycticorax nycticorax* 90–93, 110–1, 145, 157, 193, 211
Nightjars Caprimulgiformes 30–32
Nightjar *Caprimulgus europaeus* 133, 176, 192, 221, 246
Nile, Egypt 100, 102
Nornour, Scilly 79, 100, 145, 210 1, 215–7, 254
Numenius arquata Curlew 38, 70, 74, 79, 96, 111, 131–4, 151, 153, 190, 196, 217
Numenius phaeopus Whimbrel 38, 47, 92, 96, 131, 133, 153, 192, 197, 217
Numida meleagris Guineafowl 12, 13, 107, 208
Nycticorax nycticorax Night Heron 90–93, 110–1, 145, 157, 193, 211

Oligocene 18, 32–3
Olduvai Gorge 47
Oriolus oriolus Golden Oriole 62, 64, 71, 154–5, 193, 227
Oronsay 20, 57, 168–9, 205, 207, 214–5, 219, 241, 243, 248

Oscines 33
Osprey *Pandion haliaetus* 31, 35, 38, 66–7, 88, 110, 133, 137, 157, 159–60, 172–3, 176, 192, 194, 197–8, 213, 245, 250
Ossom's Cave 56, 108, 221
Ossom's Eyrie Cave 108, 111, 112, 213, 226, 227
Ostend 34, 206
Otis tarda Great Bustard 44, 50, 58, 68, 111, 145, 161, 163–7, 176, 185, 195, 215, 250, 254–5
Owl Strigiformes 6–9, 15–6, 30, 34–5, 39, 40, 45, 56, 58–60, 65–6, 68, 71, 78–9, 87, 111, 134, 137–8, 151, 154, 157, 220–1, 253

Palaeocene 18, 29, 31
Pandion haliaetus Osprey 31, 35, 38, 66–7, 88, 110, 133, 137, 157, 159–60, 172–3, 176, 192, 194, 197–8, 213, 245, 250
Papa Westray 73–4, 167–9, 207, 209–10, 213–5, 217–9, 222, 228, 234, 249
Passer domesticus House Sparrow 16, 38, 56, 96, 111, 112, 150, 158, 160–1, 180–3, 189, 194, 200, 229, 239
Passerines (Passeriformes) 6–9, 15–6, 32–4, 60, 63, 68–69, 71, 111, 133, 180–1, 225
Pavo cristatus Peacock 12–3, 101, 106–7, 113, 119, 130–1, 150, 154–5, 170, 194, 209
Peacock *Pavo cristatus* 12–3, 101, 106–7, 113, 119, 130–1, 150, 154–5, 170, 194, 209
Pelecanus crispus Dalmatian Pelican 80, 81, 87–9, 91, 93, 185, 195, 211
Perdix perdix Grey Partridge 6–7, 9, 12–3, 35, 38, 44, 50, 56, 58, 68, 70, 78, 88, 96, 111, 132, 138, 147, 189, 199, 208, 254
Peregrine *Falco peregrinus* 14, 44, 58, 66–7, 88, 95–6, 110, 133–4, 136–9, 144, 154–5, 157, 173, 177–80, 191, 198, 200, 213, 251
Pernis apivorus Honey Buzzard 66–7, 71, 129, 170, 173, 192, 197, 211
Phalacrocorax aristotelis Shag 10, 16, 39, 57–8, 73–4, 77, 79, 96, 112, 133–4, 181, 183, 189, 210
Phalacrocorax carbo Cormorant 10, 16, 34–6, 57, 73, 74, 77, 79, 87, 90, 96, 112, 132–4, 150, 154, 156, 190, 193, 210, 236
Phalacrocorax pygmaeus Pygmy Cormorant 10, 16, 90, 92, 185, 188, 195, 210, 236
Phasianus colchicus Pheasant 12–3, 101–2, 106–7, 113, 121, 124, 132, 143, 150, 156–7, 162, 170, 180–4, 187, 189, 194, 199, 204, 208, 245
Pheasant *Phasianus colchicus* 12–3, 101–2, 106–7, 113, 121, 124, 132, 143, 150, 156–7, 162, 170, 180–4, 187, 189, 194, 199, 204, 208, 246
Phoebastria anglica 33–4, 238
Pica pica Magpie 6–7, 9, 15–6, 25, 38, 45, 50, 77–8, 134, 151, 155, 157, 160, 181, 183, 188–9, 200–1, 227, 242, 254
Picus viridis Green Woodpecker 6–7, 9, 120, 150, 154, 155, 191, 222

Pied Flycatcher *Ficedula hypoleuca* 40–1, 62–4, 191, 226
Pinguinus impennis Great Auk 75, 77–8, 93, 96, 112, 133, 158, 166–70, 185, 219, 240–1, 252
Pin Hole Cave, Creswell 38–9, 145, 203, 209, 212–5, 217, 220–4, 228–9, 232
Pintail *Anas acuta* 10, 56, 57, 77, 79, 87, 101, 109, 131, 132, 192, 206
Place-names 1–2, 82, 115, 116–30, 151, 239, 241, 248
Platalea leucorodia Spoonbill 90, 92, 160, 180, 185, 193, 202, 211
Pluvialis apricaria Golden Plover 6–7, 9, 14, 16, 35, 37–8, 46, 56, 58, 69, 78–9, 95–6, 108, 110, 131, 133, 176, 190, 215, 240
Pluvialis squatorola Grey Plover 14, 38, 50, 56, 58, 74, 108, 130, 215
Podiceps cristatus Great Crested Grebe 57, 87, 132, 177, 190, 209
Point of Cott, Westray 73–5, 103, 148, 205, 207, 210, 215–6, 219, 243
Portchester, Sussex 84–5, 91–2, 107, 110, 131, 138, 205–6, 209, 212, 216–7, 219, 222–3, 230, 239
Port Eynon Cave, Gower 57, 148, 207, 209–10, 212–3, 215, 217, 219–20, 225, 228–9
Prunella modularis Dunnock 35, 38, 50, 62, 64, 77, 96, 118, 154–5, 181–2, 189, 224
Ptarmigan *Lagopus muta* 12–3, 38–40, 44, 50, 52–8, 69, 190, 195–6, 207, 246, 254
Pterocles sp. Sandgrouse 44, 220
Pterodroma feae Fea's Petrel 134, 210
Puffin *Fratercula arctica* 6–7 , 9–11, 38, 57, 75, 79, 87, 96, 110, 112, 133–4, 181, 183, 189, 196, 220
Puffinus puffinus Manx Shearwater 58, 79, 87, 96, 133–4, 181, 183, 189, 210
Pygmy Cormorant *Phalacrocorax pygmaeus* 10, 16, 90, 92, 185, 188, 195, 210, 236
Pyrrhocorax pyrrhocorax Chough 44, 45, 46, 50, 58, 97, 133, 155, 156, 157, 160, 191, 228

Quail *Coturnix coturnix* 13, 44, 95, 97–8, 109, 132–3, 150–1, 156–7, 192, 208
Quanterness 73–4, 76, 207, 208, 210, 215–9, 222, 224–6, 229, 236
Quercy 32–3

Ratites 29, 31
Raven *Corvus corax* 2, 10, 15, 32, 36, 38, 39, 40, 44, 50, 56, 58, 67, 76–79, 96–7, 108, 109, 113, 119–20, 124, 127–8, 131–4, 151, 154–58, 160, 171–3, 190, 228, 233, 248, 254
Razorbill *Alca torda* 10–1, 34, 36, 57, 58, 75, 77, 79–81, 96, 110, 112, 133, 167, 189, 196, 219
Recurvirostra avosetta Avocet 32, 109, 111, 159, 188, 192, 194, 202, 215
Red-breasted Merganser *Mergus serrator* 34–5, 44, 57, 80, 87, 109, 132, 190, 207

Red Kite *Milvus milvus* 36, 66–7, 70, 77, 87, 95–6, 108, 131–2, 137–8, 171–4, 192, 197–9, 212, 233, 246
Red Grouse *Lagopus lagopus* 12–3, 38–9, 50, 52–8, 69, 74, 79, 95, 97, 112, 134, 162, 175, 196, 199, 207, 251
Red-legged Partridge *Alectoris rufa* 13, 36, 111, 187–9, 208
Regulus regulus Goldcrest 16, 62, 64, 181–2, 184, 190, 226
Ring Ouzel *Turdus torquatus* 15–6, 36, 38, 44, 50, 55, 57–8, 69, 78, 97, 112, 133, 175, 191, 225
Rissa tridactyla Kittiwake 35, 57, 75, 79, 95, 133–4, 153, 181, 183, 189, 196–7, 219
Robin Hood's Cave, Creswell 50, 145, 203, 204, 207, 211–6, 220–2, 227–30, 236
Rock Dove *Columba livia* 44, 50, 57, 101, 105–6, 133, 183, 189, 220
Rook *Corvus frugilegus* 15, 26, 38, 58, 77–79, 87–8, 95–6, 108, 119–20, 124, 133, 155–7, 159, 181–3, 189, 228

Sandgrouse *Pterocles* sp. 44, 220
Scolopax rusticola Woodcock 14, 35, 57, 74, 77–8, 80, 96, 108, 130–3, 138–9, 147, 149–50, 157, 162, 190, 216
Shag *Phalacrocorax aristotelis* 10, 16, 39, 57–8, 73–4, 77, 79, 96, 112, 133–4, 181, 183, 189, 210
Short-eared Owl *Asio flammeus* 6–7, 9, 38, 45, 50, 75, 97, 111, 134, 173, 191, 221
Shoveler *Anas clypeata* 10, 35, 56, 87, 90, 101, 109, 133, 134, 191, 206
Shoveler *Platalea leucorodia* 90, 160
Sinornis 18, 27–8
Skaill, Orkney 75, 96–7, 100, 103, 106, 133, 205, 207, 209, 214–5, 220–1
Skylark *Alauda arvensis* 16, 21, 36, 38–9, 56, 68–70, 78, 95, 97, 111, 157, 180–3, 189, 222
Smew *Mergus albellus* 34, 36, 44, 56, 80, 87, 96, 207
Snowy Owl *Bubo scandiaca* 39, 44, 56, 59, 173, 193, 221
Soldier's Hole, Cheddar 50, 204–5, 208, 211–4, 216, 218, 221, 224–5, 227–8, 230
Song Thrush *Turdus philomelos* 15–6, 34–5, 38, 45, 50, 55, 57–8, 62, 64, 71, 77–9, 87–8, 95–7, 112, 130–1, 133, 155, 157, 181–3, 189, 196, 200, 225, 242
Sparrowhawk *Accipiter nisus* 44, 58, 65–8, 71, 124, 131, 133, 135–8, 156, 179, 190, 198, 200–1, 212, 249
Spoonbill *Platalea leucorodia* 90, 92, 160, 180, 185, 193, 202, 211
Star Carr 20, 21, 56–7, 91, 140, 145, 204, 206–7, 209, 211, 213–5, 236, 240, 243, 245, 255
Starling *Sturnus vulgaris* 16, 34–6, 38–9, 56, 58, 62, 77–9, 95, 97, 111–2, 130–1, 133, 155, 157, 159, 181–3, 188–9, 194, 228
Stone Curlew *Burhinus oedicnemus* 29, 31, 68, 79, 161, 192, 215

Stonehenge 21, 77–8, 148, 164, 247, 252
Streptopelia decaocto Collared Dove 7, 105, 159, 189, 202, 220
Streptopelia risoria Barbary Dove 105
Streptopelia turtur Turtle Dove 105, 155, 190, 220
Strix aluco Tawny Owl 16, 35, 38, 46, 56, 58, 65–6, 68, 71, 78–9, 97, 111, 138, 151, 155, 173, 190, 194, 199, 221
Sturnus vulgaris Starling 16, 34–6, 38–9, 56, 58, 62, 77–9, 95, 97, 111–2, 130–1, 133, 155, 157, 159, 181–3, 188–9, 194, 228
Suboscines 33
Swallow *Hirundo rustica* 15–6, 36, 38, 77, 79, 97, 120, 149, 151, 154, 156–8, 180–2, 190, 223
Swanscombe 21, 35, 59, 146, 204, 206–8, 210, 221, 226, 229, 243, 250
Swift *Apus apus* 30–2, 35, 44, 46, 179, 190, 221

Tachybaptus ruficollis Dabchick 6, 57, 87, 156, 209
Tarsometatarsus 4–5, 8–10, 14, 16, 50, 56, 58, 81, 86, 143, 213, 222
Tawny Owl *Strix aluco* 16, 35, 38, 46, 56, 58, 65–6, 68, 71, 78–9, 97, 111, 138, 151, 155, 173, 190, 194, 199, 221
Teal *Anas crecca* 10, 34–35, 38–9, 44, 50, 56–7, 74, 78–9, 87, 95–6, 101, 104, 109, 130–4, 191, 205–6
Tetrao tetrix Black Grouse 12–3, 38, 40, 44–5, 50, 54, 56, 58, 60, 78–9, 88, 95, 97, 102, 111–2, 132–4, 139, 175, 190, 199, 200, 208
Tetrao urogallus Capercaillie 12–3, 35, 40, 44, 56–58, 77–8, 102, 112, 121, 124, 146–7, 151, 154, 163, 170, 187, 190, 194–5, 208, 234, 238
Tetrax tetrax Little Bustard 39, 44, 215
Thatcham 56, 65, 140, 207, 214, 245
Tibiotarsus 4, 5, 8, 10, 33, 50, 111
Tornewton Cave 35–6, 39, 41, 59, 148, 204–5, 211, 215, 222–3, 230, 243
Tree Pipit *Anthus trivialis* 15, 36, 62, 69, 158, 191, 200, 223
Turdus merula Blackbird 15, 16, 34, 36, 38, 44–5, 50, 55–58, 62–4, 71, 77–9, 95–7, 112, 133, 150–1, 157, 180–4, 189, 194, 201, 225, 242
Turdus philomelos Song Thrush 15–6, 34–5, 38, 45, 50, 55, 57–8, 62, 64, 71, 77–9, 87–8, 95–7, 112, 130–1, 133, 155, 157, 181–3, 189, 196, 200, 225, 242
Turdus pilaris Fieldfare 15, 16, 38, 39, 44, 50, 58, 78, 96, 112, 150, 193, 225
Turdus torquatus Ring Ouzel 15–6, 36, 38, 44, 50, 55, 57–8, 69, 78, 97, 112, 133, 175, 191, 225

Turkey *Meleagris gallopavo* 12, 13, 97, 101, 103, 107, 157, 170, 204, 209
Turtle Dove *Streptopelia turtur* 105, 155, 190, 220
Tyto alba Barn Owl 38, 66–7, 78–9, 87, 108, 111, 132–3, 151, 173, 177, 191, 194, 220, 254

Uley, Gloucestershire 100, 110, 148, 225, 227, 230, 236
Uria aalge Guillemot 10, 11, 34, 57–8, 73, 75, 77–9, 96, 110, 112, 133–4, 167, 180–1, 183, 189, 219

Vanellus vanellus Lapwing 15–16, 38, 50, 57, 70, 74, 77–80, 95–6, 108, 120, 130, 132, 157, 189, 196, 215
Vulture 32, 44, 45, 76, 136, 154, 212

Westbury-sub-Mendip 54–5, 205, 207–8, 216, 221, 223
West Runton, Norfolk 34, 204, 205, 206, 207, 214, 217, 225, 227, 228
West Stow, Suffolk 23, 103, 130, 131, 140, 141, 204, 212, 214, 217, 237
Wetton Mill Rockshelter 56–8, 146, 225–7, 234
Whimbrel *Numenius phaeopus* 38, 47, 92, 96, 131, 133, 153, 192, 197, 217
White-backed Woodpecker *Dendrocopus leucotos* 60, 64, 222, 254
White Stork *Ciconia ciconia* 38–9, 57, 79, 82, 111, 145, 151, 185, 211
White-tailed Eagle *Haliaeetus albicilla* 12, 14, 36, 44, 50, 57–8, 66–7, 74, 76–8, 80, 87–8, 96, 108, 110, 116, 118, 120, 124–6, 133–4, 137, 148, 151, 159, 163, 172, 173, 174, 184, 187, 192, 194–8, 211, 232, 243, 246, 256
Whooper Swan *Cygnus cygnus* 10–1, 34–6, 44, 57, 74, 80, 82, 84, 86–7, 109, 130, 132–4, 153, 184, 192, 203
Willow Grouse *Lagopus lagopus* 44, 50–2, 54
Woodcock *Scolopax rusticola* 14, 35, 57, 74, 77–8, 80, 96, 108, 130–3, 138–9, 147, 149–50, 157, 162, 190, 216
Wood Pigeon *Columba palumbus* 35, 38, 44, 46, 57–8, 64, 68, 95, 101, 106, 131–3, 155, 162, 180–4, 189, 220
Wroxeter 84, 110, 111, 141, 206, 214, 221, 227, 247
Wryneck *Jynx torquilla* 63–5, 157, 188, 193–4, 202, 222

York 23, 84–5, 107, 110–1, 138, 140–3, 146–7, 149, 203–9, 219, 221–4, 227, 229–31, 249, 251